THE MOON AND THE PLANETS

A CATALOG OF ASTRONOMICAL ANOMALIES

Compiled by:

William R. Corliss

Published and Distributed by

The Sourcebook Project P.O. Box 107 Glen Arm, MD 21057

Library of Congress Catalog Number 85-61380

ISBN 0-915554-19-4

First printing: June 1985

Printed in the United States of America

TABLE OF CONTENTS

LIST OF PROJECT PUBLICATIONS

CATALOGS: The Moon and the Planets (categories AG, AH, AJ, AL, AM, AN, AP, AR, AU, AV)

Lightning, Auroras, Nocturnal Lights, and Related Luminous Phenomena (category GL)

Tornados, Dark Days, Anomalous Precipitation, and Related Weather Phenomena (category GW)

Earthquakes, Tides, Unidentified Sounds, and Related Phenomena (categories GH, GQ, GS)

Rare Halos, Mirages, Anomalous Rainbows, and Related Electromagnetic Phenomena (category GE)

HANDBOOKS: The Unfathomed Mind: A Handbook of Unusual Mental Phenomena

Incredible Life: A Handbook of Biological Mysteries

Unknown Earth: A Handbook of Geological Enigmas

Mysterious Universe: A Handbook of Astronomical Anomalies

Ancient Man: A Handbook of Puzzling Artifacts

Handbook of Unusual Natural Phenomena

SOURCEBOOKS: Strange Phenomena, vols. G1 and G2

Strange Artifacts, vols. M1 and M2

Strange Universe, vols. A1 and A2

Strange Planet, vols. E1 and E2

Strange Life, vol. B1

Strange Minds, vol. P1

NEWSLETTERS: Science Frontiers (current anomaly reports)

For information on the availability, prices, and ordering procedures, write:

SOURCEBOOK PROJECT
P.O. Box 107
Glen Arm, MD 21057

PREFACE

After more than twelve years of scouring the scientific and semiscientific literature for anomalies, my major conclusion is that this is an amazingly fruitful activity. In fact, organized science should have been doing the same searching and compiling for the past 200 years. It is simply astounding that a Catalog of Scientific Anomalies does not already exist to guide scientific thinking and research. It is at least as important to realize what is not known as it is to recognize the well-explained. With this outlook, here is the fifth volume in such a Catalog. It is largely the product of one person's library research, carried forward without grants, contracts, or donations.

Under the aegis of the Sourcebook Project, I have already published 20 volumes, totalling well over 7,000 pages of source material on scientific anomalies. (See page iv for a list of titles.) As of this moment, these 20 volumes represent only about 25% of my data base. New material is being added at the rate of about 1,200 new articles and items per year, about 300 of which are from the current literature. These rates could be easily multiplied several times over by spending more time in libraries. Even after twelve years, only the scientific journals of the United States and England have received my serious attention. There remain the English-language journals of the rest of the world, those journals in other languages, university theses, government reports, the publications of scientific research facilities, conference papers, untold thousands of books, and an immense reservoir of newspapers. The cataloging task has just begun. The anomalies residing in the world's literature seem nearly infinite in number.

Given this rough assessment of the magnitude of the anomaly literature, one can understand why the planned Catalog of Anomalies will require at least 25 volumes of about the same size as the one you now hold. I visualize a shelf of 25 volumes, with master indexes, to be only the initial step in providing scientists with ready access to what is not, in my opinion, well-explained.

Will the Catalog of Anomalies revolutionize science? Probably not---at least not immediately. Quite often the initial reaction to the books of anomalies already published has been disbelief. The data must be in error; the data are mainly testimonial; the data are too old; the supposed anomaly was explained long ago. Germs of truth reside in all these complaints. But for every anomaly or example that can be legitimately demolished, ten more take its place. Nature is very anomalous or, equivalently, Nature is not yet well-understood by science. Much remains to be done.

William R. Corliss

P.O. Box 107
Glen Arm, MD 21057
June 1, 1985

"ROUND ABOUT THE ACCREDITED AND ORDERLY FACTS OF EVERY SCIENCE THERE EVER FLOATS A SORT OF DUST-CLOUD OF EXCEPTIONAL OBSERVATIONS, OF OCCURRENCES MINUTE AND IRREGULAR AND SELDOM MET WITH, WHICH IT ALWAYS PROVES MORE EASY TO IGNORE THAN TO ATTEND TO . . . . ANYONE WILL RENOVATE HIS SCIENCE WHO WILL STEADILY LOOK AFTER THE IRREGULAR PHENOMENA. AND WHEN THE SCIENCE IS RENEWED, ITS NEW FORMULAS OFTEN HAVE MORE OF THE VOICE OF THE EXCEPTIONS IN THEM THAN OF WHAT WERE SUPPOSED TO BE THE RULES." **William James**

HOW THE CATALOG IS ORGANIZED

Purpose of the Catalog

The Catalog of Anomalies is designed to collect and categorize all phenomena that cannot be explained readily by prevailing scientific theories. Following its definition, each recognized anomaly is rated in terms of: (1) its substantiating data; and (2) the challenge the anomaly poses to science. Next, all examples of the anomaly discovered so far are noted, some of more interesting ones in more detail. Finally, all examined references are listed. Thus, the Catalog is a descriptive guide as well as a compendium of examples and references. Scientific researchers have a substantial foundation for beginning further studies of these intriguing phenomena. This is the basic purpose of the Catalog: the collection and consolidation of the unknown and poorly explained to facilitate future research and explanation.

General Plan of the Catalog

It was tempting to organize this Catalog alphabetically, making it an "encyclopedia" of anomalies." But many of the phenomena have obscure names or, even worse, no names at all. Under these circumstances, access to the data base would be difficult. Therefore, a system of classification was designed based upon readily recognized classes of phenomena and the means by which the observer detects them. Subject matter is first divided into nine general classes of scientific endeavor, as illustrated in the diagram on the following page. Few would have difficulty classifying a phenomenon as biological, astronomical, etc. The second, third, and fourth levels of classification are also based on generally recognized attributes. The similarity of this kind of categorization to those employed in natural history field guides is quite intentional. Like bird identification, phenomenon classification soon becomes second nature. In fact, many of the phenomena described in the Catalog are accessible to anyone with normal senses and, especially in astronomy, a little optical help.

Most catalogs boast numbering systems, and this one is no exception. Rather than employ a purely numerical system, the first three classification levels are designated by letters. The triplets of letters selected have some mnemonic value. Thus, an AVO anomaly is easily recognized as being in the astronomy class (A), involving the planet Venus (V), and being detected through telescopic observation (O). The number added to the triplet of letters marks the fourth classification level, so that AVO3 signifies the Venusian ashen light phenomenon, the third type of Venusian telescopic anomaly. Every anomaly type has such a unique alphanumeric code. All indexes and cross references are based on this system. Catalog additions and revisions are also made easier with this scheme.

The Catalog codes may seem cumbersome at first but their mnemonic value to the compiler has been significant. The codes are simple, yet flexible enough to encompass the several thousand anomalies identified so far in many diverse disciplines.

A glance through this volume will reveal that each example of a specific anomaly bears an X-number, and each reference an R-number. AVO3-X3 therefore specifies the third example of the ashen light phenomenon; and AVO3-R7 the seventh reference to the ashen light. Indexes and cross references can consequently be made more precise than page numbers.

How Data and Anomalies Are Evaluated

Each anomaly type is rated twice on four-level scales for data "validity" and "anomalousness," as defined below. These evaluations represent only the opinion of the compiler and are really only rough guides.

Data Evaluation Scale

1 Many high-quality observations. Almost certainly a real phenomenon.

2 Several good observations or one or two high-quality observations. Probably real.

3 Only a few observations, some of doubtful quality. Phenomenon reality questionable.

4 Unacceptable, poor-quality data. Such phenomena are included only for the purposes of comparison and amplification.

Anomaly Evaluation Scale

1 Anomaly cannot be explained by modifications of present laws. Revolutionary.

2 Can probably be explained through relatively minor modifications of present laws.

3 Can probably be explained using current theories. Primarily of curiosity value.

4 Well-explained. Included only for purposes of comparison and amplification.

Anomalies that rate "1" on both scales are very rare. Such anomalies, however, are the most important because of their potential for forcing scientific revolutions. As additional Catalog volumes are published, the relative proportion of "double-1s" will increase. especially in the fields of biology and psychology.

Catalog Coding Scheme

First-order classification	Second-order classification	Third-order classification	Fourth-order classification
A Astronomy	G Earth	B Celestial mechanics	1 Cusp phenomena
B Biology	H Mercury	E Geological phenomena	2 Ring of light
C Chemistry & physics	J Jupiter	F Radiation phenomena	3 Ashen light phenomenon
E Earth sciences	L Moon	O Telescopic phenomena	4 Phase anomaly
G Geophysics	M Mars	W Atmospheric phenomena	⋮
L Logic & math	N Neptune	X Transit & occultation phenomena	⋮
M Archeology	P Pluto	Z Magnetic phenomena	⋮
P Psychology	R Saturn		⋮
X Unclassified	U Uranus		⋮
	V Venus		10 Terminator irregularities

Anomaly Examples

Examples of anomaly types are designated by the letter X in the body of the Catalog. All examples discovered so far are listed. If the example is of the event type, time and place are specified where available. Such data are the foundations of the Time-of-Event Index, which could in principle lead to the discovery of obscure cause-and-effect relationships. Where library research has unearthed many examples of a specific type of anomaly, only the most interesting and instructive are quoted in detail. Direct quotations from eye-witnesses and scientific experts are employed frequently to convey accurately the characteristics of the phenomena.

The References and Sources

Each anomaly type and the examples of it are buttressed by all references that have been collected and examined. Since some references describe several examples, each reference includes the X-numbers of the examples mentioned. When a reference covers more than one type of anomaly, it is repeated in the bibliography following each anomaly type. Actually, there is little repetition of this sort in the Catalog.

Perusal of the Source Index will demonstrate that the great majority of the references comes from the scientific literature. Heavily represented in this volume of the Catalog are such journals as: Nature, Science, Journal of the British Astronomical Association, and Monthly Notices of the Royal Astronomical Society. Several less technical publications are also mentioned frequently: Popular Astronomy, Sky and Telescope, and the English Mechanic. The English Mechanic was for many years an important English technical magazine, with a strong astronomical content. All of the serials mentioned above are generally very reliable, though one must always be wary when unusual events are reported. In addition to these often-referenced publications, there is a wide spectrum of other journals and magazines. In this context, it should be remembered that anomalous phenomena do not seek out scientists, and that the laymen who observe many anomalies do not have access to the scientific journals.

The time span covered by the sources ranges over 150 years. Many excellent reports come from the latter half of the Nineteenth Century. Although many scientists frown on such old reports, the quality is often good, and they should not be discarded arbitrarily. Not only were the observers of a century ago competent but they were unbiased by modern dogmas about what should be observed. They also lived in a less-polluted time when observing conditions were usually better than today. It should also be mentioned here that modern astronomical literature tends heavily toward theory and mathematical modelling as opposed to eye-witness accounts of strange phenomena of the skies. A possible exception to this generalization is the Strolling Astronomer, published by the Association of Lunar and Planetary Observers. Even so, the reader will find a strong concentration of event-type observations in the 1870-1910 period.

The Indexes

Each Catalog volume concludes with five separate indexes. At first glance this may seem to be too much of a good thing. But in the context of a science-wide catalog of anomalous events and unusual natural phenomena, each index has its special utility.

The subject index is essential in any work of this type. It is placed last for easy access. The time-of-event and place-of-event indexes are analytical tools for the anomaly researcher. They help identify phenomena that are reported separately (perhaps in widely different journals) but are really different aspects of the same event. This integrating feature will become more apparent as additional Catalog volumes appear. To illustrate, the effects of large solar flares may be felt throughout much of the solar system. It is possible that judicious indexing will reveal many synchronicities. It is the intent of the Catalog effort to generate a composite set of indexes that will link astronomy to geophysics, to biology, and so on.

The source index shows immediately the dependence of this Catalog upon scientific literature rather than newspapers and other popular publications. Its real purpose, though, is the rapid checking to determine if a specific reference has or has not been caught already in the fishing nets of this Catalog project. The source index is doubly valuable because many footnotes and bibliographies in the scientific literature display sources only; that is, titles and authors are omitted entirely. The researcher also comes across many vague references to such-and-such an article by so-and-so back in 1950 in Nature. The exhaustive and rather ponderous source and first-author indexes can help pin down many references lacking specifics.

All five indexes use the catalog codes described above rather than page numbers. The codes are permanent whereas the page numbers will change as addenda and revised volumes are produced. The mnenonic value of the catalog codes is useful here, too, because the approximate nature of each index entry is readily apparent, while page numbers give only location.

Supporting Publications of the Sourcebook Project

The Catalog volumes currently being published are actually distillations of huge quantities of source material. The Sourcebook Project has already published 20 volumes of this source material, as detailed on p. iv. Phase I of the Sourcebook Project resulted in 10 looseleaf notebooks called "sourcebooks." To meet the objections of librarians, Phase II supplanted the sourcebooks with a series of 6 "handbooks," which are hardcover and much larger and more comprehensive than the sourcebooks. Phase III, now in progress, is the cataloging phase, which involves the systematization of a data base comprising some 30,000 articles. The Sourcebook Project also publishes a bimonthly newsletter, SCIENCE FRONTIERS, which informs customers about scientific anomalies appearing in the current literature.

Catalog Addenda and Revisions

Over 1,200 new reports of anomalies are collected from current and older scientific journals each year. New anomaly types and additional examples of types already cataloged are accumulating rapidly. When sufficient new material has been assembled, Catalog volumes will be revised and expanded.

Request for Additions and Corrections

The Sourcebook Project welcomes reports of new anomalies and examples of recognized anomalies not yet registered in extant Catalog volumes. Reports from scientific journals are preferred, but everything is grist for the mill! Credit will be given to submitters in revised volumes of the Catalog of Anomalies. Send data to the Sourcebook Project, P.O. Box 107, Glen Arm, MD 21057.

AG INTRODUCTION TO THE EARTH

Key to Categories

AGB ANOMALIES IN THE EARTH'S ROTATION
AGL EARTH SATELLITE PHENOMENA
AGO OBSERVATIONS OF EARTH FROM SPACE

The earth itself is an astronomical object only for astronauts and extraterrestrial beings. Nevertheless, we can make a few useful astronomical observations from the earth's surface that bear on the earth's own motion and its halo of gases. Sightings on the fixed stars have, for example, revealed the still unexplained Chandler Wobble. Through the precision tracking of artificial earth satellites, astronomers have deduced the superrotation of the earth's upper atmosphere and the presence of hitherto unappreciated drag forces. There is also observational evidence that dust and other space debris have collected at two of the earth-moon Lagrangian points, creating in effect dust-cloud satellites of the earth. Another curious aspect of astronomy restricted to the earth's environs focusses on ephemeral natural satellites of the earth. Does the earth actually have small moonlets in orbit around it? Some data say "yes", but there seems to be unusual scientific resistance to the idea.

In this volume's chapters on the moon and other planets, considerable space is devoted to the "geophysics" and "geology" of each object. In the earth's case, seeing we know so much more about our own planet, the Catalog of Anomalies includes four separate volumes on geophysics and four additional volumes on geology are projected.

AGB ANOMALIES IN THE EARTH'S ROTATION

Key to Phenomena

AGB0 Introduction

This short section is devoted to astronomical observations that provide us with knowledge about the earth's orientation in space and, in consequence, its spin rate, precession, nutation, wobble, and other dynamical data. With the advent of geodetic satellites and high-precision clocks, our scientific understanding of the earth's motion has advanced rapidly. Two decades ago, there would have been more anomalies in this category than registered below.

Finding no data to the contrary, it is assumed that the earth's precession and nutation and the forces that cause them are well understood. So too with the earth's spin rate---or, equivalently, the length of the day. Variations in the length of the day, although they are measured only in milliseconds, have been the subject of intense investigation recently. Although a few minor discrepancies still exist, it seems quite sure that these variations can be linked to meteorological variables. No significant scientific anomalies seem to exist here any more. (R4, R5, R7)

The variations of latitude measurements do not seem to be as well understood. That latitude measurements based on observations of the fixed stars vary with time has been recognized for over a century. One of these variations, the 14-month Chandler Wobble, is still the subject of much controversy. No one is positive as to which natural forces sustain this wobble and keep it from damping out. Earthquakes are widely blamed, but there remains enough doubt to list it here.

Finally, it was tempting to include as bona fide anomalies historical references to what seem to have been gross changes in the length of the day, such as the Biblical accounts of Joshua's commanding the sun to stand still and the curious tale of the Dial of Ahaz, on which the shadow of the sun behaved erratically. (R1-R3) In a similar vein, some claim that ancient gnomon records and the orientations of ancient buildings; viz., the Temple at Karnak; suggest a radical change in the orientation of the earth's axis circa 2500 B. C. (R6) However, these data are so vague that it was decided to exclude them until they could be substantiated in other historical records; say, those of China and Mesoamerica, where avid astronomers would certainly have been amazed at the purported motions of the earth.

Ancient historical records have a place in this Catalog, but here we do not have enough high quality accounts of the purported events.

References

R1. Bird, Alfred; "Joshua Commanding the Sun to Stand Still," Astronomical Register, 1:186, 1863.
R2. Garrett, E. L.; "Joshua and the Sun," Knowledge, 3:331, 1883.
R3. Morrison, J.; "The Sun Dial of Ahaz," Popular Astronomy, 6:537, 1898.
R4. Challinor, R. A.; "Variations in the Rate of Rotation of the Earth," Science, 172:1022, 1971.
R5. Gribbin, John, and Plagemann, Stephen; "Discontinuous Change in the Earth's Spin Rate Following Great Solar Storm of August 1972," Nature, 243:26, 1973.
R6. Bowden, M.; "The Recent Change in the Tilt of the Earth's Axis," Creation Science Movement, Pamphlet 236, 1983. This item claims to be based on the work of the highly regarded Australian astronomer, George F. Dodwell.
R7. Carter, W. E., et al; "Variations in the Rotation of the Earth," Science, 224:957, 1984.

AGB1 Variations in Latitude

Description. Variations in the latitudes of fixed points on the earth's surface as measured by observations of the fixed stars.

Background. The measurements of the Chandler Wobble (X2 below) actually represent changes in the angular distance between the point of measurement and instantaneous pole of rotation. A non-rigid earth is implied.

Data Evaluation. The effects noted here are supported by many thousands of latitude measurements the world over. Rating: 1.

Anomaly Evaluation. Geophysicists have not agreed upon any specific cause for the latitude variations, although earthquakes are thought by many to sustain the Chandler Wobble. It seems very likely that some prosaic geophysical or astronomical phenomenon, such as one of those listed below, will ultimately be proven to be the cause of latitude variations. In this light, a low anomaly rating seems in order. Rating: 3.

Possible Explanations. Earthquakes (GQS10); changes in liquid core motion; wind friction; solar wind interactions with the earth's magnetic field.

Similar and Related Phenomena. Earthquakes correlated with the Chandler Wobble (GQS10); the well-known precession and nutation of the earth's spin axis.

Examples

X1. Latitude variation with the moon's altitude. "Recent investigations of this laboratory (Astronomical Laboratory, Harvard) have suggested a possible connection between the variation in latitude of a given place on the earth's surface and the position of the moon in the sky at the time observations for latitude are made. An analysis of the whole series of latitude observations, which were made by Ross at Gaithersburg from 1911 to 1914, has revealed a striking correlation between the moon's hour angle and the value of the latitude obtained. The data were restricted to results obtained with the photographic zenith telescope, thus eliminating all personal equation...... The striking rise in the value of the latitude with increasing altitude of the moon is shown in the altitude-latitude curve, Fig. 2, which again was plotted from the running means. The maximum latitude occurs at altitude 30°, or when the observer was 60° from the sublunar point. It should be stated that the extreme range of variation of latitude due to this lunar effect was 0."08 for the 1913-1914 series and about 0."09 for the whole series 1911-1914. On account of the relatively greater degree of precision obtained in the 1913-1914 series and the larger number of observations included, double weight was given this series in plotting the final graph as exhibited in Fig. 2. The fact that the total variation is about twenty times the probable error of each point on the curves leaves little ground for interpreting the curve as a chance phenomenon. The curve of observations for the moon below the horizon is radically different. A marked fall in the value of latitude follows the negative altitude of 30°." The author eliminates meteorological causes for this phenomenon, stating that they were averaged out because several thousand observations were used for each point. Changes in refraction due to the lunar atmospheric tide were considered too small. (R1) No further reference to this work has been found in the literature examined to date. The moon does, of course, affect the earth's precession and nutation, but this seems to be a different phenomenon. (WRC)

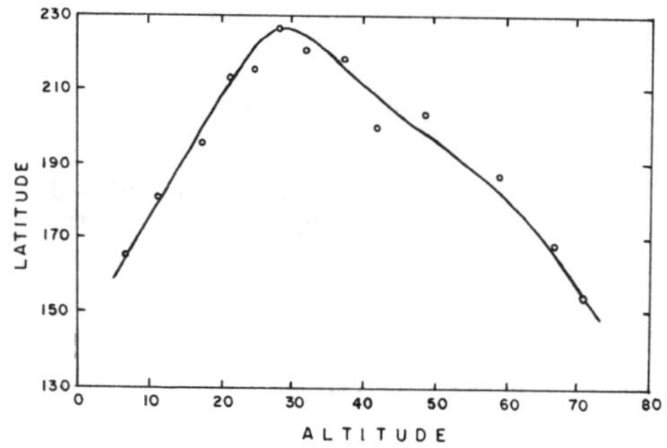

Variation of latitude with the altitude of the moon, according to H. T. Stetson's measurements. Each circle represents 600 observations. (X1)

X2. The 14-month variation of latitude or Chandler Wobble. Seth Carlo Chandler was not an accepted member of the scientific community. Because of this and the dogma that the earth was a rigid body, no one believed Chandler when he claimed that the earth's crust moved with respect to the earth's spin axis in a 14-month cycle. His analysis had to be incorrect. But Chandler persisted, ignoring the idealization of a rigid earth. He did not use a telescope, he simply collected latitude measurements, which were available in abundance. These data showed clearly that the latitude of all locations varied with time. The earth was plastic, because the angular distances between points on the earth's surface and the planet's spin axis changed as much as 0.2 arcsecond during a 14-month cycle. Reckoned in terms of distance, this is only a 21-foot change in the apparent distance to the geographical pole.

The 14-month period of the Chandler Wobble is difficult-to-explain because no natural potential driving force has a similar period. Yet, the wobble seems to have persisted for over 200 years, even though calculations show that friction should damp it out in about 23 years. The only possible external force is the solar wind, but it does not have the proper period. Looking inward, earthquakes have been a popular stimulus used to account for the Chandler Wobble. Although earthquakes do inject considerable energy into the earth's crust, there are many drawbacks to this theory. Geophysicists are now considering internal waves and convection phenomena in the earth's liquid core. (R2-R11)

For a much more detailed account of the possible connection between earthquakes and the Chandler Wobble, see GQS10 in another volume of this Catalog, entitled Earthquakes, Tides, Unidentified Sounds.

References

R1. Stetson, Harlan True; "On the Variation of Latitude with the Moon's Position," Science, 69:17, 1929. (X1)

R2. Gold, T.; "Irregularities in the Earth's Rotation," Sky and Telescope, 17:216, 1958. (X2)

R3. Manshina, L., and Smylie, D. E.; "Earthquakes and the Earth's Wobble," Science, 161:1127, 1968. (X2)

R4. Stubbs, Peter; "The Dance of the Pole," New Scientist, 41:296, 1969. (X2)

R5. Simon, Cheryl; "Chandler Wobble," Science News, 120:268, 1981. (X2)

R6. Halm, J.; "The Relation between the Periodic Changes of Solar Activity and the Earth's Motion," Nature, 61:445, 1900. (X2)

R7. Halm, J.; "The Relation between the Periodic Changes of Solar Activity and the Earth's Motion," Scientific American Supplement, 49:20315, 1900. (X2)

R8. Halm, J.; "Latitude-Variation, Earth-Magnetism and Solar Activity," Nature, 62:460, 1900. (X2)

R9. Gold, T.; "Irregularities in the Earth's Rotation---II," Sky and Telescope, 17: 284, 1958. (X2)

R10. Hughes, David W.; "Earth Wobble, Day Length and Continental Drift," Nature, 253:591, 1975. (X2)

R11. Dahlen, F. A.; "Simulation Shows Wobble Period Neither Multiple Nor Variable," Nature, 301:657, 1983. (X2)

AGL EARTH SATELLITE PHENOMENA

Key to Phenomena

AGL0 Introduction

Earth satellite anomalies fall basically into two classes. The first deals with artificial earth satellites, which can be tracked with high precision from the earth's surface. Tracking data demonstrate that the orbits of artificial satellites (and, presumably small natural earth satellites) are forever perturbed by their environment. This is hardly a surprise. Aerodynamic drag is well understood, as are the effects of solar radiation pressure and the nudgings of gravitational and electromagnetic fields, and so on. There are, though, perturbations of artificial earth satellites that have not yet been explained by appealing to known physical forces acting in orbit. These perturbations constitute the first group of anomalies.

The second class of earth satellite anomalies centers on small natural satellites of the earth. The mere existence of such satellites is questioned by many. In fact, some astronomers seem to go out of their ways to debunk those observations suggestive of their existence. One is reminded of those savants of 1800 who ridiculed the possibility of stones falling from the sky. In addition, very few astronomers dabble in this byway of science. Despite such problems, a few interesting observations have accumulated, many of which are the product of one man's research. The future of this field is very limited because man-made debris now envelopes the earth is a sort of satellite smog that hides the comings and goings of natural earth satellites.

AGL1 Slow Changes in Satellite Inclination

Description. The slow change in the plane of inclination of artificial satellites relative to the plane of the equator. The total change of inclination for a low-altitude satellite (100-200 mile altitude) is about 0.1° during its lifetime.

Data Evaluation. An immense amount of satellite tracking data confirm this effect for all low-altitude satellites. Rating: 1.

Anomaly Evaluation. High-altitude winds or atmospheric superrotation are usually presumed to be the cause of this phenomenon, but no acceptable driving force has been found for superrotation. If superrotation does not in fact exist (as some experiments show), there is no ready explanation for the inclination changes. Rating: 3.

Possible Explanations. Drag by high-altitude winds blowing west-to-east at about 100 meters/second nudge the orbital planes of satellites. These presumed winds suggest that the whole atmosphere is moving faster than the earth's surface or "superrotating". Some satellite experiments show no indications of winds of such magnitude.

Similar and Related Phenomena. The superrotation of the Venusian atmosphere (AVW1)

Examples

X1. Analysis of the orbits of nine early satellites. "Because the atmosphere rotates, a close Earth satellite is subjected to small lateral aerodynamic forces which have the effect of slightly changing the inclination of the orbit to the equator. For an eastbound satellite the inclination decreases, sometimes by as much as 0.1°, and the amount of the decrease is a measure of the angular velocity of the atmosphere at heights near that of the satellite's perigee. Such a decrease was first noted in 1958 with Sputnik 2, and it was pointed out that the decrease was rather greater than would be caused by an atmosphere rotating at the same rate as the Earth." Using Sputnik-2 data and similar measurements of eight other satellites, it was inferred that at heights of 200-300 kilometers, the atmosphere rotates west to east with a mean wind speed of 100 meters per second at midlatitudes. (R1)

X2. Analysis of satellite 1968-59A. The so-called superrotation of the earth's atmosphere was found to extend down to heights of 150-170 kilometers. (R2) Note that the basic anomaly is the change of satellite inclination, and that atmospheric superrotation is inferred. (WRC)

X3. General observations. "Observations of small changes in the orbital inclinations of artificial satellites have shown that the Earth's upper atmosphere (at altitudes of 150-400 km) is rotating about 20-30% faster than the Earth itself. This phenomenon has become known as super-rotation. Even though it was discovered by King-Hele in the early 1960s, the cause of this 100 m s^{-1} west to east wind is still a subject of considerable debate. The change of inclination because of super-rotation is only about 0.1° during the satellite lifetime, and so spatial variations and short term effects are very difficult to detect. No clear evidence has been found of year to year variations, such as might be caused by the 11-yr-periodicity in solar activity, but over daily periods the wind seems to maximise between 2100 and 2400 local time." After this preamble, various causes of superrotation were considered and found wanting. Included were: (1) Horizontal pressure gradients set up in the atmosphere by the thermospheric heating from solar radiation and energetic particle bombardment; (2) Ionospheric or magnetospheric motions that are coupled to the neutral atmosphere; (3) The moving flame effect created by the rotation of the earth with respect to solar heat input; (4) The momentum imparted by atmospheric waves; and (5) The Scott Effect in which a magnetic field applies a torque to a low-pressure gas. (R3)

Despite the general acceptance of superrotation as the cause of satellite inclination changes, the direct measurement of high altitude winds by an experiment on the Dynamics Explorer 2 indicate average velocities of only 18 meters per second west-to-east along the equator and a westward wind of the same magnitude at midlatitudes. (R4) This lack of direct evidence of superrotation may mean that some other phenomenon is changing the inclinations of earth satellites. (WRC)

References

R1. King-Hele, D. G.; "Rotational Speed of the Upper Atmosphere, at Heights of 200-300 km," Nature, 202:893, 1964. (X1)
R2. King-Hele, D. G.; "'Super-Rotation" of the Upper Atmosphere at Heights of 150-170 km," Nature, 226:439, 1970. (X2)
R3. Hughes, David W.; "Super-Rotation of the Upper Atmosphere," Nature, 249: 405, 1974. (X3)
R4. Wharton, L. E., et al; "The Earth's Superrotation from Dynamics Explorer 2," Geophysical Research Letters, 11: 531, 1984. (X3)

AGL2 Sudden Perturbations of Orbital Elements

Description. Sudden changes in the inclination, perigee, line of nodes and other orbital parameters of artificial satellites. The changes may be temporary or permanent.

Data Evaluation. While the published elements of some artificial satellites do indeed show sudden perturbations, such changes could be the consequence of faulty tracking, incorrect data processing procedures, the use of erroneous physical constants, and a variety of other human and equipment artifacts. (R3) For example, some so-called "perturbations" are suspicious because the elements return to their original values---something unlikely if there have been real orbital changes. Rating: 3.

Anomaly Evaluation. Sudden orbital perturbations could easily be attributed to collisions or near-collisions with orbital debris or even satellite propulsion system malfunctions. This phenomenon, therefore, is not highly anomalous. Rating: 3.

Possible Explanations. Collisions and near-collisions with natural and artificial orbital debris, involving actual contact, gravitational forces, electrostatic forces, and/or electromagnetic forces. Equipment malfunction could cause orbital perturbations if propellant from maneuvering engines or the attitude control system were vented.

Similar and Related Phenomena. The suddenly perturbed paths of comets and natural meteors in the upper atmosphere (ACO and AYO).

Examples

X1. Survey of published orbital elements. "From an analysis of the considerable amount of published data on the orbits of artificial satellites, I have recently found that many sudden anomalies have been occurring in the orbital elements of a large number of the artificial bodies. These changes are often quite drastic and of a semipermanent nature. To my knowledge, they have not been previously noted or discussed. Inclinations have changed by up to tenths of a degree, apogees and perigees have fallen and increased from tens to hundreds of kilometers, and the lines of nodes and apsides have changed their orientations by up to ± 50% of the normal rate. These anomalies are most easily seen by plotting the published tabular data. A typical example is shown in Fig. 1 for the orbit of Telstar 1. The curious thing about such sudden anomalies is that so far as is known over 90% have taken place at epochs when one or more of the orbits of the proposed natural satellite subgroups was intersecting the orbits of the artificial bodies affected. Under these conditions, one or more close passages of natural and artificial bodies could take place. This would be possible if the natural and artificial satellites in their own orbits were to reach the common intersection point at the same time." The author theorizes that these perturbations of artificial satellites were caused by a group of natural moonlets which broke off a larger parent body on December 18, 1955; because, when projected back in time, the moonlet orbits converge to a common point. (R1, R2) See AGL4 for possible direct observations of the proposed moonlets. (WRC)

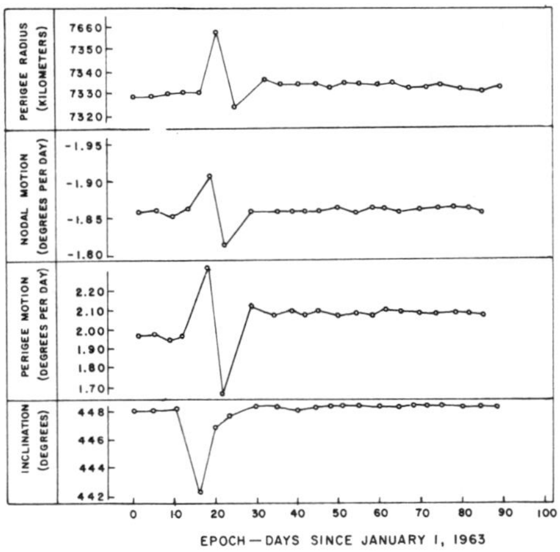

References

R1. Bagby, John P.; "Terrestrial Satellites: Some Direct and Indirect Evidence," Icarus, 10:1, 1969. (X1)

R2. "The Earth Has 'Moonlets'---and May Have Had a 'Super Moonlet' until December 1955," New Scientist, 42:223, 1969. (X1)

R3. Meeus, Jean; "Bagby's Phantom Moonlets," Icarus, 19:547, 1973.

Behavior of some Telstar satellite orbital parameters in early 1963. (X1)

AGL3 Slow, Unexplained Descent of Satellites

Description. The slow, sometimes erratic descent of artificial earth satellites at rates much faster than can be accounted for by aerodynamic drag and other known forces.

Data Evaluation. The literature reviewed to date deals only with the Lageos satellite, but the data are of high quality, coming from laser tracking stations over a period of several years. Rating: 1.

Anomaly Evaluation. The anomaly probably owes its existence to the interaction of satellites with the surrounding medium which, despite the high altitudes involved, is still present, being derived from gases escaping from the earth and the action of solar radiation and wind. There are doubtless several unappreciated aerodynamic and plasmadynamic effects at work. Rating: 3.

Possible Explanations. A so-called "plasma drag" caused by the interaction of an electrically charged satellite with the thin surrounding plasma. Other possibilities are: the presence of unexpected quantities of gases, such as helium, escaping from the top of the earth's upper atmosphere; and radiation pressure from the earth's albedo.

Similar and Related Phenomena. Lower altitude satellites always encounter appreciable aerodynamic drag, which ultimately leads to their reentry to the earth's atmosphere. The moon, however, slowly recedes from earth due to the tidal dissipation of energy.

Examples

X1. The Lageos case. Lageos (an acronym for Laser Geodynamic Satellite) was orbited in May 1976. It is a 2-foot aluminum sphere spangled with 426 mirrors that reflect laser beams aimed at it from tracking stations around the world. The core of Lageos is a 386-pound cylinder of brass. Lageos's orbit is known with great accuracy for it is intended to be an orbital benchmark. Scientists were understandably perturbed when shortly after launch, tracking stations reported that Lageos was losing altitude at the rate of 1/25 of an inch per day---roughly ten times the rate attributable to aerodynamic drag and other known forces at its altitude of about 3700 miles. The Lageos problem is made more perplexing by its changing rate of fall. From late 1978 through late 1979, its rate of descent was 60% greater than the 1982 value of 1/25 inch per day. In 1983, however, Lageos seemed to stop falling and had possibly started to regain altitude. (R1, R2)

References

R1. Maran, Stephen P.; "Fall from Space," Natural History, 91:74, December 1982. (X1)

R2. Smith, David E.; "Acceleration on Lageos Spacecraft," Nature, 304:15, 1983. (X1)

AGL4 Direct Visual Observation of Natural Earth Satellites

Description. Visual and photographic observation of small, moving points of light or nebulous luminous clouds, which can be shown to be in earth orbit from tracking data.

Background. Several types of objects may mimic natural earth satellites. Some meteors, for example, may be observed over tracks thousands of miles long, as in the case of the famous 1913 meteor procession (AYO). Today, with many thousands of artificial satellites and pieces of man-made debris in orbit, observations of natural earth satellites are difficult to separate from artificial objects, although NASA and other agencies do try to keep track of most objects in orbit.

Data Evaluation. The visual and photographic evidence for accumulations of dust and debris at two of the earth-moon Lagrangian points (L/4 and L/5) seems quite convincing. It has been possible to repeat observations of the same clouds and ascertain that they move at the same velocity as the moon itself. The existence of ephemeral earth satellites of a more palpable nature is suggested by dozens of recent observations. However, it has been difficult to convince many astronomers that these fleeting glimpses represent true earth satellites. No purported natural satellite has been observed repeatedly by a large group of observers. Rating: 2.

Anomaly Evaluation. Most astronomers will acknowledge that the solar system teems with miscellaneous debris from dust to 1000-mile chunks. It seems inevitable that some of these bits and pieces would be captured by earth in temporary orbits. Nevertheless, the reactions to observations supporting such a situation are often treated with much disdain. Although physically reasonable; small, natural earth satellites seem anomalous to some. Rating: 3.

Possible Explanations. The temporary capture by earth of space debris, with occasional gravitational assists by the moon.

Similar and Related Phenomena. The sudden perturbation of the orbits of artificial earth satellites (AGL2); radio and radar anomalies associated with natural earth satellites (AGL5); long-distance meteor processions (AYO); observations of the (supposed) intramercurial planet and other enigmatic objects (AEO).

Examples

X1. Typical older claim of observations of other natural earth satellites: "M. Petit, Director of the Observatory of Toulouse, made observations and calculations which induced him to conclude that there is at least one meteoric stone of considerable magnitude which is attached as a satellite to the earth. Its orbit is about 5,000 miles from the surface, and, therefore, 9,000 miles from the centre, or about 25 times nearer than the moon. It makes a complete revolution in three hours and 20 minutes, and, therefore, revolves around the earth about seven times per day." (R1, R2) Note that there are a great many other observations of objects which were thought at the time to be earth satellites. Usually, these were one-time sightings, and no satellite orbit could be obtained. Such observations are filed in Category AEO, in another volume of this Catalog, along with other "enigmatic objects". (WRC)

X2. Cloud satellites in lunar orbit. For two centuries, astronomers have realized that there are several points in space where the gravitational fields of the earth and moon combine in such a way that small particles can collect at these points and remain in fairly stable orbits around the earth. These points are called Lagrangian points. Some have supposed that dust and rubble might collect at these points and be visible from earth, even though perturbations from the sun's gravitational field might eventually smear out such concentrations into bands.

The Polish astronomer K. Kordylewski was apparently the first to actually observe these so-called cloud satellites. "Beginning in October, 1956, on a number of occasions Dr. Kordylewski saw with the naked eye slight brightenings in the sky near the L_4 and L_5 points. These difficult sightings were made in the very dark, pure sky of high mountain stations in Czechoslovakia, on Mts. Lomnica and Kasprowy Wierch, and at Skalnate Pleso. The luminous patches appeared to be at least two degrees in diameter, and were a magnitude or two fainter than the central part of the gegenschein. In a few cases, the Polish astronomer could observe such a patch on consecutive nights, confirming that it was moving at the same rate as the moon." Kordylewski also photographed these faintly luminous patches. (R3-R5)

Kordylewski's discovery was confirmed in great detail photographically from aircraft by J. W. Simpson and his colleagues at Lockheed. However, as Simpson testified, many uncertainties remained: "At present we have a considerable number of confirmed multiple observations of both L/4 and L/5. We have found that they do not fall precisely at the lunar sextile libration points but are displaced from these points by varying distances and directions. We do know that they can not be seen at the time of the new moon or when in conjunction with the earth and moon. This reaffirms the previously stated opposition or phase-angle criterion. It is quite possible that the orbit of the libration points around the earth is not identical with that of the moon, and orbit analysis is now underway... The size of the particles and the dynamics of the individual areas are presently unknown. These properties are of interest because they will determine what mechanisms could inject particles into the cloud and how stable they are. The answers to these questions will in turn determine to what extent the libration areas are reservoirs of interesting cosmic rubble... At present, however, despite educated guessing, the true nature of the clouds remains to be determined." (R7, R8)

Despite the publicity given these data and the apparent general acceptance of the reality of the cloud satellites, some astronomers were unconvinced. The supposed observations of cloud satellites might in reality be sightings of clouds of particles not in lunar orbit at all. "The observational problem is that of detecting a cloud of dust particles at a libration point against a background brightness arising from a large and bright zodiacal 'cloud' of particles distributed throughout the solar system. This cloud causes the zodiacal light, probably also the gegenschein, and in fact produces diffuse brightness over the entire sky. There is no reason to expect that this zodiacal cloud will be completely homogeneous; bright patches of the size that have been observed might be seen as a result of several causes. Although much more work must be done on this subject, we suggest as one possibility that, when seen 'end on,' the orbits of many old comets may still contain enough diffuse matter to produce enhanced sky brightness. Also interaction of the interplanetary gas and dust grains with the filamentary solar wind might produce observable effects." (R9) To which the Lockheed group responded: "Apparently the Texas-group investigation covers areas of 'patchy brightness' over the 'entire sky.' Our own investigation is concerned only with those areas of 'patchy brightness' in or near the plane of the lunar orbit, at or near the libration points, moving at a rate approximating that of the moon. In the cases of Kordylewski, his ten scientist associates, the writers and NASA personnel, the clouds have been observed with their expected departure from their predicted position. In numerous cases we have observed their movement from night to night at a daily rate closely approximating the daily motion of the moon when it occupied that portion of its orbit during the same revolution. Airglow patches and loosely distributed material along the lunar path would hardly be expected to have this daily motion, which have all observed for the libration clouds." (R9)

A recent telescopic search for "objects" as opposed to "dust" in the vicinities of the Lagrangian points yielded negative results. (R15)

In sum, nebulous luminous patches seem to have been observed visually and photographically near the Lagrangian points L/4 and L/5, although they are sometimes displaced from the theoretically ordained points, they move as if they were in lunar orbit. The data are confused, however, by the frequent appearance of similar luminous patches of various pedigrees. (WRC)

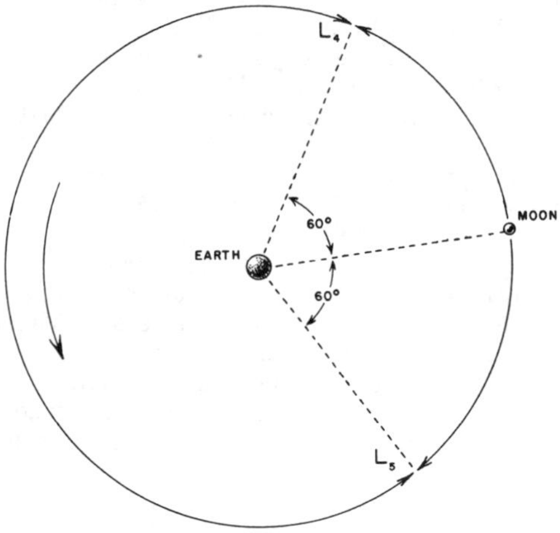

The L4 and L5 Lagrangian points form equilateral triangles with earth and moon. Kordylewski's clouds were observed near L5. (X2)

X3. The Glenn Effect. A field of luminous particles observed by John Glenn in 1962 while in orbit was publically attributed to material from the space capsule itself. However Glenn's testimony is so at variance with this interpretation that this event must be added here as possible evidence of an orbiting dust cloud. "The strangest sight of all came with the very first ray of sunrise as I was crossing the Pacific toward the U. S. I was checking the instrument panel and when I looked back out the window I thought for a minute that I must have tumbled upside-down and was looking up at a new field of stars. I checked my instruments to make sure I was right-side-up. Then I looked again. There, spread as far as I could see, were literally thousands of tiny luminous objects that glowed in the black sky like fireflies. I was riding slowly through them, and the sensation was like walking backwards through a pasture where someone had waved a wand and made all the fireflies stop right where they were and glow steadily. They were greenish yellow in color, and they appeared to be about six to 10 feet apart. I seemed to be passing through them at a speed of from three to five miles an hour. They were all around me, and those nearest the capsule would occasionally move across the window as if I had slightly interrupted their flow. On the next pass I turned the capsule around so that I was looking right into the flow, and though I could see far fewer of them in the light of the rising sun, they were still there. Watching them come toward me, I felt certain that they were not caused by anything emanating from the capsule. I thought perhaps I'd stumbled into the lost batch of needles the Air Force had tried to set up in orbit for communications purposes. But I could think of no reason why needles should glow like fireflies, nor did they look like needles. As far as I know, the true identity of these particles is still a mystery." (R6, R16)
If the "fireflies" truly came from Glenn's capsule, why did he think he was flying through an immense field of them spread out ahead of the capsule? And why are there not many similar observations from other manned space flights? (WRC)

X4. Modern visual sightings of possible ephemeral earth satellites. Transitory moving luminous objects resembling earth satellites are occasionally picked up by astronomers and other sky observers. As mentioned above most of these are seen so briefly that true satellite status cannot be confirmed and orbits computed; they thus end in in Category AEO, enigmatic objects. John P. Bagby, however, has tried to correlate some of these observations and translate them into evidence for temporary natural earth satellites. Data from three of his papers are summarized below.

In preparation for the launching of the first American satellites in the mid-1950s, the Smithsonian Institution established Project Moonwatch. Groups of amateur astronomers formed teams and set up special telescopes all across the country to help compute the orbits of the first artificial satellites. Scanning the sky frequently, the Moonwatch teams inevitably saw unidentified moving points of light. For example, in 1960 six Moonwatch teams from New York to California tracked a mysterious object across the continent. From the telescopic fixes it seems to have been a small natural satellite in retrograde orbit. Bagby has collected additional evidence for a family of such retrograde satellites orbiting beyond the moon. He also suspects that these objects may have produced the many observations of Vulcan, the lost intramercurial planet, recorded in the nineteenth century. (R14)

Next, quoting Bagby's 1966 paper in Nature: "During the mid-fifties, I analyzed evidence of two swarms of such objects in retrograde orbits. However, it has never been conclusively proved that such an ephemeral satellite, with a perigee in the upper atmosphere, has been observed while in orbit and before its final revolution. I now suggest that the Earth has had such a satellite and that it was observed at least eight times during the period 1956-65. Fortunately, these eight known observations were recorded with sufficient accuracy to permit the determination of an orbit. These observations are listed in Table I. The first three observations were accidentally made by Metcalfe and forwarded to me as possible evidence of an inner planet. The brilliance of the objects to the naked eye caused most persons contacted to doubt that they were astronomical despite the fact that this was confirmed with optical aids. However, when Kayser and I made the fourth observation (also with optical aid) I computed an approximate orbit, assuming all four observations to be of an Earth satellite. Continued surveillance over the next 60 days during every possible evening twilight period resulted in the fifth and sixth observations being made. The fifth observation was made with optical aid, but the sixth was by naked eye only. These further observations made it possible to improve the determination of the orbits, and osculating orbits were distributed to about twenty individuals and astronomical agencies. During most of 1965, no one else reported to me that they had seen the object. Then Hartmann and I made the seventh observation, and this further extended the orbit description. It also led to another period of evening twilight surveillance, which resulted in an eighth observation being made and further orbital refinement." Bagby believes the object may have reentered in January 1966. (R13)

Bagby's third set of visual observations of possible ephemeral satellites is presented in the accompanying table. Bagby believes that these are direct observations of the the parent satellite and its fragments discussed in AGL2-X1, where unaccounted for perturbations of artificial satellites are offered as evidence. (R10) Meeus, in his critique of Bagby's work, claims that the observations Bagby employed are insufficient for orbital computations. He ventures further that objects in the orbits Bagby claims would be bright enough near perigee to attract wide attention---which they didn't. (R12)

Table 1

Long.	Lat.	Date	Time	Azimuth	Elevation	Magnitude	Observer
87.7 W.	39.6 N.	Nov. 17	01:20	248	20		E. Metcalfe
		1956	01:33	242	0		E. Metcalfe
87.7 W.	39.6 N.	May 24	01:20	291	20		E. Metcalfe
		1957	01:37	291	0		E. Metcalfe
87.7 W.	39.6 N.	Nov. 10	23:30	287	40		E. Metcalfe
		1957	23:58	287	0		E. Metcalfe
118.5 W.	34.0 N.	Dec. 9	03:32	233	19		B. Kayser
		1964	03:35:40	260	3.5	3	J. P. Bagby
118.5 W.	34.0 N.	Dec. 29	02:20:39	250	12.5	2	J. P. Bagby
		1964	02:22:24	280	5	2	J. P. Bagby
118.5 W.	34.0 N.	Jan. 10	02:16:48	278.5	5.5	3	J. P. Bagby
		1965	02:16:56	280	5	3	J. P. Bagby
118.5 W.	34.0 N.	Oct. 25 1965	02:08	231	23	2	R. M. Hartmann
			02:11	281	4.5	2	J. P. Bagby
118.5 W.	34.0 N.	Dec. 14	02:57:09	303.5	30	3.5	J. P. Bagby
		1965	03:00:10	296	13	3.5	J. P. Bagby

References

R1. Proctor, Richard A.; "Intralunar Satellite," English Mechanic, 20:94, 1874. (X1)

R2. King, S.R.; "Has the Earth More Than One Satellite?" English Mechanic, 61: 350, 1895. (X1)

R3. "Earth's Natural Satellites," Science

Table 1. Direct Observations of Proposed Natural Satellites[a]

Observation:	I	II	III	IV	V	VI
Year	1947	1952	1956	1956	1957	1957
Date, UT	Sep 23 1	Jan 21	Jan 31	Jan 31	May 4	May 4
Universal Time (hr)	06.0 1.0	15.78	11.63	11.73	03.83	03.83
Latitude, N (degrees)	47.58	33.96	42.33	42.33	42.03	42.03
Longitude, W (degrees)	122.33	118.35	83.12	83.12	87.75	87.75
Right ascension (degrees)	304.5	302.76	180.51	180.56	194.3	197.5
Declination (degrees)	-24.7	-20.04	-5.55	-5.57	17.5	14.3
Position angle (degrees)	No record	57.8 5	45 8	45 8	Ambiguous	Ambiguous
Rate of motion (deg/min)	2.0 1.0	0.5 0.1	1.6 0.3	1.6 0.3	0.9 0.1	0.9 0.1
No. of objects	1	1	6	4	4	4
Aperture of instrument (cm)	8.9	10	15.2	15.2	8.5	8.4
Observer	J. M. Hammond	T. A. Cragg	D. Craig	D. Craig	J. P. Bagby	J. P. Bagby

[a] The position angle (of travel) is the stellar direction in which the objects appeared to be moving, where stellar "north" would be 0 degrees and stellar "east" would be 90 degrees, etc.

News Letter, 80:82, 1961. (X2)

R4. "New Natural Satellites of the Earth," Sky and Telescope, 22:10, 1961. (X2)

R5. "More about the Earth's Cloud Satellites," Sky and Telescope, 22:63, 1961. (X2)

R6. Ley, Willy; "What Was the Glenn Effect?" Science Digest, 52:5, July 1962. (X3)

R7. "'Rubble' Orbits Earth," Science News Letter, 90:232, 1966. (X2)

R8. Simpson, J. Wesley; "Dust-Cloud Moons of the Earth," Physics Today, 20:39, February 1967. (X2)

R9. Roosen, R. G., et al; "Doubt about Libration Clouds," Physics Today," 20:9, May 1967. (X2)

R10. Bagby, John P.; "Terrestrial Satellites: Some Direct and Indirect Evidence," Icarus, 10:1, 1969. (X4)

R11. Bagby, John P.; Terrestrial Satellites: Further Evidence and Possible Terrestrial Effects," Privately published report, 1973.

R12. Meeus, Jean; "Bagby's Phantom Moonlets," Icarus, 19:547, 1973. (X4)

R13. Bagby, John P.; "Evidence of an Ephemeral Earth Satellite," Nature, 211:285, 1966. (X4)

R14. Bagby, J.P.; "Natural Earth Satellites," British Interplanetary Society, Journal, 34:289, 1981. (X4)

R15. Valdes, Francisco, and Freitas, Robert A., Jr.; "A Search for Objects near the Earth-Moon Lagrangian Points," Icarus, 53:453, 1983. (X2)

R16. Glenn, John; "If You're Shook Up, You Shouldn't Be There," Life, 52:29, March 9, 1962. (X3)

AGL5 Radio Propagation and Natural Earth Satellites

Description. Periodic enhancements of radio signal strengths that can be correlated with the regular perigee appearances of hypothesized natural earth satellites.

Data Evaluation. A single series of regular enhancements of signal strength has been reported. (More cases may exist in the unexplored literature of radio propagation.) Rating: 3.

Anomaly Evaluation. Radio propagation anomalies of various sorts are rather common; and temporary natural earth satellites are not unreasonable, given the large population of debris in the solar system. Finally, meteor trails in the upper atmosphere are known to reflect radio waves. This phenomenon is in itself more of a curiosity than an anomaly, but if verified does support the existence of natural earth satellites. Rating: 3.

Possible Explanations. Ion mirrors created by the perigee passages of natural earth satellites through the upper atmosphere periodically enhance the propagation of some radio signals. Periodic variations in the height or density of the ionosphere might also explain the effect.

Similar and Related Phenomena. Radio whistlers caused by the reflection of radio waves from the ionized trails of meteors.

Examples

X1. Occasional occurrence. John P. Bagby, a hunter of natural earth satellites, has found several cases where anomalous enhancements of radio signals seem to have been caused by ion mirrors created by small satellites at the perigees of their orbits. For example, the high frequency timing signals broadcast by station CHU, in Ottawa, at 7.33 megahertz, were regularly enhanced at his receiver in Pacific Palisades, California in the early 1960s. Bagby thinks that the patterns of increased signal amplitudes can be best explained by assuming that a natural satellite created ion mirrors when it periodically dipped into the fringes of the upper atmosphere between Ottawa and Pacific Palisades. (R1)

References

R1. Bagby, John P.; "Radio Anomalies Associated with an Ephemeral Satellite Still in Orbit," Nature, 215:1051, 1967. (X1)

AGL6 Correlation of Geophysical Events with Perigee Passages of Natural Earth Satellites

Description. The synchronicity of geophysical events, such as earthquakes, volcanic eruptions, and variations in the Chandler Wobble, with close approches of natural earth satellites.

Data Evaluation. Although the dates of major geophysical events are well known, the data supporting the existence and orbital parameters of natural earth satellites is weak and not widely verified. Rating: 3.

Anomaly Evaluation. The ability of a dimensionally small natural earth satellite to trigger major geophysical events would be remarkable, for gravitational and other effects would appear to be negligible. Indeed, to overcome these kinds of objections, the analyst reporting this correlation had to invoke a mini-black hole in orbit to establish a cause-and-effect link. Rating: 2.

Possible Explanations. Orbiting mini-black holes gravitationally or electrostatically trigger earthquakes and other geophysical events.

Similar and Related Phenomena. Lunar tidal triggering of earthquakes and the terrestrial tidal triggering of moonquakes.

Examples

X1. Statistical analysis. Abstract. "At the 1972 and 1974 annual meetings, I presented evidence of correlations between the orbit elements of certain natural earth satellites and the Chandler Wobble's variation, and also with the times of triggering of earthquakes and volcano eruptions. The latter two correlations were of a 'last straw' type of influence on stresses already near the point of strain release. When the satellites are considered to consist of meteoric material, their angular size in transit across the sun or moon and their magnitude in the dark sky both suggest dimensions around 300 m. Although massive, this dimension has caused some to doubt that the satellites are large enough to cause the observed correlations. Additionally, the amount of electric charge implied by correlations given in Nature and Icarus, regarding ionospheric and other effects, has been criticized. All of these objections can easily be satisfied by considering some or all of the satellites to be mini-black holes, surrounded by ephemeral clouds of particles picked up during each perigee passage through the ionosphere. The apparent size is thus due to the cloud of particles and the radar cross section would vanish. This might explain the large number of visual and photographic observations relative to the few radar ones, as only two objects are carried regularly by the NASA Goddard 'Satellite Situation Report'. Recent confirmation of my orbit computational method and observational evidence has been obtained through one of the tracking agency's good offices. These and other reasons supporting the mini-black hole hypothesis are explored." (R1)

References

R1. Bagby, J. P.; "An Explanation for Certain Dochotomies Associated with the Observational Evidence for, and Terrestrial Effects of, Recently Discovered Natural Earth Satellites," Eos, 58:374, 1977. (X1)

AGO OBSERVATIONS OF EARTH FROM SPACE

Key to Phenomena

AGO0 Introduction

Observing earth from a great distance, as if it were one of the other solar system planets, is a relatively unexplored branch of astronomy. We assume that by living on the earth's surface and sending up satellites with cameras we see all there is to see. But we have seen in the cases of Mercury and Venus that what one sees through a distant telescope is often a bit different from what we see close up. Does the earth, for example, display a phase anomaly like Venus, or cusp blunting, or linear markings that do not seem to be correlated with "ground truth" surveys? Research in this area is almost nonexistent, with just one potential anomaly to contribute to the Catalog.

AGO1 Periodic Changes in Earth's Brightness

Description. A 6-day variation in the brightness of the earth as measured by spacecraft instruments.

Data Evaluation. An experiment on the Mariner 2 spacecraft bound for Venus. Rating: 3.

Anomaly Evaluation. A daily variation in the earth's brightness, when seen from afar, is easily attributable to the rotating cloud cover; but there is no obvious explanation for a 6-day period in albedo. Rating: 3.

Possible Explanations. An unappreciated cyclic change in cloud cover, associated perhaps with the natural generation of storm systems at certain times of the year.

Similar and Related Phenomena. None.

Examples

X1. Mariner 2 data, 1962. "Mariner 2 kept an electronic eye on the earth as it flew past Venus back in 1962. In order to keep its orientation in space, the satellite eye reported via radio the brightness of the earth's image at increasing distances. Meaningful signals came back to earth for 52 days in the middle of the flight. Analysis of this data revealed two cycles in the earth's brightness. One cycle is daily, the other requires five or six days. In the analysis which was published in the Journal of Geophysical Research, Robert L. Wildey of the Mount Wilson and Palomar Observatories in California says there is no obvious explanation for that six-day cycle." (R1)

References

R1. "Our Blue Planet's Light Rhythm," Cycles, 24:143, 1972. (X1)

AH INTRODUCTION TO MERCURY

Key to Categories

Mercury is an unusual planet in many ways. Its small size and heavily cratered surface give it a distinctly moon-like appearance. Yet, its high density (5.44) is radically different from that of our moon (3.34), rather it is more like that of earth (5.53). Mercury's orbit is scarcely better behaved. Only Pluto's orbit is more eccentric and more highly inclined. These facts suggest to some that Mercury may not have been formed in the orbit it now occupies. The curious 2:3 resonance between Mercury's yearly revolution around the sun and its axial spin only strengthens the suspicion concerning the planet's pedigree. One of the greatest surprises about Mercury has been its apparently intrinsic magnetic field. It is the only inner planet besides the earth to have a self-generated magnetic field. Given Mercury's small size and low density, most astronomers dismissed the possibility of a magnetic field before the 1974 Mariner 10 flyby.

With Mariner 10's close-up photographs and experimental data, few scientists attend to the many telescopic anomalies of Mercury, which keep on accumulating. Over the years, Mercury has displayed---to some observers at least---blunted cusps, irregular terminators, spots, systems of lineaments, and a ring of light around its dark limb. And during transits, bright spots and halos appear. These observations often have highly subjective content, because some astronomers see phenomena that others cannot. No wonder modern planetologists prefer to work with the sharp pictures and hard numbers from the Mariner 10 flight. Through the telescope, one requires an experimental psychologist to help unravel what is real and what is illusory.

AHB MERCURY'S ORBITAL AND SPIN ANOMALIES

Key to Phenomena

AHB0 Introduction

Mercury is a wayward planet. Its orbit is uncomfortably eccentric; its spin is locked in a curious way to its yearly voyage around the sun; and its perihelion advances at a rate that is at variance with Newtonian celestial mechanics and, possibly, relativistic celestial mechanics, too. Most critical to science is Mercury's potential for undercutting the General Theory of Relativity, in contrast to its present fame as one of the "convincing proofs" of Einstein's famous theory. Beyond this, Mercury's orbital anomalies may indicate past solar system upheavals involving our moon and perhaps Venus.

AHB1 The Residual Advance of Mercury's Perihelion

Description. The difference between the observed advance of Mercury's perihelion and the value predicted by prevailing theory; i.e., the General Theory of Relativity.

Background. That a discrepancy exists between observations of Mercury's perihelion advance and that predicted by Newtonian celestial mechanics has been recognized for over 150 years. The widely published value of 43.0 arcseconds per century for the discrepancy or residual is believed by most scientists to be fully and triumphally accounted for by Einstein's Theory of General Relativity. Recent measurements of solar oblateness (departure from a uniform sphere) substantially alter the value of the residual of 43.0 arcseconds per century, which was computed assuming a perfectly spherical sun. In other words, the additional advance in Mercury's perihelion caused by an oblate sun changes the value of the residual so much that General Relativity's prediction may be far off the mark. Newtonian celestial mechanics might, in fact, completely explain Mercury's orbit around an oblate sun!

Data Evaluation. The modern value of 574.1 arcseconds per century for the advance of Mercury's perihelion is considered very accurate. The measurements of solar oblateness, however, are still highly controversial. Since these are crucial in determining whether a real anomaly exists here, the overall rating for the data is low. Rating: 3.

Anomaly Evaluation. If solar oblateness is large enough to change the residual so that the predictions of the General Theory of Relativity are far off the mark, one of the major supporting pillars of General Relativity will be shattered. Rating: 1.

Possible Explanations. Solar oblateness may be sufficiently great to explain the residual advance of Mercury's perihelion with Newton's Law of Gravitation.

Similar and Related Phenomena. All of the planets, especially the inner ones, display perihelion advances with residuals, although they are less than Mercury's.

Examples

X0. General observations. The discrepancy between observed and computed values for the advance of Mercury's perihelion has been explained in several ways in addition to relativistic effects. (R4) Some of these are now listed as a matter of general interest: (1) The exponent of distance in Newton's Law of Gravitation differs slightly from -2; viz., -2.0000001574. (R1-R3); (2) The magnetic field of the sun is the cause of the residual. (R1); (3) Perturbations of an intramercurial planet or dust ring. (R3, R9; see AEI0.); and (4) Electrostatic force between the sun and Mercury in effect changes the distance exponent in Newton's Law of Gravitation. (R3)

X1. Permanent feature. Solar oblateness is sufficient to reduce the residual to a value significantly below that accounted for by the General Theory. In 1967, R. Dicke of Princeton observed solar flattening which would reduce the residual to 39.6 arcseconds per century. (R5, R6) Even this change of 3.4 arcseconds is considered a serious blow to the General Theory. (WRC)

X2. Permanent feature. Measurements of the sun's flattening by P. Goode and colleagues at the University of Arizona indicate that the prediction of the General Theory is off by about 1%---but this is felt to be significant. (R7)

X3. Permanent feature. Using estimates of the sun's internal rotation based upon global solar oscillations, the quadrupole moment of the sun is large enough to explain Mercury's perihelion advance more accurately than the General Theory of Relativity. (R8)

References

R1. Mumford, N.W.; "Mercurial and Lunar Perturbations," Popular Astronomy, 18: 319, 1910. (X0)

R2. Turner, Arthur B.; "Two Old Unsolved Problems of Astronomy," Royal Astronomical Society of Canada, Journal, 7: 276, 1913. (X0)

R3. Gerjuoy, E.; "Feasibility of a Nonrelativistic Explanation for the Advance of th the Perihelion of Mercury," American Journal of Physics, 24:3, 1955. (X0)

R4. Morton, Donald C.; "Relativistic Advances of Perihelions," Royal Astronomical Society of Canada, Journal, 50:223, 1956. (X0)

R5. Holcomb, Robert W.; "Experimental Tests of Relativity," Science, 169:40, 1970. (X1)

R6. Trefil, James S.; "Was Einstein Wrong?" Astronomy, 8:67, January 1980. (X1)

R7. "Reputed Mistake Found in Einstein's Theory of Relativity," Baltimore Sun, April 6, 1982, p. A9. (X2)

R8. Campbell, L., et al; "The Sun's Quadrupole Moment and Perihelion Precession of Mercury," Nature, 305:508, 1983. (X3)

R9. "Mercury's Perihelion: From Le Verrier to Einstein," Sky and Telescope, 66:220, 1983. (X0)

AHB2 Mercury's High Eccentricity and Inclination

Description. Mercury's orbital eccentricity is 0.206; its inclination to the plane of the ecliptic is 7°. These values, especially the eccentricity, are anomalous in the inner solar system. Only Pluto has a higher eccentricity, and it is thought by many to be an escaped satellite of Neptune rather than a bona fide planet.

Data Evaluation. Orbital data for Mercury are very accurate. Rating: 1.

Anomaly Evaluation. Except for Mercury and Pluto, the other planets have orbits of low eccentricity and low inclination. Some astronomers therefore conjecture that Mercury's orbit was somehow disturbed in the past, breaking the pattern it once shared with the other planets. This conjecture is grounded in the belief that Mercury was formed in the same way and at the same time as the other planets (except Pluto). Since Mercury's orbit implies such substantial solar system disturbance, the anomaly rating here is high. Rating: 2.

Possible Explanations. (1) Mercury could have condensed in its present orbit, although most theories of solar system formation do not favor orbits with high eccentricities and inclinations; (2) Mercury could be an interloper from another part of the solar system that was captured in its present orbit; (3) Mercury's orbit could have been disturbed when the earth captured the moon from a position between Mercury and Venus (R1) or inside Mercury's orbit (R2); (4) Mercury was once a satellite of Venus and escaped into solar orbit as a consequence of tidal interaction with Venus (R3).

Similar and Related Phenomena. Pluto's orbital eccentricity and inclination are also high and are taken as an indicator of past disturbance in the outer solar system.

Examples

X0. Permanent feature. Mercury's orbital elements are to be found in many reference works. The figures quoted above come from R4.

References

R1. Bailey, J. Martyn; "The Moon May Be a Former Planet," Nature, 223:251, 1969. (X0)

R2. Cameron, A. G. W.; "Orbital Eccentricity of Mercury and the Origin of the Moon," Nature, 240:299, 1972. (X0)

R3. Van Flandern, Thomas C., and Harrington, Robert S.; "A Dynamical Investigation of the Conjecture That Mercury Is an Escaped Satellite of Venus," Icarus, 28:435, 1976. (X0)

R4. "Mercury," McGraw-Hill Encyclopedia of Science and Technology, 8:292, 1977. (X0)

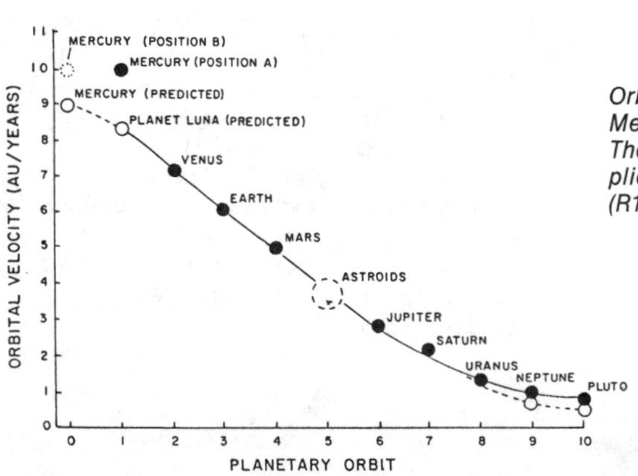

Orbital velocities of the planets, illustrating how Mercury defies the trend at A, its present position. The fit is better if Mercury is located at B, which implies the past existence of another planet (luna). (R1)

AHB3 Short Transit Times across the Sun

Description. Shorter than computed solar transit times.

Data Evaluation. This phenomenon has been reported for only one transit, although many observers confirmed the anomaly. It should be remembered in this context that the times of transit commencement and termination are very difficult to pin down precisely. Rating: 3.

Anomaly Evaluation. Given our precise knowledge of Mercury's orbit and the sun's diameter, large discrepancies in transit times are very surprising, providing they cannot be written off as due to observational errors. Rating: 2.

Possible Explanations. Observational errors in timing.

Similar and Related Phenomena. The so-called "black drop" phenomenon (AHX0), which illustrates the difficult of timing transits.

Examples

X1. November 11-12, 1940. Over 200 observers cooperated in studying the transit of Mercury. "They found that the transit happened 36 seconds later than expected and lasted 18 seconds less. The tardiness is assumed to be caused by irregularities in the rotation of the earth, which lead to errors in the reckoning of astronomical time. The accuracy of the observed time deviation in this instance could be checked by observations of the moon, which yielded a value of the 'fluctuation' of 35.3 seconds, in good agreement with the Mercury transit value. The shortness of the transit is not wholly accounted for. Observational difficulties may readily account for part of Mercury's apparent haste, but the discordance between observation and theory is too large to be entirely attributed to this cause." (R1)

References

R1. Hoffleit, Dorrit; "Tardy Mercury Hurried," Sky and Telescope, 1:17, August 1942. (X1)

AHB4 Mercury's Spin Resonance

Description. The sidereal period of Mercury's revolution around the sun is 87.969 earthdays, whereas the planet's spin period is 58.646 days, exactly 2/3 of its sidereal year.

Background. Mercury's spin period was long recorded in the textbooks as 88 days; that is, a 1:1 resonance with its revolution around the sun, like the earth's moon in its motion around the earth. Details of Mercury's surface are difficult to follow through the telescope. Most observers who thought they caught glimpses of planetary detail proclaimed that Mercury always kept the same face toward the sun. Because Mercury orbits so close to the sun, this situation seemed reasonable and was universally adopted. The 88-day period was not questioned until 1965, when radar measurements found the spin to actually be 58.646 days.

Data Evaluation. The radar measurements of Mercury's spin period are considered very reliable and accurate. Rating: 1.

Anomaly Evaluation. Ordinarily, astronomers would expect a 1:1 resonance rather than the observed 2:3 resonance. However, the seriousness of the situation is mitigated by Mercury's highly eccentric orbit and the possibility that the planet may be a prolate rather than an oblate sphere. Rating: 3.

Possible Explanations. Mercury's anomalous orbit and possible prolate shape may be sufficient to account for the 2:3 resonance, although no proofs of this have been found as yet.

Another possibility is that Mercury is a recently escaped satellite of Venus. Calculations indicate that at its hypothetical escape from Venus Mercury's spin period would be something over 40 days---not too different from its present spin period. (R1)

Similar and Related Phenomena. The resonance of Venus with the earth (AVB1) and the many other solar system resonances (AB); Mercury's possible prolate shape (AHE2).

Examples

X0. Permanent feature. Mercury's spin period of 58.646 days came from R2.

X1. Permanent feature. In 1927, the famed planetary astronomer E. M. Antoniadi completed a long series of Mercury observations, particularly the motion of the markings. His conclusion: Mercury definitely has a period of 88 days, thus confirming the results of many other astronomers. (R3) Even the best astronomers have had their problems with Mercury's elusive markings. (WRC)

References

R1. Van Flandern, Thomas C., and Harrington, Robert S.; "A Dynamical Investigation of the Conjecture That Mercury Is an Escaped Satellite of Venus," Icarus, 28:435, 1976.

R2. "Mercury," McGraw-Hill Encyclopedia of Science and Technology, New York, 8:292, 1977. (X0)

R3. "Mercury Has a 2112-Hour Day," Science News Letter, 12:217, 1927. (X1)

AHE GEOLOGY AND FIGURE OF MERCURY

Key to Phenomena

AHE0 Introduction

Mercury is a difficult object through the telescope. It is small and hugs the sun. Planetary surface markings are vague at best. Before March 1974, when the spacecraft Mariner 10 swept past Mercury taking close-up pictures and making other measurements, astronomers knew virtually nothing of Mercury's geology. In fact, the surface markings are so poorly defined that those who studied them declared, almost-to-a-man, that Mercury turned once on its axis every 88 days. The correct figure is about 59 days. Obviously, Mercury's great distance and the vicissitudes of the terrestrial atmosphere also hid the fact that Mercury has a heavily cratered surface---an asymmetrically cratered surface at that. Mercury's craters were a great surprise to almost everyone.

It is pertinent to mention early in this book, which deals heavily in elusive phenomena, that some astronomers see planetary markings that other astronomers, equally capable, cannot discern at all. Sometimes these controversial apparitions turn out to be physically real, sometimes not. In the case of Mercury, the rather colorful astronomer T. J. J. See reported seeing craters on Mercury in June 1901. His drawing resembles the photos Mariner 10 snapped as it flew past 73 years later. Until Mariner 10, See's observation was never taken seriously. It was well-known that the surface details of Mercury were beyond the reach of telescopic observation. The precocious observations of Martian craters were pooh-poohed for similar reasons. (R1-R3)

References

R1. See, T. J. J.; "Origin of the Lunar Terrestrial System by Capture, . . . ," Popular Astronomy, 18:106, 1910.

R2. Young, Andrew T.; "Mercury's Craters from Earth," Icarus, 34:208, 1978.

R3. Baum, Richard; "Historical Sighting of the Craters of Mercury," Strolling Astronomer, 28:17, 1979.

AHE1 The Asymmetry of Mercury's Topography

Description. One hemisphere of Mercury is characterized by the presence of large basins, similar to the lunar mares, but the opposite hemisphere is devoid of these features and pockmarked with many craters.

Data Evaluation. High-quality, close-up pictures from the Mariner 10 1974 flyby. Rating: 1.

Anomaly Evaluation. The asymmetry of Mercury's topography calls attention to the postulated ancient bombardment of the inner solar system by meteors that somehow afflicted only one hemisphere of Mercury, Mars, our moon, and perhaps the earth as well. Since the asymmetric bodies occupy different locations in the solar system and possess radically different rates of rotation on their axes, the nature of the hypothetical bombardment is difficult to model. If it lasted for more than a few hours, rapidly spinning Mars would have been uniformly cratered. Rating: 2.

Possible Explanations. If Mercury had been a moon of Venus during the hypothetical meteor bombardment, Venus might have shielded it from the bulk of the meteor flux. The same argument is employed to explain the moon's asymmetry. (R2-R5)

Similar and Related Phenomena. The asymmetric hemispheres of the moon (ALE1) and Mars (AME1).

Examples

X1. Permanent features. A description based on the Mariner 10 flyby. "The terrain viewed on the inbound segment resembles lunar highlands and is marked by an absence of large basins, Gault said. There is some evidence of faulting or other deformational activity, but not of terrestrial, global plate motion. No evidence of Mars-like shield volcanism was seen during the Mariner 10 flyby.

Outbound, Mariner 10 saw a totally different terrain. There is a strong suggestion of albedo variation, and at least three large circular basins are visible, the largest about 1,300 km. (806 mi.) in diameter, on the scale of the moon's Mare Imbrium. The smallest of the three, approximately 500 km. (310 mi.) across, is peculiarly marked by a dark ring that surrounds the lighter center material. There definitely is mare-like filling in the largest basin, suggesting flooding, Gault said. Mare-like material extends about 1,000 km. (620 mi.) beyond the basin rim.

Two prominent ray craters were observed on the outbound flight---one in the northern hemisphere, surrounded by a dark halo, which Gault compared to the moon's Tycho, and one in the southern hemisphere, surrounded by a bright halo.

The terrain differences observed on Mercury are comparable to the differences observed between the near and far sides of the moon, the northern and southern hemispheres of Mars, and even the Atlantic and Pacific basins on earth, Gault said. These phenomena indicated that planets---or at least the inner, or terrestrial ones---tend to evolve asymmetrically. The formation of large basins also seems to be a basic phenomenon of planetary formation, Gault said." (R1)

References

R1. Elson, Benjamin M.; "Planetary Formation Ideas Questioned," Aviation Week, April 15, 1974, p. 17. (X1)

R2. Van Flandern, Thomas C., and Harrington, Robert S.; "A Dynamical Investigation of the Conjecture that Mercury is an Escaped Satellite of Venus," Icarus, 28:435, 1976.

R3. "Was Mercury a Moon of Venus?" New Scientist, 71:641, 1976.

R4. Metz, William D.; "Mercury; More Surprises in the Second Assessment," Science, 185:132, 1974.

R5. Murray, Bruce C., et al; "Mercury's Surface: Preliminary Description and Interpretation from Mariner 10 Pictures," Science, 185:169, 1974.

AHE2 Mercury's Possible Prolate Shape

Description. Mercury is a prolate sphere; i.e., football-shaped.

Data Evaluation. No direct measurements of Mercury's prolateness have been found. Rather, this unusual geometry is inferred from the existence of the 2:3 resonance between the planet's spin period and the length of its year. (AHB4) This type of resonance, rare among the denizens of the solar system, is favored by objects with certain moments of inertia, such as that possessed by a prolate sphere. (R1) Rating: 3.

Anomaly Evaluation. All rotating, planet-sized objects should respond to centrifugal force and be slightly flattened at the poles; that is, they should be oblate rather than prolate spheres. Since Mercury may not conform to this expectation, it may have a different origin than the rest of the planets. Rating: 2.

Possible Explanations. Mercury might have been a close satellite of some other, much larger planet (Venus?). Locked in a tight orbit with one face perpetually facing this planet, its shape could have been gravitationally drawn out into a prolate shape.

Similar and Related Phenomena. None.

Examples

X0. Permanent feature. Mercury's prolateness is only an inference. See Data Evaluation above. (R1)

References

R1. Van Flandern, Thomas C., and Harrington, Robert S.; "A Dynamical Investigation of the Conjecture that Mercury is an Escaped Satellite of Venus," Icarus, 28:435, 1976. (X0)

AHE3 Swirl Markings

Description. A variety of ribbon-like patterns, open loops and closed loops, and swirls of bright and dark material. The swirls may be nearly 50 kilometers across.

Background. Swirl patterns were first identified on the moon (ALE5), where they are associated with concentrations of remanent magnetism. (ALZ3)

Data Evaluation. The photographs of Mercury's surface do not have the high resolution of those taken of the moon. Nevertheless, the gross characteristics of Mercury's swirls are very similar to those on the moon. No magnetic exploration of Mercury's surface has been made, so we do not know if this planet's swirls are colocated with magnetic patches. Rating: 2.

Anomaly Evaluation. Swirl patterns are rare in geology; and generally indicate fluid action of some kind---liquid, gaseous, magnetohydrodynamic. Even with this hint, the origin of the lunar swirls and those analogous structures on Mercury remains enigmatic. Rating: 2.

Possible Explanations. The magnetohydrodynamic action of a colliding comet is the most popular explanation of the moon's swirls. Other suggestions are: nue ardente deposits, volcanic sublimates, secondary impact effects, albedo effects modified by local magnetism, antipodal effects of impact basin formation.

Similar and Related Phenomena. Lunar swirl patterns. (ALE5)

Examples

X1. Permanent features. In reference to lunar swirl patterns: "Similar patterns are recognized on Mercury near latitude 20°N, 47°W and 4°N, 35°W. Because different phase angles and high-resolution images of these regions are unavailable, it is impossible to verify this identification in detail. Nevertheless, the bright loops and swirls noted at these locations are very similar to the gross patterns found on the moon." (R1)

References

R1. Schultz, Peter H., and Srnka, Leonard J.; "Cometary Collisions on the Moon and Mercury," Nature, 284:22, 1980. (X1)

AHF INTRINSIC RADIATION SOURCES

Key to Phenomena

AHF0 Introduction

Mercury is seen by reflected sunlight. If intrinsic sources of visible light exist, no one has reported them. But Mercury's magnetic field, which apparently is derived from an active core dynamo, suggests that volcanic action and lava flow, along with the visible radiation they might produce, are not impossible. Mercury's magnetosphere, detected by Mariner 10, may generate radio signals; but, here too, nothing has been detected at earth. The only possible intrinsic source of radiation (other than infrared) was a fleeting ultraviolet source detected by Mariner 10. This observation is remarkable and rather suspect. If confirmed, it would certainly be an important anomaly.

AHF1 Extreme-Ultraviolet Emissions from Mercury

Description. Extreme-ultraviolet radiation emitted by Mercury.

Data Evaluation. A single observation of a transitory nature. Rating: 3.

Anomaly Evaluation. No natural sources of extreme-ultraviolet radiation seem likely on Mercury's surface or in its very tenuous atmosphere. Rating: 2.

Possible Explanations. Instrumental error, or perhaps a strong stellar ultraviolet source in the same direction as Mercury.

Similar and Related Phenomena. None.

Examples

X1. March 27, 1974. Two days before Mariner 10 flew past Mercury, one of the ultraviolet detectors on the spacecraft began registering strong radiation at short ultraviolet wavelengths coming from Mercury's dark side. The next day, the emissions had disappeared. However, after the flyby, when the instruments were looking back at Mercury, similar ultraviolet radiation was detected. The latter signals were ultimately attributed to a star located in almost the same direction as Mercury. However, the first observations have never been explained, except in terms of possible instrument malfuntion. (R1, R2)

References

R1. "Mercury's Moon That Wasn't," New Scientist, 63:602, 1974. (X1)

R2. Broadfoot, A. L., et al; "Mercury's Atmosphere from Mariner 10: Preliminary Results," Science, 185:166, 1974. (X1)

AHO ANOMALOUS TELESCOPIC OBSERVATIONS

Key to Phenomena

AHO0 Introduction

Telescopic Mercury is quite different from the planet photographed by the space probe Mariner 10. Rather than a finely etched profusion of craters, the telescope reveals only a tiny crescent of light shimmering in the eyepiece. The surface features are vague; some people see markings that others cannot no matter how hard they study the planet. Chief among the telescopic anomalies are Mercury's blunted southern cusp and terminator irregularities. These anomalies come and go, tantalizing the observer at the limits of his seeing. Certainly the Mariner 10 pictures show no planetary features that can be associated with the cusp and terminator phenomena. Even more puzzling are the systems of lineaments claimed by a few observers of Mercury. Like the Martian canals, these lines are simply not on the close-up photos. The anomalies in this section are all frustrating to the telescopic observer. Perhaps some or all of them are illusory; that is, subjective in the main. The reader will find that all of the planets and the earth's moon as well are plagued like Mercury with elusive phenomena that depend on good seeing, a good eye at the telescope, and not infrequently a touch of imagination.

AHO1 Mercury's Occasional Blunted Cusps

Description. The frequent blunting of Mercury's southern cusp. The amount of blunting varies; and not all observers perceive the phenomenon equally. In contrast, the northern cusp is invariably sharp. Cusp blunting is often associated with terminator irregularities (AHO2) and may actually be a localized example of this more general phenomenon.

Background. Some nineteenth century astronomers thought the blunting to be periodic and therefore a measure of the length of Mercury's day. From careful observations of the blunting's waxing and waning, Mercury was estimated to spin once on its axis every 24 hours, 50 minutes. (R1) The modern, radar-derived figure is about 59 days.

Data Evaluation. Many observations of Mercury's blunted southern cusp, including photos, are on file (R4). We have here an objective phenomenon in which visual acuity and seeing conditions play key roles. Apparently, not all astronomers can discern the blunting. Rating: 1.

Anomaly Evaluation. The frequent blunting is almost certainly an optical effect, involving shadows and/or surface reflectivity. In this sense, blunting is a curious but rather mild anomaly. The Mariner 10 pictures give no hint as to why the southern polar region might be different from the rest of the planet. Rating: 3.

Possible Explanations. Since Mercury also displays terminator anomalies and brightness variations outside the south polar region, it is probable that some unappreciated reflectivity or perhaps fluorescent factors are to blame. Beyond these thoughts the literature is mute. Many observers comment that the region surrounding the blunted cusp is unusually bright; so contrast phenomena cannot be ruled out.

Similar and Related Phenomena. Terminator irregularities (AHO2); blunted cusps of Venus (AVO1-X2).

Examples

X1. Frequent occurrence. April 19, 1864: southern horn blunted, northern horn singularly sharp. Early May 1923: blunted southern cusp associated with terminator irregularities; viz., an extremely bright bulge near the southern cusp, making the entire terminator look like a reversed S. (R3)

X2. General observations. "For many years observers have reported that the south cusp of the crescent Mercury occasionally appears blunted as opposed to the usually sharp north tip. Schroeter noted this in 1800 and 1801; and, according to Antoniadi, the effect was confirmed by Burton, Noble, Franks, Trouvelot, Denning, and Antoniadi. The Jarry-Desloges observers (V. and G. Fournier) also noted this aspect from time to time, as well as small terminator deformations. Of this phenomenon, Schiaparelli says, 'I must confess...that I have always seen the entire southern cusp very well though its light was weaker (than the north); and only once (June 5, 1882) I found that cusp so little luminous, that from time to time, during the least distinct moments of vision, one could suppose it truncated. That unequal splendor of the polar regions is the real cause of the apparent truncation of the southern cusp...' Because of the

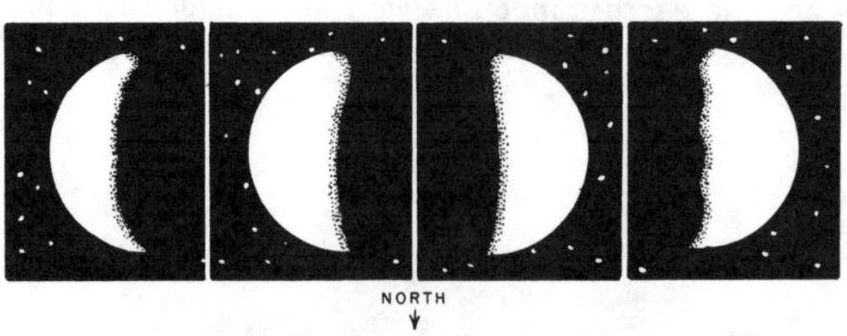

Four drawings of Mercury, all showing blunted southern cusps. Some terminator irregularities are also shown. (X2)

brightness diminution at the cusps approximately according to the Lambert cosine law of diffuse reflection, usually imperfect seeing conditions, and general low surface surface brightness of the image, one expects to find the cusp tips on the limit of visibility." (R4)

X3. 1963."Fig. 1 (not reproduced) shows four drawings of Mercury made in a nine-day period in 1963. Drawing A was made with a 36-in. reflector and the remainder with a 12-in. reflector, all in fairly good seeing conditions. Between the first and last drawings in Fig. 1, a darkening at the south cusp is apparent. Later drawings made as the crescent waned show the south cusp darkening further until it was the darkest region on the planet." (R5) Solar-wind-induced fluorescence was suggested as the cause of these variations.

X4. March 20, 1912. "There was a very brilliant area extending some distance along the limb from the S. cusp, which appeared to project considerably, the north one being normal." In the accompanying drawing, the terminator forms a forward S, with the southern cusp extended abnormally. The terminator line was displaced appreciably from the theoretical terminator in the direction of the sun-lit side. (R2) This observation is obviously the opposite of cusp-blunting, underscoring the great variability of reflection from Mercury's surface.
See the preceding illustration, sketch 2, for the more usual reversed-S terminator phenomenon. (WRC)

References

R1. "Mercury," English Mechanic, 10:397, 1870. (X1)
R2. McHarg, John; "Mercury," English Mechanic, 95:205, 1912. (X4)
R3. McHarg, John; "Mercury," English Mechanic, 117:205, 1923. (X1)
R4. Cruikshank, Dale P.; "Mercury, Part I. The Blunted Cusp Effect and Terminator Irregularities," Strolling Astronomer, 17:129, 1963. (X2)
R5. Cruikshank, Dale P.; "Possible Luminescence Effects on Mercury," Nature, 209:701, 1966. (X3)

AHO2 Terminator Irregularities

Description. Bulges, cavities, and sundry irregularities along Mercury's terminator (the line separating night from day). Mercury also displays occasional departures from theoretical phase; that is, the entire terminator leads or lags the theoretically computed line dividing night from day. The planet's frequently blunted southern cusp is treated separately (AHO1), although it is a terminator irregularity in a sense.

Data Evaluation. Few papers on terminator irregularities have been uncovered in the scientific literature if the blunted cusps are excluded. However, such "miscellaneous" effects are rarely reported formally, even though noticed rather regularly. For example, the records of the Association of Lunar and Planetary Observers (ALPO) contain about as many cases of terminator irregularities as blunted cusps. (R4) Rating: 1.

Anomaly Evaluation. As in the blunted-cusp phenomenon (AHO1), surface reflectivity and fluorescence may play important roles. Surface reflectivity may vary with the sun-Mercury-earth angle. Fluorescence stimulated by the solar wind would change with solar activity. Attractive as these possibilities are, they are merely guesses. Rating: 3.

Possible Explanations. See above.

Similar and Related Phenomena. Mercury's blunted cusp (AHO1); the phase dichotomy of Venus (AVO4).

Examples

X1. April 19, 1864. A gentle reversed-S terminator with a blunted southern cusp. (R1)

X2. March 20, 1912. Significant departure of actual phase from theoretical phase. See AHO1-X4. (R2)

X3. Early May 1923. A gradual change over a period of five days from a nearly straight terminator to a reversed-S. (R3)

X4. General observations. Many deformed terminators recorded in the records of the Association of Lunar and Planetary Observers. Several drawings from these records are shown in the illustration in AHO1. (R4)

References

R1. "Mercury," English Mechanic, 10:397, 1870. (X1)

R2. McHarg, John; "Mercury," English Mechanic, 95:205, 1912. (X2)

R3. McHarg, John; "Mercury," English Mechanic, 117:205, 1923. (X3)

R4. Cruikshank, Dale P.; "Mercury, Part I. The Blunted Cusp Effect and Terminator Irregularities," Strolling Astronomer, 17:129, 1963. (X4)

AHO3 White Projections and Spots

Description. White spots on the golden, sunlit side of Mercury. Sometimes these spots project beyond the planet's disk. The shapes and positions of the spots change slowly.

Data Evaluation. A single series of observations by an amateur astronomer. Rating: 3.

Anomaly Evaluation. Similar spots appear on Mars, where they are readily identified as clouds. Given Mercury's very thin atmosphere, clouds are most unlikely. Rating: 2.

Possible Explanations. None.

Similar and Related Phenomena. White spots and projections on Mars.

Examples

X1. July 17-25, 1885. A series of observations from Rolfontein, South Africa. See illustration. "July 17. Observed Venus and Mercury soon after sunset, before 'glow' appeared, and as soon as eye settled to eyepiece, I thought to see a small white, horn-like point, near or on bright limb of Mercury, some 30° from (inverted) south cusp on terminator. Thinking it an illusion, I observed bright limb all around, but no other similar object was seen; slightly averted vision also always distinctly revealed it. A more careful examination placed it at about 55° from (inverted) south cusp. A few minutes later fancied to see a whitish patch deeper on bright limb, and perhaps a little further S. (inverted); it narrowed towards bright limb, and widened slightly on the bright disc. Planet's colour a reddish golden yellow." Similar observational notes follow for six more days, during which the spot changed shape. (R1, R2)

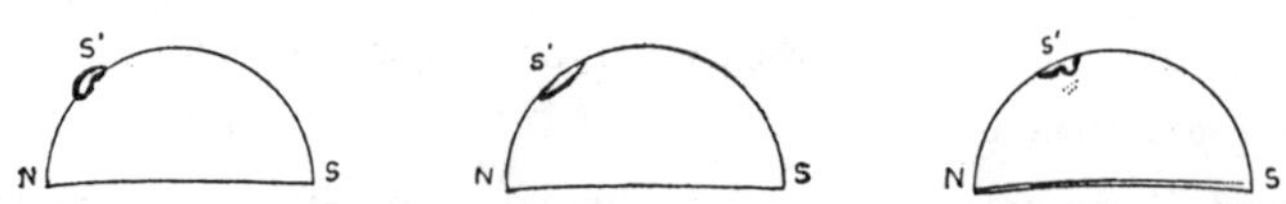

Series of sketches showing the development of a white spot on July 17, 19, and 21, 1885. (X1)

References

R1. Ballot, John; "Suspected White Markings on Mercury," English Mechanic, 42:139, 1885. (X1)

R2. Ballot, John; "Suspected White Markings on Mercury," English Mechanic, 42:199, 1885. (X1)

AHO4 Dark Linear Markings

Description. Long, dark streaks running in various directions and often intersecting in knot-like spots.

Background. Some nineteenth and early-twentieth century astronomers claimed that they saw networks of streaks and lines on Mars and many other solar system objects. Giovanni Shiaparelli and Percival Lowell are the best-known of the streak observers.

Data Evaluation. Apparently only Shiaparelli, Lowell, and Poor (also at Lowell Observatory) discerned extensive networks of lineaments on Mercury. A few others saw vague streaks; most observers saw no linear features. One cannot contest that linear patterns were thought to have been seen on Mercury, for the honesty and ability of the observers is beyond reproach. Whether the lines are physically real is another question. See below. Rating: 1.

Anomaly Evaluation. Systems of lineaments have been reported on several planets by many honest, qualified astronomers; but they are not verified by most other observers and, even more importantly, by spacecraft. This is an important perceptual anomaly, because science is based on all observations being accessible to all normal people. Rating: 1.

Possible Explanations. A slight possibility exists that Mercury possesses some subdued surface patterns that some people can discern while others cannot. Mercury is a difficult telescopic object; and observer's eyes do vary considerably when it comes to slight shadings and patterns. But it seems more likely that the lineaments are generated mentally; that is, they are illusory. Why some people see such apparitions is an unanswered question---a psychological anomaly.

Similar and Related Phenomena. The Martian "canals" (AMO1); Venus' "spoke marks" (AVO6).

Examples

X1. General observations. "Schiaparelli has discovered many marks upon the disk of the planet which had not been noticed before, and he has made a little map or diagram which shows that these marks strikingly resemble some of the features discovered in recent years on Mars. They are elongated streaks running in various directions, and frequently presenting at their points of junction the appearance of an enlargement or knot. Similar streaks on Mars have been assumed to be long narrow seas or watercourses. The geometrical figures formed by the intersection of these streaks on Mercury strikingly resemble those on Mars. In one place there is a shape of this kind that roughly resembles a huge figure 5, covering a quarter of one hemisphere of the planet." (R1)

X2. September 23, 1896. "In 1896 Professor Lowell published in the Monthly Notices (vol. lvii, p. 148) a series of drawings of Mercury, showing extremely curious markings on the planet, which he speaks of as much more distinct than the Martian canals, and, in fact, quite easy to see. They appear to have been observed when the planet's surface was not more than half illuminated, and the drawings show them as linear in nature, though otherwise they bear little resemblance to the canals of Mars. I give as an example Professor Lowell's drawing of September 23, 1896, in which the markings appear mostly as stripes projecting from the terminator or running parallel with it. There

Surface markings on Mercury as seen by Percival Lowell. (X2)

is also a slight shading off of the cusps." (R2)

X3. General observations. A drawing of Mercury prepared by Professor Poor, at Lowell Observatory, is reproduced. It portrays many long and slightly arcuate lines, many intersecting in common junctions. "It will be noticed that many of the markings on Mercury---their straightness and the oases---have a striking resemblance to those on Mars. In some cases the lines are exactly parallel to the parallels of latitude, and equally distant." (R3) Drawing not reproduced. (WRC)

References

R1. "Wonderful News about Mercury," Scientific American, 62:211, 1890. (X1)
R2. Wesley, W. H.; "Markings on Mercury and Venus," Knowledge, 4:228, 1907. (X2)
R3. Watson, H.; "Mercury: Its Markings," English Mechanic, 90:40, 1909. (X3)

AHO5 Anomalous Brightness Temperatures

Description. Unexpectedly low radio brightness temperatures of Mercury coupled with a lack of variation with Mercury's phase. Brightness temperatures of about 200°K were measured rather than the expected 500°K.

Data Evaluation. A single series of observations over a three-month period. Confidence in these measurements is increased by the fact that observations of the brightness temperatures of other planets were consistent with those of other observers. Rating: 3.

Anomaly Evaluation. Reasonable models of Mercury's surface and atmosphere do not come close to explaining these data. Rating: 2.

Possible Explanations. None.

Similar and Related Phenomena. None.

Examples

X1. July 16 through October 17, 1965. "Radio observations of Mercury at 3.4 millimeters from July to October 1965 showed, contrary to expectation, brightness temperatures of only about 200°K, even when major fractions of Mercury's illuminated hemisphere were observed. There was no significant variation with phase." (R1)

References

R1. Epstein, Eugene E.; "Mercury: Anomalous Absence from the 3.4-Millimeter Radio Emission of Variation with Phase," Science, 151:445, 1966. (X1)

AHO6 Ring of Light around Mercury's Dark Side

Description. Dark side of Mercury surrounded by a halo or aureole of light.

Data Evaluation. One occurrence by two different observers. Rating: 3.

Anomaly Evaluation. Optical refraction seems excluded by the planet's negligible atmosphere. Again, we must fall back on psychological effects, which are themselves poorly understood. Rating: 2.

Possible Explanations. A contrast effect. The type of explanation would also apply to the similar phenomena seen around Venus and the moon. Unfortunately, these rings of light or "annular phases" are not always observed under identical conditions.

Similar and Related Phenomena. Annular phase of Venus (AVO2); the moon seen as a ring (ALO9); the ashen light of Venus (AVO3); the illumination of Mercury during transit (AHX3); the ring around Mercury during transit (AHX2).

Examples

X1. May 18, 1896. "Observing Mercury on the 18th May, between 22^h and 23^h G.M.T., I was astonished to see not only spots.... but even the dark side surrounded by an aureole, just like the appearance of Venus in July 1895. Fearing to be the victim of an optical illusion, I tried various eyepieces (powers 146, 196, 242, 310, 410), changed the position of the planet in the field and shook the telescope, but both phenomena remained unchanged (respectively dancing with the illuminated disk in the same manner), so that there remained no doubt. Besides, after having made the drawing (at 22 $1/2^h$) I called Mrs. Manora, who believed for the first moment that it was Venus, as the appearance was so similar. She pronounced the dark side and the aureole to be very conspicuous objects, saying that she saw them at the first look, whilst she saw the spots on the illuminated disk later. The dark side was darker than the sky, just as I (with one single exception) have always found it in the case of Venus, so that I share now M. Flammarion's views on the explanation of this strange phenomenon." (R1)

References

R1. Brenner, Leo; "Visibility of the Dark Side of Mercury," British Astronomical Association, Journal, 6:387, 1896. (X1)

AHX ANOMALIES OBSERVED DURING TRANSITS

Key to Phenomena

AHX0 Introduction

The transits of Mercury are much more frequent than those of Venus. They are to be observed only in May and November, a fact suggestive of orbital resonance between Mercury and earth (AB). Mercury, as it creeps across the sun's face, is usually a vivid black, but on occasion a bright spot appears just south of the planet's center. A halo of bright light around the black disk is sometimes seen, too. The halo may accompany the bright spot, but not always, and vice versa. These optical and/or psychological phenomena have not been reported in the general scientific literature for many decades. Presumably, they still occur but no one bothers with them. They are merely miscellaneous data.

The "black drop" effect is a transit phenomenon occurring with both Mercury and Venus. It is not considered anomalous but should be mentioned here. As Mercury begins its transit of the solar disk, its black circle invades the sun's bright limb, but does not quickly and

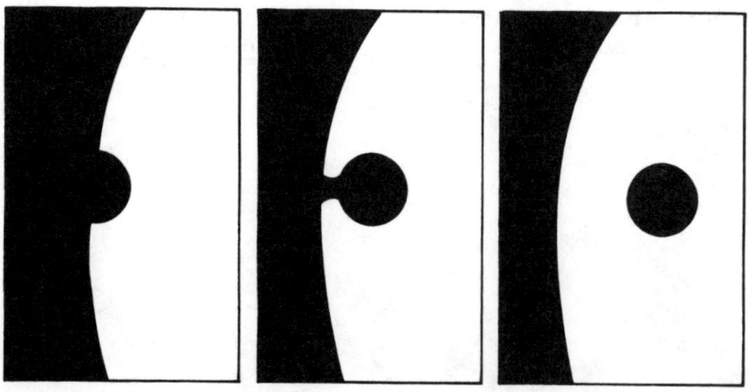

sharply detach itself from the inner edge of the sun's disk. A black ligament usually extends, for a few moments, from Mercury's black disk to the edge of the sun. The ligament quickly disappears as Mercury further invades the sun's disk. This is evidently a trick of optical diffraction that is analogous to the ligament formed when one's thumb and forefinger are held almost touching in front of a bright light. The black drop phenomenon seems to be closely related to the observations of bright stars "hanging" on the moon's limb (ALX8).

AHX1 Bright Spots on Mercury during Transit

Description. One, rarely two, small bright points of light seen on the black disk of Mercury as it transits the sun. Sometimes, the spot appears grayish. It is rarely centered, being mostly south of the center of the disk. The spot(s) does not appear at every transit.

Data Evaluation. Mercury's bright spot has been observed many times by many different astronomers utilizing various telescopes. The abundancy and consistency of these reports, coupled with multiple observers and instruments, argue against telescopic ghosts and other instrumental aberrations. Rather strangely, the last observation of the spot recorded in the literature examined occurs in 1914, although it was reported often before that date. Rating: 1.

Anomaly Evaluation. Although some astronomers have joked about Mercury "having a hole in it", the appearance of the bright spot is genuinely puzzling. As mentioned above, instrumental effects seem to be ruled out. Neither does ordinary refraction seem to be the culprit, for the spot is invariably off-center and sometimes doubled. Rating: 2.

Possible Explanations. Mercury surface phenomena can probably be eliminated; that is,fluorescence and white mineral deposits can be ruled out. Rather, some sort of optical effect seems indicated. Refraction by Mercury's miniscule atmosphere does not appear to be involved. Perhaps most likely of all is optical illusion. The small black disk against the blindingly bright sun may trigger some contrast effect, or something akin to those optical illusions where gray images appear out of nowhere amid geometrical designs.

Similar and Related Phenomena. The halo around Mercury during some transits (AHX2); lineaments seen on Mercury (AHO4).

Examples

X1. November 1697. First record of the bright spot found so far. (R1)

X2. May 1799. Schroeter and Harding at Lilienthal, and Koehler at Dresden, saw a small luminous spot on the dark disk. The spot was not stationary, for Harding saw it change its position, and later in the day Schroeter saw it sometimes on one part of the disk, sometimes on another. Other observers saw, not one, but two small spots of a greyish colour." (R5)

X3. 1802. Spot observed by Schroeter and Harding. (R1)

X4. May 5, 1832. "In the transit of May 1832 Professor Moll observed a spot, the periphery of which was not well defined, but was always situated in the same position, a little south of the center of the planet, and preceding the centre." (R1, R5)

X5. November 8, 1848. A single grayish spot. (R5)

X6. November 11, 1861. "J.W. Jeans saw a slight ashy light on the eastern limit or following the centre of the planet on its way to perihelion." (R5)

X7. November 4, 1868. Spot recorded. (R1, R5, R7)

X8. May 6, 1878. "A bright spot was seen on Mercury's disk. It appeared to me perfectly central and of sensible magnitude. My eldest daughter, who observed with me, described it as a mere point, and quite central, as if the disk were a round piece of black card, and the bright spot were a hole pierced through with the compass point in striking out its circular outline. I noticed one feature in the bright spot which seemed to me decisive as to its subjective nature; when a small cloud passed over part of the sun's face, nearly the whole of which was in the field of view, the bright spot perceptibly waned in brightness, though not crossed by the cloud. This I noticed distinctly three times. Another feature---perhaps a mere illusion---was that it seemed to me, as the spot thus waxed and waned in brightness, that it was triangular in shape. I could not distinctly recognize this peculiarity when the luster of the spot was steady. The aspect of the spot was not perceptibly modified when the telescope was released from the driving clock and Mercury allowed to approach the edge of the rather wide field of view." (R2-R4, R7)

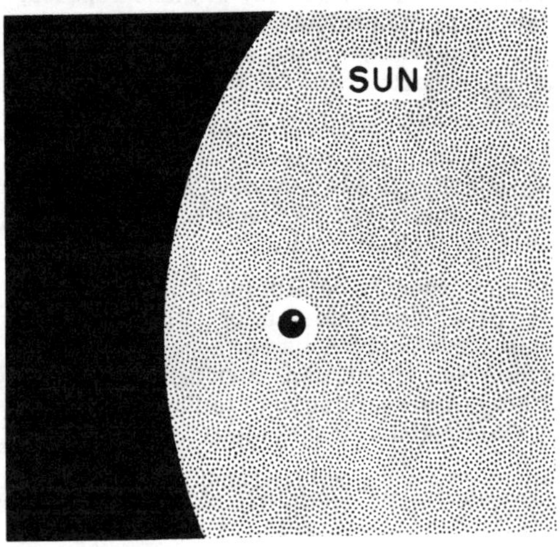

Mercury in transit with bright halo and off-center white spot. (X10)

X9. November 7, 1914. White spot seen during transit. Attributed to the telescope optics. (R8) A hasty conclusion. (WRC)

X10. General observations. A summary. "1st. That in the May transits, when Mercury is near its aphelion, the luminous spot is in advance of the planet, preceding the centre; in the November transits, when Mercury is near its perihelion, the luminous spot follows the planet. 2nd. The luminous spot has never been seen at the centre, but always south of it, and therefore cannot be due to diffraction, as Professor Powell suggests. 3rd. Sometimes in the same transit two spots have been seen close together, where shortly before only one was observed." (R5) The account in X8 contradicts the claim that the spot is never centered. (WRC)

References

R1. "Mercury," English Mechanic, 10:397, 1870. (X1, X3, X4, X7)

R2. English Mechanic, 27:261, 1878. (X8)

R3. "Transit of Mercury," Scientific American Supplement, 6:2099, 1878. (X8)

R4. "Is There a Hole through Mercury?" Scientific American, 38:392, 1878. (X8)

R5. Jenkins, B.G.; "The Luminous Spot on Mercury in Transit," Royal Astronomical Society, Monthly Notices, 38:337, 1878. (X2, X4-X7, X10)

R6. "Observations of the Transit of Mercury, 1878, May 6, Made at the Royal Observatory, Greenwich," Royal Astronomical Society, Monthly Notices, 38: 397, 1878. (X8)

R7. Johnson, P.H.; "Markings on Mercury," English Mechanic, 117:94, 1923. (X7, X8)

R8. Hollis, H.P.; "The Alleged White Spot on Mercury," English Mechanic, 117: 120, 1923. (X9)

AHX2 Ring around Mercury during Transit

Description. A bright or dark ring sometimes seen around the black disk of Mercury during transit. The bright rings are brighter than the adjoining surface of the sun. Dark rings are nebulous and occasionally tinged with purple. Double concentric rings of different brightnesses have been recorded. A bright spot on the planet itself often accompanies the ring phenomenon (AHX1).

Background. In the late nineteenth century, astronomers interpreted the bright ring seen during transit as indicative of a substantial planetary atmosphere. The Mariner 10 flyby in 1974 found no atmosphere of consequence.

Data Evaluation. Many old observations, often by noted astronomers, exist up to and including the transit of 1878. No records have been found since then. As with the bright spot observed during transit, we are dealing here with a phenomenon that is (was?) observed widely through a variety of telescopes. Instrumental effects seem precluded. Rating: 1.

Anomaly Evaluation. Since instrument effects and atmospheric optical phenomena, such as diffraction and refraction, are apparently eliminated, a psychological contrast phenomenon may be to blame. However, we have found no laboratory experiments to bolster this supposition. Even though the cause of the ring may be psychological, it is still an anomaly. Rating: 3.

Possible Explanations. A psychological contrast effect.

Similar and Related Phenomena. The so-called "annular phase" of Venus (AVO2); and the moon seen as a ring (ALO9).

Examples

X1. May 1707. The Astronomer Royal saw "Mercury in transit encompassed by a thick haze or atmosphere." (R4)

X2. November 1736. A bright ring was observed surrounding Mercury. (R1, R4)

X3. May 1753. Another bright ring. (R1, R4)

X4. May 1786. Bright ring. (R1, R4)

X5. November 1789. A ring of unspecified type. (R4)

X6. May 1799. A dark or nebulous ring, with a purplish tinge toward the planet. (R1, R4)

X7. November 1802. A ring of unspecified type. (R4)

X8. May 5, 1832. "In May 1832 the planet was seen at the Royal Observatory surrounded by a dusky tinge. In this transit also the ring has been described as of 'a violet hue, the colour being strongest near the planet.'" (R1, R4)

X9. November 4, 1868. An aureola of light a little brighter than the sun's disk. (R4, R5)

X10. May 6, 1878. "A bright halo of somewhat irregular outline, and having a breadth of 3" to 4" was seen round the planet with an inner and much brighter ring about 1" in breadth. The halo was much brighter than the ordinary surface of the Sun. It may have been an effect of contrast. It was last seen when the light was reduced to extreme faintness by turning the Nicol (sic) of the direct-vision solar eye-piece." (R5, also R2, R3)

References

R1. "Mercury," English Mechanic, 10:397, 1870. (X2-X4, X6, X8, X9)

R2. English Mechanic, 27:261, 1878. (X10)

R3. "Transit of Mercury," Scientific American Supplement, 6:2099, 1878. (X10)

R4. Jenkins, B. G.; "The Luminous Spot on Mercury during Transit," Royal Astronomical Society, Monthly Notices, 38: 337, 1878. (X1-X9)

R5. "Observations of the Transit of Mercury, 1878, May 6, Made at the Royal Observatory, Greenwich," Royal Astronomical Society, Monthly Notices, 38: 397, 1878. (X10)

AHX3 Illumination of Mercury's Disk during Transit

Description. The very faint illumination of the entire disk of Mercury just as it encroaches on the sun's disk at the beginning of transit.

Data Evaluation. Only two observations uncovered so far. In one of these cases, an independent observer using the same telescope failed to see the effect. Rating: 3.

Anomaly Evaluation. Once again we have a probable contrast phenomenon; again substantiating laboratory experiments are wanting. The low anomaly rating assumes a psychological rather than a physical cause. Rating: 3.

Possible Explanations. Mercury's atmosphere is so slight that atmospheric refraction seems a very unlikely source of the planet's illumination. Mercury possesses a magnetosphere, but one cannot specify a mechanism by which it could radiate visible light. More reasonable is an optical illusion where the eye-brain combination responds to the intense contrast by brightening the encroaching planetary disk.

Similar and Related Phenomena. The black drop effect (AHO0); bright spots seen during transit (AHX1); ring or halo seen during transit (AHX2); the ashen light of Venus (AVO3).

Examples

X1. May 6, 1878. Philadelphia Observatory. "I wish to acknowledge the kind attention of astronomers, observing the last transit of Mercury, to the phenomenon of the complete illumination of the planet's disk when half of it was projected on the chromosphere. The observations as a whole, while not as decisive as could be wished, seem to point to the reality of the phenomenon as originally observed by me May 6, 1878. The illumination mentioned seems to be a very delicate phenomenon, and perhaps visible only at a critical moment. I should therefore deem it worth the while of observers to make the attempt to see this phenomenon at future transits of the planet Mercury..... While the planet was approaching the solar edge I noticed a grayish illumination of the following semidisk of Mercury, and, without describing the phenomenon, asked Dr. Heyl to see if he noticed any difference of illumination in the planet. He reported no difference, and at subsequent trials could not see what I saw then, and subsequently. My faith in the reality of the semi-illumination of the planet while approaching the solar edge was somewhat staggered by this negative decision of my friend, and hence I made up my mind that I would make no report of the phenomenon, although I saw this grayish illumination change position as the planet approached the limb of the Sun.

On reading the detailed description of the observations of Professor G. A. Hill, as reported by Dr. W. S. Eichelberger in No. 601 of the Astronomical Journal, I decided to state that I had verified, in its essential elements, Professor Hill's physical observations of the dull illumination of the following half of the planet early in the approach to the limb, and the change of this illumination to the preceding side. It was all so evident and yet so puzzling to me, and so completely unverified by the colleague who so kindly agreed to try to verify anything seen, but that for the excellent report of Dr. Eichelberger the appearance would never have been mentioned." (R1)

References

R1. Snyder, Monroe B.; "Physical Phenomena of the Transit of Mercury," Popular Astronomy, 16:390, 1908. (X1)

AHZ MERCURY'S ANOMALOUS MAGNETIC FIELD

Key to Phenomena

AHZ0 Introduction

Mercury's magnetic field is weak compared to the earth's; but it is there, and it is intrinsic. This was a great surprise in 1974, when it was first measured by Mariner 10. In those days, Mercury was not expected to have any field at all.

We have now sent spacecraft magnetometers to Mercury, Venus, and Mars; and we can discern no consistency in what they measure. Venus and Mars have no magnetic fields of consequence; that of Mercury is weak, the earth's is strongest of all. If these four planets all formed in the same way, at the same time, one would expect field-generating core dynamos to be operating in Venus and perhaps Mars. Venus is comparable to the earth in size and orbits between two planets with intrinsic fields. Conclusion: the inner planets may not have formed at the same time or in the positions we now see them. This would be a radical departure from prevailing thoughts about solar system evolution.

AHZ1 Mercury's Unexpected Magnetic Field

Description. The Mariner 10 spacecraft measured a maximum magnetic field strength of 98γ at its closest approach to the planet (704 kilometers altitude).

Data Evaluation. The only direct measurements come from a single spacecraft that made only a single pass. Rating: 2.

Anomaly Evaluation. Since Mercury appears moon-like, rotates slowly, and emits no radio signals detectable at earth, the pre-1974 expectation was that the planet had no magnetic field. In the context of prevailing wisdom, the discovery of what is believed to be an intrinsic magnetic field adds another anomaly to Mercury's list. Given the planet's small size and very slow rotation, conventional dynamo action---which itself is not really understood---is difficult to imagine. Mercury's core under present solar system scenarios should have solidified long ago. If the field is a remanent one, where did it come from? Rating: 2.

Possible Explanations. (1) Mercury"s magnetic field is not intrinsic; i.e., dynamo-generated; but is created by some unappreciated solar plasma induction process; (2) Contrary to expectations Mercury still has a fluid core and active dynamo; (3) Mercury's field is remanent or frozen-in-place.

Similar and Related Phenomena. The earth has a much stronger intrinsic magnetic field. A typical value for the earth's field at the surface, in the United States, is 5500 γ.

Examples

X1. Permanent feature. "The data still do not force one to conclude that there is an intrinsic magnetic field, but the case is strong. Mariner 10 found that the solar wind passing Mercury formed a detached bow shock. If it is not the result of a complicated process that induces a magnetic field around the planet, then Mercury has a significant intrinsic field. That field is apparently not tilted more than 10° away from the pole, but it seems to be offset by 47 percent of the radius of the planet. Perhaps it is the remanent of an extinct dynamo." (R1)

"The magnetic field intensity on Mercury's surface is approximately 1% of Earth but because of its small size, the magnetic moment of $4.9 \pm 0.2 \times 10^{22}$ G cm^3 is only 6×10^{-4} of the Earth's magnetic moment. There is no question that the observed magnetic field is intrinsic to the planet, but its origin is uncertain. Possible sources are either remanent magnetization after cooling, with the magnetizing field due to either an internal dynamo or an external source, or a present day active internal dynamo such as on Earth and Jupiter. Which of these sources is indeed the main source, depends upon the thermal history of the planetary interior. It is not possible to uniquely distinguish between the two mechanisms, from either the available magnetic field data or from models of the planetary interior." (R3, R4)

References

R1. Metz, William D.; "Mercury: More Surprises in the Second Assessment," Science, 185:132, 1974. (X1)
R2. Ness, N. F., et al; "Magnetic Field Observations near Mercury: Preliminary Results from Mariner 10," Science, 185:151, 1974.
R3. Ness, Norman F.; "Mercury: Magnetic Field and Interior," Space Science Reviews, 21:527, 1978. (X1)
R4. Ness, N. F., et al; "Observations of Mercury's Magnetic Field," Icarus, 28: 479, 1976. (X1)

AHZ2 Mercury's Offset Magnetic Field

Description. Mercury's magnetic field is offset from the planet's center by 47% of the planet's radius.

Data Evaluation. The only direct measurements come from a single spacecraft that made only a single pass. Rating: 2.

Anomaly Evaluation. If Mercury's field is generated by a core dynamo, one would expect that the field would be roughly centered on the planet due to convection-cell symmetry. Thus, some asymmetry in the planet's interior or mode of formation is implied. Rating: 2.

Possible Explanations. Mercury's interior may be nonuniform in composition, and the core dynamo correspondingly offset. On the other hand, if the field is actually the product of some unknown plasma induction effect, its offset character would be easier to understand because of deformation by the outwardly streaming solar wind.

Similar and Related Phenomena. The axis of the earth's magnetic field does not coincide with the axis of rotation (EM); Jupiter's offset magnetic field (AJZ1).

Examples

X1. Permanent feature. See summary description in AHZ1-X1. (R1)

References

R1. Metz, William D.; "Mercury: More Surprises in the Second Assessment," Science, 185:132, 1974. (X1)

AJ INTRODUCTION TO JUPITER

Key to Categories

The giant planet of the solar system is Jupiter. The naked eye sees Jupiter as a bright jewel in the night sky; some observers claim they can even see the four Galilean satellites without a telescope. With a little optical help these four large moons become bright planets circling a miniature sun---a solar system within the solar system.

The mysteries of Jupiter are many. Is it a huge sphere of ices or an aspiring star with an internal energy source? What is the Great Red Spot, and does its character and position vary with solar activity? How do Jupiter's satellites modulate radio emissions from the planet? What are the lineaments on Europa and the strange grooved terrain of Ganymede. Terrestrial radio telescopes and the modern visits of the Pioneer and Voyager spacecraft have helped understand some of these problems, but they have also found new anomalies. There are many perplexities left over from the days when the telescope was king. The frequent observation of double satellite shadows on Jupiter's disk does not yet have a good explanation. Apparent physical distortion of satellites in transit and the hanging of stars on Jupiter's limb during occultation are typical of the unsolved optical enigmas. Jupiter is truly an alien planet ---a huge reservoir of anomalies, most of which we probably don't even know about yet.

AJB JOVIAN ORBITAL ANOMALIES

Key to Phenomena

AJB0 Introduction

The only celestial mechanics anomaly associated with Jupiter's orbit around the sun is a cyclic discrepancy between computed and observed positions. Other astrodynamic problems are found in predicting occultations of Jupiter's satellites (AJL15).

AJB1 Cyclic Disturbances of Jupiter's Orbit

Description. The cyclic discrepancy between Jupiter's calculated and observed positions. On occasion, Jupiter is ahead of its computer position; at other times, behind; with a maximum angular discrepancy of 0.0007°. The discrepancy is cyclic with a 12.4-year period.

Data Evaluation. Astronomical data from the last 160 years confirm the phenomenon. However, the effect is so small that some systematic error might creep in. Despite recent spacecraft missions to Jupiter, the phenomenon has not been mentioned in the literature surveyed since 1959. Rating: 2.

Anomaly Evaluation. Assuming there is no observational error, this small error implies that Newton's Law of Gravitation needs a minor modification. Rating: 2.

Possible Explanations. Systematic errors in observations. Since relativistic effects are not important here, Newton's Law of Gravitation may be slightly in error.

Similar and Related Phenomena. Whenever astronomical situations like this occur, Newton's Law is always blamed; viz., the advances of the perihelions of the inner planets (AHB1, AVB2).

Examples

X1. General observations. "Jupiter sometimes appears to be ahead of where it should be, sometimes behind. The difference changes regularly with time and goes through a complete cycle once every 12.4 years. The magnitude of the effect is small; Jupiter never gets out of place in its orbit by more than 600 miles. Jupiter is so far away that this 600-mile shift, when observed by astronomers here, puts the planet out of position by only 0.00007 degrees, an angle barely measurable..... The cycle in which Jupiter departs from its predicted position has repeated every 12.4 years for the past 160 years. This is slightly longer than the 11.9 years required by Jupiter to revolve once around the sun. Since these two time periods are nearly equal, the error may be considered due to a steady but gradual change in the shape and orientation of Jupiter's orbit. However, there is at present no known reason for such a gradual change to occur." (R1)

References

R1. "Query Newton's Theory," Science News Letter, 75:291, 1959. (X1)

AJF INTRINSIC RADIATION FROM JUPITER

Key to Phenomena

AJF0 Introduction

Jupiter is an alien planet in comparison to the inner "terrestrial" planets. Its density is not much higher than that of water. Instead of rock, we must deal with swirling gases, liquids, and ices. At the center of Jupiter, where pressures and temperatures may be very high, we may have to core of a star that could not ignite because of its small size. We do not really know what lies far beneath the colorfully banded surface of this giant planet. The gist of these observations is that bizarre physical and chemical processes may make Jupiter the source of a variety of radiation---electromagnetic and particulate.

Two types of Jupiter-generated radiation are considered anomalous at this writing: (1) the totality of the planet's electromagnetic radiation, which exceeds that received from the sun several-fold; and (2) the apparent modulation of Jupiter's decametric radio noise by the orbital motions of its large Galilean satellites. While the latter phenomenon does not turn out to be particularly anomalous, the former implies an internal energy source of some kind; perhaps nuclear or chemical or gravitational, or some combination thereof. There is something cooking at Jupiter's core.

AJF1 Jupiter's Intrinsic Radiation

Description. The radiation by Jupiter of more energy than it receives from the sun.

Data Evaluation. Modern analyses of Jupiter in the infrared and close-up observations by the Pioneer spacecraft. Rating: 2.

Anomaly Evaluation. The nature of Jupiter's implied internal energy source is unknown. Rating: 2.

Possible Explanations. Astronomers have long speculated that Jupiter might be a star that failed to make the grade; i.e., ignite thermonuclearly. However, there are other possibilities beyond a smoldering nuclear core, such as gravitational energy, phase-change energy, and even chemical energy.

Similar and Related Phenomena. Saturn also radiates more heat than it receives from the sun (ARF2); also Neptune (ANF1).

Examples

X1. September 27, 1879. Most spectra of Jupiter so closely resemble those of the sun that significant intrinsic radiation from the planet seems unlikely. However, on September 27, 1879, a spectrogram of the equatorial region showed considerable absorption of incident sunlight and its subsequent reradiation with a spectrum different from that of sunlight. (R1)

X2. A study of planetary brightness. This study showed that the brightness of Jupiter was much greater than could be accounted for by incident solar radiation. (R2) This, though, was a subjective analysis with no instrumental measurements of brightness. (WRC)

X3. Brightness of satellite shadows. If Jupiter is self-luminous due to incandescence, the shadows of its satellites should be relatively bright when photographed in the red region of the spectrum. This does not happen; rather, the satellite shadows seem unusually bright in the ultraviolet.

The author also comments that if Jupiter were self-luminous, its satellites would be illuminated by this light when in Jupiter's shadow and all sunlight cut off. In fact, Jupiter's satellites become totally invisible when they pass into the planet's shadow. (R3)

X4. Infrared analysis. "Abstract. The most accurate infrared photometric observations (8 to 14 microns) to date of the average limb darkening of Jupiter have been combined with the most refined deduction of jovian model atmospheres in which flux constancy has been closely maintained in the upper regime of radiative equilibrium and a much more accurate approximation of the 10- and 16-micron vibration-rotation bands of ammonia has been incorporated. The theoretically predicted emergent specific intensity has been multiplied by the spectral response function and folded (mathematically convolved---intersmeared) with the spatial response function of the atmosphere-telescope-photometer combination. The resulting comparison indicates that Jupiter is radiating from three to four times as much power as the planet is receiving from the sun." (R4, R5)

X5. Computation of core temperature from theory. The temperature of Jupiter's nucleus came out to be 196,000°K, which would result in a surface thermal flux comparable to that measured during the flybys of Pioneers 10 and 11. (R6)

References

R1. Draper, Henry; "On a Photograph of Jupiter's Spectrum, Showing Evidence of Intrinsic Light from That Planet," American Journal of Science, 3:20:118, 1880. (X1)

R2. Biggs, A. B.; "Is Jupiter Self-Luminous?" English Mechanic, 44:255, 1886. (X2)

R3. "Is Jupiter Self-Luminous?" British Astronomical Association, Journal, 38: 104, 1927. (X3)

R4. Trafton, Laurence M., and Wildey, Robert L.; "Jupiter: His Limb Darkening and the Magnitude of His Internal Energy Source," Science, 168:1214, 1970. (X4)

R5. "Jupiter's Radiation Mysteries," Science News, 97:577, 1970. (X4)

R6. "Does Gas Pressure Keep Jupiter Apart?" New Scientist, 77:796, 1978. (X5)

AJF2 Variations in Jovian Decametric Radiation

Description. The correlation of the intensity of Jupiter's decametric radio emissions with the positions of its Galilean satellites. The innermost satellite, Io, exerts the greatest influence. The frequencies involved run from about 8 to 40 MHz.

Data Evaluation. Jupiter's decametric radiation has been monitored by terrestrial receivers for over 20 years. Rating: 1.

Anomaly Evaluation. Although the details of the physical interactions of the Galilean satellites with Jupiter's magnetosphere still need to be worked out, there does not seem to be any serious confrontation with electromagnetic theory. Rating: 3.

Possible Explanations. When the Galilean satellites disturb Jupiter's magnetosphere, decametric radiation generated within the magnetosphere is modulated. The satellite disturbances are mechanical and/or electromagnetic, with the latter probably being the more important. Since Io is known to be surrounded by a cloud of electrically conducting sodium, it would naturally be more effective electromagnetically than the other satellites.

Similar and Related Phenomena. The correlation of solar activity with planetary positions (ASF); the correlation of terrestrial radio propagation with planetary positions (GER11).

Examples

X1. Analysis of radio signals received from Jupiter. In 1964, E.K. Bigg discovered a striking dependence of the intensity of the decametric radio emission from Jupiter on the angular position of Io, the closest of the Galilean satellites. He reported that he had not yet found similar correlations for the other three Galilean satellites. (R6, R1) This correlation was confirmed in 1965 by G.R. Lebo and his colleagues, although they also found that Europa and Ganymede did exert some influence, too. (R1)

Io's control is limited. "The stimulation by Io of some fraction of the jovian decametre-wavelength radio emission was recognized in 1964. Since then many observations have indicated that strong Io control is limited to frequencies between about 22 and 40 MHz. At lower frequencies, that is 8-10 MHz, the degree of Io modulation declines markedly. Recent low-frequency RAE 1 satellite, Voyager 1, and ground-based studies have shed some doubt on the conclusion that Io control predominates only at high frequencies. Data from the Earth-orbiting RAE 1 satellite revealed a strong dependence of emission detection probability on Io orbital phase at frequencies in the range 6.5-22 MHz. These apparently conflicting results lead to two different conclusions: (1) Io control decreases markedly in the 10-20 MHz regime but reappears near 6 MHz, or (2) observations above 8 MHz require that the emission intensity be taken into account. We demonstrate here that the latter is the case and offer a reinterpretation of these data which is consistent with the recent low-frequency satellite and ground-based results." (R5)

X2. Profferred explanations. Io disturbs the orbits of particles trapped in Jupiter's magnetosphere. (R2) The Galilean satellites have magnetic properties that distort Jupiter's magnetic field, so that charged particles are trapped in these distorted regions and emit synchrotron radiation. (R3) If Io has a conducting ionosphere, a large voltage would be generated as it moves through Jupiter's magnetic field, and this voltage could drive a plasma process which generates the observed radio noise. The sodium cloud around Io provides just such a conductor. (R4)

References

R1. Lebo, G.R., et al; "Jupiter's Decametric Emission Correlated with the Longitudes of the First Three Galilean Satellites," Science, 148:1724, 1965. (X1)

R2. Dulk, George A.; "Io-Related Radio Emission from Jupiter," Science, 148:1585, 1965. (X2)

R3. Burns, J.A.; "Jupiter's Decametric Radio Emission and the Radiation Belts of Its Galilean Satellites," Science, 159:971, 1968. (X2)

R4. Metz, William D.; "Moons of Jupiter: Io Seems to Play an Important Role," Science, 183:293, 1974. (X2)

R5. Desch, Michael D.; "Io Control of Jovian Radio Emission," Nature, 287:815, 1980. (X1)

R6. Bigg, E.K.; "Influence of the Satellite Io on Jupiter's Decametric Radiation," Nature, 203:1008, 1964. (X1)

AJL JUPITER'S REMARKABLE GALILEAN SATELLITES AND RING

Key to Phenomena

AJL0 Introduction

Jupiter is orbited by four planet-sized moons---the Galilean satellites---named after their discoverer. The other Jovian satellites are much smaller and, as far as we can tell at present, not nearly as interesting as Io, Europa, Ganymede, and Callisto, the four Galilean satellites in order of increasing distance from Jupiter. When the Voyager spacecraft neared the Jovian system, planetologists expected the photographs radioed back to earth to reveal heavily cratered spheroids much like Mercury and the earth's moon. The expected did not occur. The Galilean satellites turned out to be quite different from one another, and all were apparently cast in different molds than Mercury and the moon. There are craters, it is true, but Ganymede also boasts a tapestry of meandering systems of grooves quite unlike anything seen elsewhere in the solar system. There are curious lineaments marking Europa's icy crust that bear an amusing resemblance to the Martian canals drawn by Lowell. Then there is Io with its huge clouds of sodium and sulphur, and powerful volcanos spewing out still more sulphur. The word "bizarre" is employed often in this book, but it can be applied to Io with emphasis.

AJL1 Pre-Voyager Sightings of Jupiter's Ring

Description. The pre-Voyager detection of a ring of material surrounding Jupiter using telescopic data.

Background. The scientific literature generally attributes the discovery of Jupiter's ring to the Voyager flights to that planet. Yet, the work of S.K. Vsekhsvyatskii, pre-dating Voyager by some 20 years, clearly anticipates the ring photographed by the Voyager cameras.

Data Evaluation. S.K. Vsekhsvyatskii, and possibly others, collected evidence from telescopic observations that Jupiter possesses a ring---all well before the Voyager flights. The literature is emphatic on this point. Of course, the Voyager data are superior. Rating: 1.

Anomaly Evaluation. There is no scientific anomaly here. Researchers often ignore the early literature, especially if it is in a foreign language. Rating: 4.

Possible Explanations. None needed.

Similar and Related Phenomena. Pre-Mariner sightings of craters on Mars (AMO2) and Mercury (AHO0).

Examples

X1. Pre-Voyager evidence for Jupiter's ring. "A ring of comets and meteorites probably circles around the planet Jupiter, a Russian scientist reports. A dark band girdling Jupiter at its equator is actually the shadow cast by the huge halo, Dr. S.K. Vsekhsvyatskiy, astronomer at Kiev State University in Russia, emphasizes. The 'shadow' shifts position as the planet rotates around the sun in a manner expected for a shadow, Dr. Vsekhsvyatskiy notes. The ring of comets and meteorites supposedly formed many millions of years ago when an explosion on Jupiter sent huge masses of rock hurtling into space." (R1) The above material was published in the United States in 1961.

This from England, 1964: "Vsekhsvyatsky has made a close study of drawings of Jupiter done by experienced observers in many countries, and believes that a very thin, broken streak in the region of the planet's equator indicates the existence of a ring. He considers that it has been formed as a result of gigantic outbursts in Jupiter itself, and that its composition is identical with that of the ring-system of Saturn. These conclusions are likely to be hotly challenged. There is no theoretical bar to a Jovian ring, but if it exists it is extremely elusive, and confirmation is likely to prove a very difficult matter." (R2)

X2. General observations. The great preponderance of literature about Jupiter attributes the discovery of the ring to the flights of the Voyager Program of NASA. "A ring was discovered surrounding Jupiter." (R3) So goes the "official" Voyager documentation. (WRC)

Further information on the work of Vsekhsvyatskii (spelling varies) may be found in Kronos, a journal supporting the hypotheses of Velikovsky. The work of Vsekhsvyatskii has been of great interest to supporters of Velikovsky because Vsekhsvyatskii, like Velikovsky, believes that great eruptions have occurred on Jupiter. (R4)

References

R1. "Jupiter's Dark Band Is Shadow of Comet Halo," Science News Letter, 80:56, 1961. (X1)
R2. "A Ring round Jupiter?" New Scientist, 21:363, 1964. (X1)
R3. Smith, Bradford A., et al; "The Jupiter System through the Eyes of Voyager 1," Science, 204:951, 1979. (X2)
R4. Vsekhsvyatskii, S.K.; "The Ring of Comets and Meteorites Encircling Jupiter," Kronos, 5:29, Summer 1980. (X2)

AJL2 Io's Bizarre Physical Makeup

Description. The presence of large quantities of unusual materials, such as sodium and sulphur, on the surface and in the atmosphere of Io. Not only are the high concentrations of these materials anomalous for planet-sized objects in the solar system, but their ejection into Io's atmosphere makes this moon unique in the solar system.

Data Evaluation. The Voyager encounters with Jupiter and recent telescopic study. Rating: 1.

Anomaly Evaluation. The simple presence of sodium and sulphur is not surprising because these elements are rather common in carbonaceous chondrites. It is the segregation and concentration of these materials that is hard to explain. Rating: 2.

Possible Explanations. The sodium and sulphur may be the remnants of water-leached carbonaceous chondritic material---the basic stuff of the primordial planets. Since Io has no water now, the leaching process, which is not well-defined, must have transpired long ago. Io's unique atmosphere is probably created when sodium is sputtered by solar particulate radiation and when Io's volcanos spew out sulphur compounds.

Similar and Related Phenomena. None.

Examples

X1. The sodium cloud. "The strange cloud of sodium that envelopes Jupiter's moon Io has fascinated planetologists ever since it was first detected in spectral studies several years ago. Apparently sputtered up from the surface by the impacts of charged particles riding in on the lines of the Jovian magnetic field, it is invisible to the eye, in large measure because the much brighter light reflected from Io and Jupiter simply drowns it out. Thus it has heretofore been recorded only as bright lines through the narrow slits of spectrographs and spectrometers. But now, apparently for the first time, a group of researchers has taken its picture.....The photo and the map show only a portion of the cloud about six times the width of Io's disk. The entire cloud is believed to be as large as the whole Jovian system, but most of it is too dim, relative to the sunlight reflected from Jupiter, to show up in such an image." (R3, R4)

X2. The exotic surface materials. "There are several other puzzling questions regarding the surface of Io, a principal one being of course why it has such strong sodium emissions. Also, why is Io as bright as if it were covered by ice, yet shows no ice absorption features in its spectrum? And why has Io dark poles? Fanale et al. (Science, 186, 922; 1974) suggest that Io's surface composition involves evaporite salt deposits which are rich in sodium and sulphur. (This would both provide a source

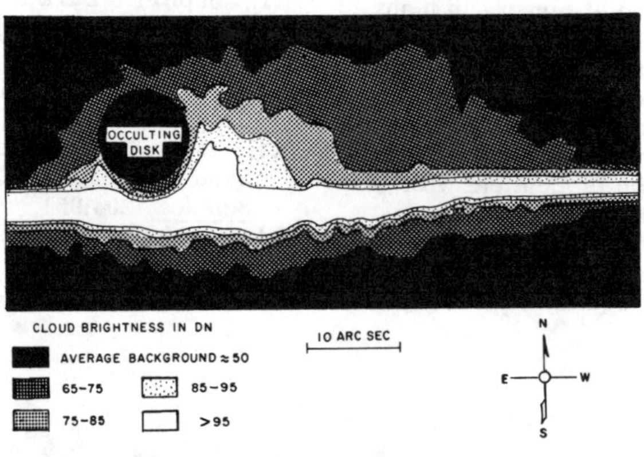

Map of sodium cloud surrounding Io, which is located behind the occulting disk. (X1)

of sodium and explain the brightness of the planet.) They believe that unlike the other Jovian satellites, Io never had large amounts of ice, yet it apparently was not totally devoid of water like our Moon. As a result of internal degassing, much of Io's water may have seeped to the surface, where it would quickly evaporate as a result of the satellite's proximity to Jupiter and its low surface gravity.... Salt-rich assemblages that Fanale et al. propose for the surface of Io are easily derivable from the leaching of carbonaceous Chondritic material." (R2) Reflectivity measurements also suggest considerable sulphur on Io's surface. (R1)

X3. General observations. "Io's unique character among the planet-sized bodies of the solar system was eminently demonstrated during the Voyager encounters with the Jupiter system. Embedded deep within Jupiter's magnetosphere, bombarded by intense radiation, and characterized by bizarre surface properties, Io is the source of remarkable extended clouds of sodium, sulfur, oxygen, and other species. These atomic and ionic clouds extend great distances from Io, and their appearance and behavior provide useful insights into the physical conditions of its surface, atmosphere, and plasma environment." (R5)

References

R1. "Brimstone in Orbit about Jupiter," New Scientist, 62:376, 1974. (X2)

R2. "Io, the Anomaly of the Solar System," Nature, 253:587, 1975. (X2)

R3. Eberhart, Jonathan; "The First Actual Look at Io's Cloud," Science News, 111: 155, 1977. (X1)

R4. Smith, Bradford A., et al; "The Jupiter System through the Eyes of Voyager 1," Science, 204:951, 1979. (X1)

R5. Goldberg, Bruce A., et al; "Io's Sodium Cloud," Science, 226:512, 1984. (X3)

AJL3 Io's Anomalously Energetic Volcanos

Description. The ejection from Io's volcanos of material at velocities of up to 1 kilometer/second. Thermal heat sources may not be adequate to produce such velocities.

Data Evaluation. Voyager photographs of the volcanos in action. Rating: 1.

Anomaly Evaluation. The high ejection velocities derived from the heights of the volcanic plumes seem to rule out thermal propulsion. Only exotic energy sources remain. Rating: 2.

Possible Explanations. The supposed great age of Io requires a perpetual energy source. Tidal flexing of Io's crust does not seem to generate enough heat through friction to create the required temperatures. More promising is electrical heating, created when Io orbits in Jupiter's powerful magentic field.

Similar and Related Phenomena. None.

Examples

X1. Energetic volcanism on Io. "Probably the most spectacular discovery of the Voyager mission has been the existence of active volcanoes on Io, erupting materials to heights of several hundred kilometers above the surface. The first discovery of an active volcanic eruption is described by Morabito et al; it appeared as an enormous umbrella-shaped plume rising 270 km above the bright limb. Since this discovery, six additional volcanic plumes have been found; most have been seen several times. In the likely case that the trajectories are ballistic, the altitudes measured on images taken in the clear filter, imply eruption velocities of several hundred meters to $\sim$ 1 km/sec.... What causes such violent volvanic activity? (R3)

X2. Possible explanations. Heat from radioactivity can probably be ruled out because Io has probably been active throughout geological time, and an unreasonable amount of long-lived radionuclides would be required. (R3) Tidal flexing as Io orbits in Jupiter's powerful gravitational field could generate an enormous amount of frictional heat in Io's interior. This perpetual heat source could power Io's volcanos. (R3, R4)

Gold, however, has rejected tidal heating as inadequate. He calculates that to produce such high velocities (up to 1 kilometer/second) in the expelled material, temperatures around 6,000°K would have to be generated, assuming the plumes are sulphur and/or sulphur compounds. Instead, Gold proposes electrical heating, caused when electrically conducting Io orbits through Jupiter's strong magnetic field. Such electrical heating could attain much higher temperatures than tidal heating. (R1, R2, R4)

References

R1. Drobyshevski, E. M.; "Magnetic Field of Jupiter and the Volcanism and Rotation of the Galilean Satellites," Nature, 282:811, 1979. (X2)

R2. Gold, Thomas; "Electrical Origin of the Outbursts on Io," Science, 206:1071, 1979. (X2)

R3. Smith, Bradford A., et al; "The Jupiter System through the Eyes of Voyager 1," Science, 204:951, 1979. (X1, X2)

R4. Hunt, Garry E.; "Io: The Electrified Satellite," Nature, 283:815, 1980. (X1, X2)

AJL4 Ganymede's Grooved Terrain

Description. A mosaic of systems of grooves criss-crossing Ganymede. Each system consists of several grooves, 5-15 kilometers apart, all running approximately parallel to each other, for distances of 10-1000 kilometers. Most systems are arcuate, but there are some fans and wedges of grooves. The width of the typical groove systems varies from 10-100 kilometers.

Data Evaluation. Voyager photography. Rating: 1.

Anomaly Evaluation. Ganymede's systems of grooves seem to be unique in the solar system. Obviously, terrestrial analogs are absent, forcing scientists to guess at what happened long ago to Ganymede's icy crust. Why do the grooves form families; and how did the complex interweaving of systems originate? It is probably only a matter of working out fault patterns and interplays of tectonic forces. Rating: 3.

Possible Explanations. Various types of faulting have been proposed; and one can conceive of Ganymede's crustal plates, like the earth's, colliding and pushing up wrinkles.

Similar and Related Phenomena. Systems of ridges on Mars (AME7); many terrestrial ranges of mountains consist of roughly parallel ridge systems.

Examples

X1. General observations. "Most of the grooved terrain is a mosaic of discrete systems of grooves, with the grooves of one system ending abruptly at the boundary of an adjacent system. In some places one system transects another. Where the grooved terrain meets the older cratered terrain, the grooves are parallel to the lines of contact. Each system contains a few to several tens of grooves. Systems are ~10 to ~100 km wide and range from ~10 to ~1000 km in length. Individual grooves are a few hundred meters deep, as determined by measurement of shadows near the terminator, and ~5 to ~15 km wide. They are generally arcuate in plan and some have sharp bends; therefore the grooves and intervening ridges cannot have been produced by strike-slip displacement of the crust. Also where one groove system transects another, grooves of the older system do not appear to be laterally displaced along grooves of the younger." (R1)

Another overview. "In detail, grooved terrain consists of alternating ridges and grooves, spaced 3 to 10 km apart, and typically 300 to 400 m deep. In near-terminator pictures, ridges appear to be sharp-crested and groove floors appear to be wide, flat bottomed, and somewhat hummocky. Ridges and grooves occur in straight or curvilinear sets. Some sets are narrow and long; others are short and wide; some are shaped like fans, bundles, or wedges. A few ridge-and-groove sets are nearly smooth and bounded by deeper grooves." (R4)

Systems of grooves on Ganymede. (X1) (Courtesy NASA)

References

R1. Smith, Bradford A., et al; "The Jupiter System through the Eyes of Voyager 1," Science, 204:951, 1979. (X1)

R2. Smith, Bradford A., et al; "The Galilean Satellites and Jupiter: Voyager 2 Imaging Results," Science, 206:927, 1979. (X1)

R3. Sutton, Christine; "Jupiter's Enigmatic Variations," New Scientist, 81:21, 1979. (X1)

R4. Lucchitta, Baerbel K.; "Grooved Terrain on Ganymede," Icarus, 44:481, 1980. (X1)

R5. Maxwell, T.A.; "Grooved Terrain on Ganymede: Characteristics and Origin of Transverse Grooves," Eos, 62:318, 1981. (X1)

R6. Squyres, Steven W.; "The Topography of Ganymede's Grooved Terrain," Icarus, 46:156, 1981. (X1)

R7. Golombek, Matthew P., and Allison, M. Lee; "Sequential Development of Grooved Terrain and Polygons on Ganymede," Geophysical Research Letters, 8:1139, 1981. (X1)

AJL5 Europa's Lineaments

Description. Lineaments, tens of kilometers wide and thousands long, appearing in the icy shell encasing Europa. These lines may be straight, curved, or segmented. They appear to be extremely shallow, perhaps only 100-200 meters deep. Some show white lines running down their centers.

Data Evaluation. Voyager photography during Jupiter encounters. Rating: 1.

Anomaly Evaluation. Europa's ice lineaments are apparently unique in the solar system; there seem to be no analogous phenomena in terrestrial ice fields. Since they are incompletely mapped, we can only guess their origin. This anomaly rating has to be a guess for the same reason. Rating: 2.

Possible Explanations. Ice cracks due to tidal and/or tectonic stresses; the upwelling and subsequent freezing of water from Europa's interior, perhaps along cracks in the crust beneath the ice.

Similar and Related Phenomena. None.

Examples

X1. General observations. "Making a map of a map would seem to be a redundant endeavor, and although the object in question is certainly no mere map---it is an ice-clad ball of rock nearly as large as earth's moon---it does have that appearance. Like a particularly convoluted roadmap, its light-colored surface is laced with a tangle of dark lines, lone ones and short, some straight, others curved or segmented, some embellished with light streaks that run along their centers. And, maplike, the lines are flush enough with the terrain to appear almost printed on, departing no more than 100 to 200 meters from the skin of a globe more than 3,100 kilometers through.... Close looks at Voyager photos indicate that some features cross others at characteristic 45° angles, suggesting that they resulted from torsional stresses, says Lawrence Soderblom of the U.S. Geological Survey, while in other cases this aspect is lacking. Also he says, from the ways in which some of the streaks cross over and under one another, 'it's clear that you can divide these things into families with different ages.'" Many of these lines appear to be tens of kilometers wide. They appear to be cracks in the icy crust of Europa. One theory is that they were created by tidal stresses; another by the upwelling and freezing of water from Europa's interior. (R2, R1)

References

R1. Smith, Bradford A., et al; "The Galilean Satellites and Jupiter: Voyager 2 Imaging Results," Science, 206:927, 1979. (X1)

R2. Eberhart, Jonathan; "Europa: Tracings in the Ice," Science News, 117:283, 1980. (X1)

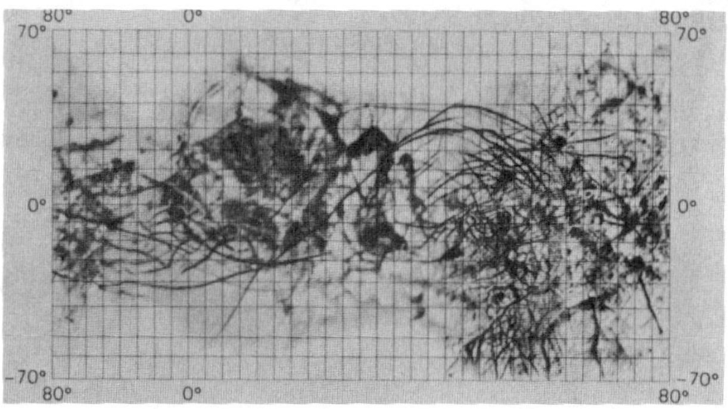

Map of the lineaments seen on Europa from R1. (X1)

AJL6 Temporary Disappearance of Ganymede

Description. The sudden, brief disappearance of Ganymede. Period of disappearance: less than a second.

Background. Ordinarily, such an obscuration would be classified with the "enigmatic objects" (AE), for an unrecognized astronomical object is the likely cause. In Ganymede's case, however, we have a history of occasional jet-black transits of Jupiter. Could black transits and Ganymede's temporary disappearance have the same cause?

Data Evaluation. A single example is on record. Rating: 3.

Anomaly Evaluation. Two obvious explanations, birds in the earth's atmosphere and more distant solid objects (asteroids?), present difficulties, as mentioned below. But there are many other possibilities for such a poorly defined, very brief, one-of-a-kind event; viz., dust clouds and negative hallucinations. In this context, a high anomaly rating is out-of-the-question. Rating: 3.

Possible Explanations. An occultation by a bird would probably have been much more brief. Any distant, solid body capable of occulting Ganymede would probably have been visible by reflected light. Other possibilities: low-reflectivity dust clouds in space, meteor trails in the earth's atmosphere, negative hallucinations (P), etc. The intriguing question, though, is whether this phenomenon is related to dark transits.

Similar and Related Phenomena. Dark transits of Ganymede (AJX3).

Examples

X1. December 1, 1915. "...at 8 hrs. 37 min. C.S.T., the aspect of Jupiter and his satellites was about as on the inclosed sketch. The distance from III to Jupiter was estimated by eye to be about 3 1/2 diameters of the planet, which would make it about 140", and from II to III was about 15" to 20". Within a very few seconds of 8 hrs. 37 min. 15 sec. C.S.T., as the satellites were being watched with 300 power on our five-inch refractor, with Jupiter himself in the field, III entirely or almost entirely winked out. II and Jupiter were not affected. The disappearance and reappearance seemed gradual, and extended over a space of at least half a second; more probably 3/5 or 3/4 second---much longer than it takes to wink the eye. The seeing was very good---9 on my scale." (R1) III is Ganymede.

References

R1. Truman, O.H.; "Peculiar Disappearance of Jupiter's Satellite, III," Popular Astronomy, 24:263, 1916. (X1)

Unexplained eclipse of Ganymede.

AJW JOVIAN ATMOSPHERIC PHENOMENA

Key to Phenomena

ALW0 Introduction

The variable features of Jupiter---the cloud patterns, ephemeral spots, changing colors---confirm for us watching from earth that Jupiter has a form of "weather", alien weather perhaps, but gross atmospheric changes analogous to the earth's. Scientists do not understand all the dynamics of the Jovian atmosphere, but even terrestrial weather cannot be predicted with certainty after centuries of close observation. We don't know how well the laws of terrestrial atmospheric physics will work on Jupiter. The strong Jovian magnetic field may play an important role, as might also bizarre chemical reactions. From all the telescopic and spacecraft observations of the restless atmosphere of Jupiter, only one class of data seem worth focussing on at the present time: the apparent cyclic behavior of the planet's surface features, particularly the size, color, and motion of the Great Red Spot. Many have suggested that solar activity modulates Jovian weather. The solar effect is hard enough to discern in terrestrial weather; at Jupiter's distance a strong solar effect would indeed be surprising. Perhaps there is an unrecognized force that affects both Jupiter and the sun and forces them to dance to the same tune.

AJW1 Periodicities of Jovian Atmospheric Features

Description. The cyclic behavior of: (1) General activity on the visible disk of Jupiter (the appearance of new spots, color changes, etc.); (2) The size and brightness of the Great Red Spot; (3) The brightness of the planet-as-a-whole; (4) The physical motion of the Great Red Spot.

Data Evaluation. Much of the data involve subjective determinations of "activity" and "prominence". Moreover, the correlations with solar activity and other phenomena are often weak. In general, the data are abundant but lack precision. Rating: 3.

Anomaly Evaluation. Considering the weakness of solar effects at Jupiter's distance, the strong correlation of Jupiter's "activity" with solar activity would be startling. Rating: 2.

Possible Explanations. Solar activity (ultraviolet intensity, particulate radiation flux, etc.) may only trigger Jovian atmospheric processes rather than force changes directly. Since Jupiter's orbital period and the length of the solar cycle are rather close, the supposed correlations with solar activity may be merely a consequence of Jupiter's changing distance from the sun. But here, too, the changes in solar effects seem much too small to control Jupiter's activity. Finally, Jupiter and the sun may both be affected by an unrecognized external force that controls both their 'activities'.

Similar and Related Phenomena. The correlation of terrestrial weather with solar activity (GWS); the correlation of solar activity with the positions of the planets (AS).

Examples

X1. Analysis of the periodicity of the surface markings of Jupiter. M. Lamey, ".... from a series of 583 drawings of this planet, covering a space of six years, concludes a periodic variation of about five and one-half years. As in the case of sun spots, the bands of Jupiter are subject to the same law of distribution in latitude. These bands---ordinarily two in number---are to be found near the equator and near each other at an epoch which seems to precede the maximum of activity. Then follows a separation, and the formation of numerous secondary bands, until the motion of the bands in latitude leads to their disintegration, and, particularly in the northern hemisphere, to an absence at times of almost all markings. The period in its oscillation about a mean, is expressed in years as 5.43 ± 0.07, just as in the case of the sun we have 11.11 ± 0.287. The last equatorial concentration appears to have attained its maximum the twenty-third of March 1885." (R2)

However, others found the period of activity to be considerably longer. "In the Monthly Notices, vol. lix, p. 76, there is an important paper by Mr. W. F. Denning dealing with the subject of periodically recurrent disturbance in the region of the north temperate belt of Jupiter. In 1880, and again in the years 1891-92, there occurred great outbreaks of spots on this belt, perhaps the most remarkable fact connects with them being the enormous velocity of the spots relative to other markings, this velocity being very considerably greater than that of the great equatorial current. In the paper above cited Mr. Denning discusses some previous occasions upon which spots had been observed in former years in this region, and comes to the conclusion that outbreaks of the kind recur at intervals of a little more than 10 years, and calls the attention of observers to the importance of examining Jupiter in February 1901, in order to ascertain if there is any trace visible of a similar outbreak of spots." The author, A. S. Williams, then asserts that his analysis points to a 12-year period. (R5) W. F. Denning reponds with data that supports his estimate of a 10-year period. (R6)

A more recent study of Jupiter's color variations. "When the short-period variations were compared with sunspot activity, it was found that when sunspots were numerous, the belts were generally a dark brown; when they were not numerous, the belts had a tendency to be orange. Because of the similarities of sunspot activity to the short-period color fluctuations, it is suspected that the color changes may be influenced by solar corpuscular emissions, the effects of which have been detected by other investigators at radio wavelengths." (R8)

Correlation of solar activity and the rotation of surface features. Abstract. "The relation between solar activity and the rotational periods of surface features on Jupiter were investigated over the period 1880-1968. Chree's superposition analysis shows that the mean rotational periods of the features display a double-maximum relationship in the course of the 11-year solar cycle." (R16)

Correlation of Jupiter's activity with its orbit. "Abstract. Analysis of features lying between the South Temperate Belt and the North Temperate Belt shows that the atmospheric activity of Jupiter is higher near and past the perihelion of its orbit than at the aphelion." (R19)

X2. Correlation of solar activity and the brightness of the Great Red Spot. "Interest in Jupiter has been stimulated by recent investigations. A correlation between solar activity likely to affect the ultraviolet radiation in the solar system and the activities of the Jovian great red spot would be of some significance for those interested in the planet. In fact, examination of the sun-spot cycle and the relative brightness of the Jovian red spot reveals just such a correlation. Fig. 1 (not reproduced) shows the curves for both Zurich sunspot numbers and the relative brightness of the Jovian red spot between 1892 and 1947. The data for the relative brightness of the red spot have been taken from Peek and his techniques for indicating the intensity of the spot, based on observations, has been adopted. Fig. 1 shows there is a pronounced correlation between the cyclic maxima and minima of the two curves during the period for which data were compared. In conclusion, it may be worth noting that a maximum in the current sunspot cycle is now anticipated, and also that recent observations have revealed a high intensity of the Jovian red spot." (R10) In reponse to this paper, E. Argyle calculated the actual, rather than apparent, correlation presented in R10, concluding that it was not significant and that a causal relationship between the two phenomena had not been established. (R11) Other investigators confirmed Argyle's conclusion. (R12)

But a more refined analysis by Basu reversed the situation. "Examination of the curves for the visibility of Jupiter's red spot and the solar cycle given by Graf et al. shows no apparent correlation because of the presence of some irregular fluctuations in the visibility function. In the analysis here the irregular fluctuations have been removed by evaluating 5 yr moving averages of the values of V. To achieve uniformity in data the values of R_Z were also subjected to the 5 yr moving average. The modified curves for the entire period 1894-1965 are shown in Fig. 1. A striking feature of the curves is the close correspondence between the two phenomena from 1894 to 1945. For this period the correlation coefficient (with moving average values is $r = 0.5714$ and the standard error for the 52 values compared is $\sigma_r = 0.14$, which shows that r is more than four times the value of σ_r. This correlation is clearly significant and indicates that the variation of the visibility of Jupiter's red spot does have a component dependent on solar activity. It is interesting, however, that subsequent values of V exhibit a completely different correspondence which does not fit into the previous period and consequently the correlation coefficient for the whole period 1894-1965 is reduced to $r = 0.2433$. Nevertheless, considering the behaviour of the red spot from 1894 to 1945, as revealed by the present analysis, it is difficult to conclude that there is no correlation between Jovian and solar activities. The apparent discontinuity around 1947 in the cyclic variation of the 5 yr moving agerages of V raises a second problem. Because the values of V provided by Peek (1892-1947) and Reese (1948-67) have been scaled by the same criteria, it must be admitted that Jupiter's red spot begins to show a clear inconsistency around 1947. It is interesting to note in this connexion that this is also the time when the Sun shows an unexpected large enhancement in sunspot numbers. It is perhaps worth investigating whether the two might have a common cause external to both, such as local variation in density of interstellar medium." (R13)

A confirming analysis. "<u>Abstract</u>. A new

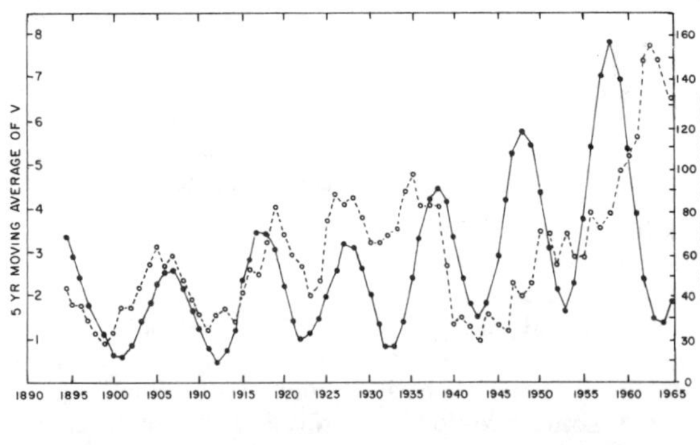

Sunspot numbers (solid line) versus Red Spot prominence index (dashed line) from R13. (X2)

inquiry has been made into the question of whether Jupiter's Great Red Spot shows a solar activity dependence. From 1892 to 1947 a clear correlation was present. A dearth of sightings in the seventeenth century, along with the Maunder Minimum, further supports the relation. An anticorrelation, however, from 1948 to 1967 removed support for such an effect. The old observations have been reexamined and recent observations have also been studied. The author reexamines this difficult question and suggests a possible physical mechanism for a Sun-Jovian weather relation. Prinn and Lewis' conversion reaction of Phosphine gas to triclinic red phosphorus crystals is a reaction dependent upon solar UV radiation. It may explain the dependence found, as well as the striking appearance of the Great Red Spot in the UV." (R20, R21) The Maunder Minimum was a period when few if any sun spots were apparent. (WRC)

X3. Correlation of Jupiter's brightness with solar activity. "Conclusions. The available data strongly imply the existence of a real relationship between the relative brightness of Jupiter and the relative sunspot number. ....The observed relative brightness variations of Jupiter may possibly be directly related to variations in the sun's ultraviolet energy output, but this possibility appears unlikely because of the small amount of energy available. The relative brightness variations cannot be accounted for by variations in the total energy output of the sun, since variations of Jupiter's distance from the sun appear to have little effect on the relative brightness of Jupiter." Some sort of triggering effect of solar radiation is suggested. (R7)

X4. Periodic motion of the Great Red Spot. "One of the most notable properties of the behaviour of the Red Spot and other features in Jupiter's atmosphere has been their complete unpredictability. It is thus both something of a surprise and a significant observational breakthrough that Dr. H. E. Solberg, working at the Observatory of New Mexico State University, has found a periodic oscillation in the longitude of the Red Spot. This motion has remained regular for over five years. The programme on which Solberg is working began in 1963, and has produced 1200 photographic plates of Jupiter on which the position of the Red Spot is shown with unprecedented accuracy. Earlier observations had shown the variation in position of this feature without being able to reveal any periodicity in its motion. The new plates, however, clearly show that the Red Spot undergoes semi-regular oscillations in its longitude, with an average amplitude of 0.8° and period close to 90 days. The mechanism causing this motion is not known, but its observation is certainly one of the most significant events in the entire history of Jovian observation, and a rash of theories attempting to explain it will inevitably follow shortly." (R14, R15)

More correlations of Spot motion. "Conclusions. The observed long and short term fluctuations in longitude of the Red Spot are indeed the actual phenomena, as well as the fluctuations of the B-term and of cometary discoveries. However, we leave the realm of strict reality and enter into that of probability if we show several relationships between them: (a) The quasi-simultaneity of rises and falls of long term fluctuations of the R. S., of terrestrial rotation and of cometary discoveries with periods approaching that of the period of great aurorae. (b) The coincidence of the period and the phase of the three-monthly oscillations of the R. S. with inferior conjunctions of Mercury and the indirect relationship of their amplitude with the degree of the alignment of Mercury on the Sun-Jupiter axis. All of these multiple coincidences may be purely fortuitous even if the probability of that is very small. Alternatively there may be an internal link, perhaps the solar activity, but we cannot propose provisorily any mechanism of this action. This lack, however, should not impede us from drawing attention to these facts and related speculations." (R18)

References

R1. Proctor, Richard A.; "Something Wrong with Jupiter," English Mechanic, 13:80, 1871. (X1)

R2. "Periodic Variations of the Spots of Jupiter," Franklin Institute, Journal, 123: 339, 1887. (X1)

R3. Waugh, W. R.; "The Periodicity of Jupiter's Markings," Observatory, 18:311, 1895. (X1)

R4. Noble, William; "The Periodicity of Jupiter's Markings," Observatory, 18: 337, 1895. (X2)

R5. Williams, A. Stanley; "Periodic Disturbances in the Northern Hemisphere of Jupiter," Observatory, 23:176, 1900. (X1)

R6. Denning, W. F.; "Periodically Recurrent Disturbances in the North Temperate Belt of Jupiter," Observatory, 23: 215, 1900. (X1)

R7. Shapiro, Ralph; "A Planetary-Atmospheric Response to Solar Activity," Journal of Meteorology, 10:350, 1953. (X3)

R8. Wegner, Gary; "Color Variations of the Great Planets," Astronomical Society of the Pacific, Publications, 74:413, 1962. (X1)

R9. Peek, B. M.; "Sudden Changes in the Motion of Jupiter's Great Red Spot during 1962," Royal Astronomical Society Monthly Notices, 130:423, 1966. (X4)

R10. Graf, E. R., et al; "Correlation between Solar Activity and the Brightness of Jupiter's Great Red Spot," Nature, 218:857, 1968. (X2)

R11. Argyle, Edward; "Correlation between Solar Activity and Brightness of Jupiter's Great Red Spot," Nature, 219:474, 1968. (X2)

R12. Solberg, H. G., jun., and Chapman, C. R.; "Correlation between Zurich Sunspot Number and Prominence of Jupiter's Red Spot," Nature, 221:352, 1969. (X2)

R13. Basu, D.; "Relation between the Visibility of Jupiter's Red Spot and Solar Activity," Nature, 222:69, 1969. (X2)

R14. "Jupiter's Great Red Spot Swings Up and Down," New Scientist, 43:632, 1969. (X4)

R15. Solberg, H. Gordon, Jr.; "A 3-Month Oscillation in the Longitude of Jupiter's Red Spot," Planetary and Space Science, 17:1573, 1969. (X4)

R16. Krivsky, L., and Pokorny, Z.; "Solar Activity and the Rotation of Jupiter," Astrophysical Letters, 12:173, 1971. (X1)

R17. Banos, Cosmas J.; "Contributions to the Study of Jupiter's Atmosphere," Icarus, 15:58, 1971. (X2)

R18. Link, F.; "Some Peculiarities Affecting the Movements of Jupiter's Red Spot," Planetary and Space Science, 23:805, 1975. (X4)

R19. Favero, Giancarlo, et al; "Periodicity in the Activity of Jupiter's Atmosphere," Strolling Astronomer, 27:240, 1979. (X1)

R20. Schatten, K. H.; "A Jovian Great Red Spot-Solar Cycle Variation," Eos, 60: 307, 1979. (X2)

R21. Schatten, Kenneth H.; "A Great Red Spot Dependence on Solar Activity?" Geophysical Research Letters, 6:593, 1979. (X2)

AJX TRANSIT AND OCCULTATION ANOMALIES

Key to Phenomena

AJX0 Introduction

Strange things happen when the large Galilean satellites slip behind Jupiter or pass across its face. Most of the odd phenomena transpire during transits. Some satellites, particularly Ganymede, seem grossly distorted as they cross Jupiter's disk. Instead of perfect circles, they may appear elliptical, with a strong tendency toward doubling. In this vein, the shadow of Io, when cast upon the planet below, has occasionally been seen doubled. All the satellites cast shadows; and they are usually black and circular as expected, but sometimes they are definitely gray. A gray shadow implies that Jupiter's surface is emitting intrinsic radiation, not reflected sunlight. Back to the satellites themselves during transit; one would expect them to be seen as bright against the planet's background due to reflected sunlight---they usually are; but rarely they will be jet black. Just why no one knows.

The Galilean satellites also pass behind Jupiter in eclipses and occultations. As they touch Jupiter's limb at the commencement of an occultation, they sometimes hang there or are projected onto the limb for several minutes. The same thing may happen with occulted stars, just as it does when our own moon occults stars. An intriguing eclipse phenomenon is the erratic brightening of Io as it emerges from Jupiter's shadow. Some astronomers have suggested that when the planet's shadow cools Io, some white chemical, perhaps SO_2, covers Io with a temporary layer of strange snow.

AJX1 Distorted Shapes of Galilean Satellites in Transit

Description. The occasional elliptical or dumbbell appearance of the Galilean satellites. Ganymede is often seen to be elliptical; Io has been observed doubled in transit.

Data Evaluation. Many noted astronomers have remarked on the elliptical tendencies of the Galilean satellites, particularly Ganymede. Io has been seen double only once. Rating: 2.

Anomaly Evaluation. Contrast effects (see below) seem reasonable explanations of Ganymede's occasional elliptical shape. The doubling of Io, however, is more mysterious, particularly because its shadow on Jupiter is sometimes doubled (AJX4), while Io itself remains single. Rating: 2.

Possible Explanations. The contrasting dark and light areas of a satellite, especially Ganymede, may make it appear irregularly shaped against the backdrop of Jupiter. In the case of Io, its emissions of sulphur and sodium may create clouds that appear dark in transit, thus distorting the satellite's figure. Another possibility is that Io's shadow on Jupiter's face may overlap the image of the satellite itself, although an experienced astronomer should not be fooled by this.

Similar and Related Phenomena. Double shadows of the Galilean satellites as they transit Jupiter (AJX4).

Examples

X1. September 8, 1890. "On Sept. 8, 1890, with the 12-inch this satellite (Io) presented a remarkable aspect while in dark transit. I noticed that it appeared elongated in a direction nearly perpendicular to the belts of Jupiter. With high powers (500 and 700) and perfect definition the satellite appeared distinctly double, the components clearly separated. At my request Mr. Burnham kindly examined the satellite with me, and we both distinctly saw the phenomenon of apparent duplicity (see Ast. Nach. 2995). In reference to the appearance of the satellite---whatever may be the explanation---Mr. Burnham has no hesitation in stating that it was as distinctly double as any double star that he has seen. The distance between the centres of the two images was about 1", and the position angle at transit 173° - . The south component was very slightly the smaller." (R2, R1, R3)

X2. March 23, 1895. Io was observed in transit to be obviously elliptical, with the long axis perpendicular to Jupiter's equator. (R4)

X3. April 17, 1896. Antoniadi confirms W. Pickering's 1893 claim that Ganymede's shape changes in transit. "On 1896, April 17, I accidentally had the opportunity of confirming the accuracy of the American astronomer's observations. The third satellite appeared, in spite of myself, and at the first glance, with a tendency to duplicity. The longest dimensions made, when I started observing, an angle of some 45° with the planet's orbital plane; but the value of this angle was slowly and steadily decreasing. This appearance lasted for about an hour, when the elongation gradually subsided into the circular form. At first I thought I was witnessing an occultation of a satellite by another, but reference to the 'Connaissance des Temps', showed that this was not the

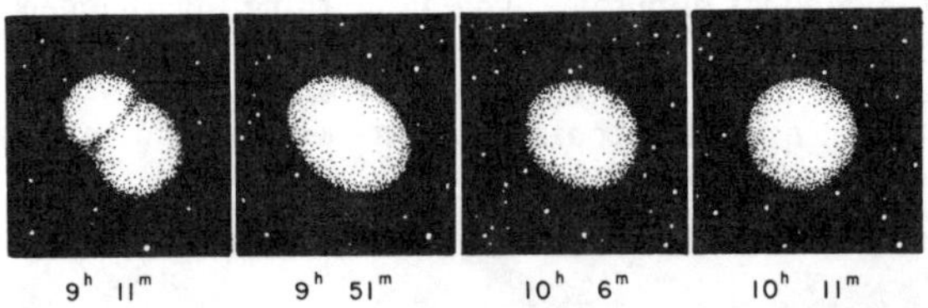

Antoniadi's drawings of Ganymede in transit on April 17, 1896. (X3)

case; so I thought myself victim of some illusion, though of what kind, of course, I could not say." (R5)

X4. May 24, 1913. J. Guillaume, at the Lyons Observatory, reports that Ganymede in transit has an irregular shape. (R6)

X5. General observations. "The Occasional Elongated Appearance of the IIIrd Satellite (Ganymede) and Its Cause.---This phenomenon must have been remarked by Herschel in 1797, when he suggested that the bodies of the Jovian satellites might not be spherical. On one occasion in 1850, Lassell found the IIIrd satellite certainly not round, and this was noticed also by Secchi, Burton and many others since. But, in 1921, the Rev. T. E. R. Phillips solved admirably and completely the mystery, by demonstrating that the apparent distortion was simply due to the spots of the satellite, and not to a really flattened figure, still less to a swarm of meteorites, as had been so startlingly asserted beyond the Atlantic." (R8)

Ganymede has often been described as elliptical during transit. The shadows of the Galilean satellites also appear elliptical on occasion. (R7)

W.H. Pickering maintained that the Galilean satellites were in fact elliptical, as demonstrated by his micrometer measurements from 1892 on. These measurements were not made during transits. (R9)

X6. Radar observations of the Galilean satellites. The effective radar cross sections---and therefore the strengths of the echos---of these satellites vary with time. This does not necessarily imply an actual change in physical cross section; rather a change in reflectivity as the satellite rotates. (R10)

References

R1. English Mechanic, 54:38, 1891. (X1)
R2. Barnard, E. E.; "Observations of the Planet Jupiter and His Satellites during 1890," Royal Astronomical Society, Monthly Notices, 51:543, 1891. (X1)
R3. Barnard, E. E.; "On the Phenomenon of the Transit of the First Satellite of Jupiter ....," Royal Astronomical Society, Monthly Notices, 52:156, 1892. (X1)
R4. Brenner, Leo.; "The I. Satellite of Jupiter Elliptical ," English Mechanic, 61:170, 1895. (X2)
R5. Antoniadi, E. M.; "Diplopia and Egg-Shaped Jovian Satellites," British Astronomical Association, Journal, 9:86, 1898. (X3)
R6. "A Curious Aspect of Jupiter's Third Satellite," Nature, 91:460, 1913. (X4)
R7. Barker, Robert; "Jupiter's Satellites," English Mechanics, 3:35, 1927. (X5)
R8. Antoniadi, E. M.; "On the Markings of the IIIrd Satellite of Jupiter in Transit Across the Planet," British Astronomical Association, Journal, 48:275, 1938. (X6)
R9. Ashbrook, Joseph; "W. H. Pickering and the Satellites of Jupiter," Sky and Telescope, 26:335, 1963. (X5)
R10. "The Mystery of Jupiter's 'Expanding' Moons," Star & Sky, 2:12, November 1980. (X6)

AJX2 Hot Satellite Shadows

Description. The occasional enhancement of radiation from the shadows of the Galilean satellites cast on Jupiter, as determined by the lightness of the shadows and/or infrared measurements of the shadows.

Background. The shadows of the Galilean satellites on Jupiter are normally inky black, so that a light or invisible shadow is thought to indicate high energy production in the region of the shadow.

Data Evaluation. The data are very skimpy. The visual observations are always compromised by contrast effects. Rating: 3.

Anomaly Evaluation. Are temporary enhancements of radiation from Jupiter's visible surface anomalous? In the context of our general ignorance of the physical and chemical processes underway in Jupiter's atmosphere, the answer has to be yes. Rating: 2.

Possible Explanations. Contrast effects against Jupiter's variegated background might account for some light shadows, but enhanced intrinsic radiation is also a reasonable explanation.

Similar and Related Phenomena. Dark transits of the Galilean satellites (AJX3); the post-eclipse brightening of Io (AJX6).

Examples

X1. October 18, 1880. Io and Europa were both in transit; both cast shadows on Jupiter. The shadow of Europa was dark brown while that of Io was jet black. (R1)

X2. November 23, 1905. The shadow of Io in transit was practically invisible, with only a shred of it being seen during moments of good definition. (R2)

X3. October 26 and December 15, 1962. "Abstract. On the evenings of 26 October and 15 December 1962, while the disk of Jupiter was being scanned for thermal emission in the 8- to 14-micron wavelength region, a large enhancement was discovered in the emission from shadows cast on Jupiter by the Jovian satellites Ganymede and Europa. However, on the evening of 14 December 1964, the shadow of Io was observed and no enhancement was detected. The effect is thus variable with time." (R3)

References

R1. Noble, W.; "Note on a Phenomenon of Jupiter's Satellites," Royal Astronomical Society, Monthly Notices, 40:41, 1880. (X1)

R2. Allison, F. B.; "Jupiter---Shadowless Transit of Satellite I," English Mechanic, 82:383, 1905. (X2)

R3. Wildey, Robert L.; "Hot Shadows of Jupiter," Science, 147:1035, 1965. (X3)

AJX3 Dark Transits of Galilean Satellites

Description. The occasional dark, sometimes inky black, appearances of the Galilean satellites Io, Ganymede, and Callisto, during transits across Jupiter. When not in transit these satellites are quite bright. But during transit, Io is often gray; Ganymede is usually white but can be jet black; Callisto often begins white but changes to black. Europa is always white.

Data Evaluation. Dark transits have a substantial literature, although it is generally from the last century. Nevertheless, no systematic study of this engaging phenomenon has appeared, even though it is unquestionably real. Rating: 1.

Anomaly Evaluation. No good explanation of this peculiar phenomenon has yet been found. Rating: 2.

Possible Explanations. Albedo and contrast phenomena have been invoked. Unfortunately, the same satellites have been seen both white and black, during different transits, over the same region of Jupiter's disk. Changes in Jupiter's brightness have also been proposed.

Similar and Related Phenomena. Post-eclipse brightening of Io (AJX6); hot shadows of the Galilean satellites (AJX2); multiple shadows of the Galilean satellites (AJX4); anomalous occultations and eclipses of the Galilean satellites (AJX5); the temporary disappearance of Ganymede (AJL6).

Examples

X1. January 28, 1848. Bond observed Io, Europa, and Ganymede all in transit. Ganymede itself was as black as the shadows of Io and Europa. (R5)

X2. May 18, 1848. Bond saw Ganymede begin transit as very bright, 20 minutes later it was hardly perceptible, then it was a perfectly black, round spot for 2 1/2 hours. (R5)

X3. December 30, 1871. Lassell observed Callisto to be as dark as its shadow while transiting Jupiter. (R1)

X4. February 18, 1872. Callisto began its transit as a dusky spot, but gradually became darker as it progressed across the disk. It did not fade at the other limb. (R1)

X5. March 25, 1873. The transit of Callisto was remarkable for its absolute blackness. The phenomenon was observed widely. (R2, R5)

X6. March 25, 1874. A dark transit of Ganymede. (R3, R4)

X7. Some time in 1884. "At the Almanac time of II occultation disappearance (8h 52m L. M.T.), Mr. W.H. Lowden took the instrument to observe the phenomenon, and he at once announced that there was a shadow of a satellite on the planet. Thinking he must be mistaken, I again looked and found, as he had said, a black spot 'as black as a drop of ink,' and I then noticed that this spot occupied about the position of the satellite then in transit. I then thought I must have made an error, but upon referring to the Ephemeris, I found that no shadow should be on at that time, and that consequently it must be the fourth satellite projected on the disk as a black spot instead of the usual bead of light. This phenomenon so occupied our attention that the occultation disappearance was allowed to pass unnoticed. Thereafter we watched the spot, at intervals for nearly an hour, during which time it remained absolutely black." (R5) Callisto is Jupiter's fourth Galilean satellite. (WRC)

X8. July 21, 1890. "On the evening of July 21, 1890, I observed the most intensely black transit of a Jovian satellite that it has ever been my fortune to witness. Both satellite III and its shadow were on the face of the planet, the former showing a round disc of dense blackness, its limb sharp and well defined, while the shadow was less black in hue, being brownish in tint, and apparently not exactly circular in form, with definition of the limb somewhat less distinct." (R8) See AJX1 for noncircular apparitions of Jupiter's satellites. (WRC)

X9. September 2, 1890. Another black transit of satellite III (Ganymede). "I turned my 5-inch telescope with power of 120 upon Jupiter and saw what appeared to be the shadow of a satellite on the lower belt and near the center of the disc of the planet, and on referring to the ephemeris I found it to occupy the position of satellite No. 3. A steady look showed it to be a jet black dot which I was unable to distinguish by appearance from a shadow. On applying the 200 power E.P. fifteen minutes later the image appeared elongated transversely with the belts (north and south). I observed it closely during the remainder of the transit, could detect no change during the entire time until within eight minutes of egress when the inky blackness became less intense. On the satellite's emerging and with the dark sky background, it appeared with its usual brightness." (R7)

X10. September 8, 1890. Dark transit of Io, during which it appeared elongated. (R10)

X11. December 27, 1894. Ganymede was of a dusky hue during its transit and elliptical in shape. Its shadow was intense black. (R11)

X12. February 14, 1921. A dark transit of Callisto. It appeared as a dark, round spot even when crossing the dusky north polar region. (R14)

X13. March 22, 1920. Ganymede during its transit was intensely and uncompromisingly black. (R12, R13)

X14. General observations. Comments of the reknowned E. E. Barnard: "In reference to the abnormal or black transits of the third and fourth satellites, from an experience of many years in actually observing the phenomenon I should feel much hesitation in endorsing the theory that they are due alone to contrast. I have seen these satellites cross the same regions of Jupiter's disc in bright transit and again like drops of ink. Assuming the albedo of the satellite to be approximately constant, that of Jupiter would need to vary so vastly that it must rank as a variable star." (R10)

An overview, circa 1888, by E.J. Spitta" "It appears that notably the fourth---the farthest from its primary---as it approaches the disk of Jupiter, becomes rapidly and increasingly fainter until it arrives at contact. When once on the limb, it shines with a moderate brilliancy for about ten or fifteen minutes, then becomes suddenly lost to view for another period of about the same duration, and lastly reappears, but as a dark spot which grows darker and darker until it equals the blackness of its own shadow on the planet. The appearance presented by the second satellite (Europa), however, is entirely different, for it never seems to have been seen otherwise than pure white during transit, whereas the first and third (Io and Ganymede) differ yet again from the preceding two. The former is sometimes a steel gray, and at others a little darker, whereas the latter has been seen perfectly white, and yet so black as to be mistaken for the fourth." Spitta believes the different albedos of the four satellites may be the cause. (R6)

References

R1. "Dark Transits on Jupiter," English Mechanic, 14:608, 1872. (X3, X4)
R2. Proctor, R. A.; "Is Jupiter a Miniature Sun?" English Mechanic, 18:186, 1873. (X5)
R3. Durrad, J. W.; "Jupiter," English Mechanic, 19:72, 1874. (X6)
R4. English Mechanic, 19:247, 1874. (X6)
R5. Sidereal Messenger, 3:126, 1884. (X1, X2, X5, X7)
R6. "On the Appearances Presented by the Satellites of Jupiter during Transit," Scientific American Supplement, 25:10294, 1888. (X14)
R7. Barnes, Willis L.; "Black Transit of Jupiter's Third Satellite," Sidereal Messenger, 9:426, 1890. (X9)
R8. Swift, Lewis; "Black Transit of Jupiter's Third Satellite," Sidereal Messenger, 9: 474, 1890. (X8)
R9. Tebbutt, John; "Note on Dark Transits of Jupiter's Satellites," Astronomical Society of the Pacific, Publications, 3:221, 1891.
R10. Barnard, E. E.; "Observations of the Planet Jupiter and His Satellites during 1890," Royal Astronomical Society, Monthly Notices, 51:543, 1891. X10, X14)
R11. Martin, E. S.; "An Interesting Transit of Jupiter's Third Satellite," Astronomical Society of the Pacific, Publications, 7:39, 1895. (X11)
R12. Porthouse, William; "A Black Transit of Jupiter's Third Satellite," English Mechanic, 111:119, 1920. (X13)
R13. Goodacre, W.; "Jupiter---Dark Transit of Sat. III," English Mechanic, 111:141, 1920. (X13)
R14. Atkins, E. L.; "Dark Transit of IV," English Mechanic, 113:58, 1921. (X12)

AJX4 Double Shadows of Io

Description. The rare, transitory appearance of a second shadow of Io on Jupiter's disk. The second shadow is often very dark and similar to the true shadow in size and shape.

Background. Multiple shadows are common on earth when two or more light sources are in operation. With Jupiter's moons, however, the sun is the only light source available.

Data Evaluation. A handful of reports from the last century, augmented by a more recent set of photographs showing the transitory phenomenon clearly. Rating: 1.

Anomaly Evaluation. All of the explanations presented below have weaknesses. Nevertheless, one anticipates that some simple optical effect will ultimately account for the doubling of Io's shadow. Rating: 2.

Possible Explanations. The two-cloud-layer theory has the second shadow appear on a lower deck of clouds, the necessary light having been transmitted through an upper layer upon which the first shadow appears. As pointed out in X4, this explanation has difficulties. Io is known to be surrounded by clouds of sodium and sulphur, which might cast separate shadows. But one would expect such shadows to be quite different from the satellite's true shadow. Actually, the true and second shadows are usually similar. Some sort of birefringence or optical ducting in the Jovian atmosphere might cause a double shadow.

Similar and Related Phenomena. The doubling of Io itself (AJX1).

Examples

X1. April 24, 1877. Observations of M. Trouvelot. "The shadow of the first satellite had entered Jupiter's disc 39 minutes before at 14h. 58m. 42s., and had not quite described a quarter of its course on the planet's disc. The shadow was black and sensibly elliptical in form, and nearly touched by its most northern point the northern edge of the rose-coloured equatorial zone. It was preceded on its western side by a dark spot of exactly the same shape, which was separated from the shadow by an interval not exceeding one third of the equatorial diameter of the shadow. This singular spot was not exactly on the same horizontal line with the shadow of Jupiter's first satellite, but was situated a little south, about one-third of the vertical diameter of the shadow. At 16h. 6m. the shadow and the spot arrived at the central meridian of Jupiter; nothing was changed in its appearance. M. Trouvelot observed the phenomenon for the last time at 16h. 45m. At that moment the shadow had described three quarters of its course on the Jovian disc, and was always preceded by its companion. The relative distance of the two objects had not varied during the observation. The most natural way of explaining this phenomenon is by supposing that the second spot was the continuation of the primary shadow formed on a layer of Jupiter's surface sufficiently thin and transparent to allow it to filter through its interstices, to pass then through a transparent medium, and to fall finally on a second deeper and solid surface of Jupiter's globe." (R1)

X2. September 29, 1891. "Satellite I (Io) was in transit, showing a well-defined white disk against the darker background of the southern equatorial belt. Following this was the customary shadow, and following this at an equal distance was a second shadow. This secondary shadow was a trifle smaller and considerably fainter than the true shadow and seemed to be surrounded by a faint penumbra. This secondary shadow disappeared simultaneously with the egress of the true shadow. The seeing was very good and the air very steady." (R3) This observation was connected to Barnard's report of Io's doubling on September 8, 1890, in an attempt at explaining the phenomenon. (R4) See AJX1-X1 for Barnard's observation. (WRC)

X3. November 14, 1891. "On the night of Nov. 14, after the egress of Jupiter's Satellite I (Io), I was watching the shadow as it hung on the edge of the great red spot, and while intently observing, the round shadow seemed to become oblong, in which form it stayed for at least thirty minutes and then the shadow seemed to open and presented a double shadow, both round and black, the one much smaller than the other." (R2)

X4. October 26, 1963. P.R. Glaser took three photographs of Jupiter on this date. The first shows Io's shadow black and round, with good planetary detail. The second photograph, taken 16 minutes later, reveals a clearly doubled shadow. Again, definition was good so that instrument bumping can be eliminated. Three minutes later, the third photo was taken; but it showed no shadow doubling at all.

Acknowledging previous observations of this phenomenon, the author doubts that the favorite explanation of two shadows on two cloud layers can hold up under scrutiny. If a cloud layer is dense enough to form a dark shadow, it will not transmit enough light to form a second dark shadow. (R5)

Commenting on the double shadow photograph, K. J. Delano suggests that Io's surrounding clouds might cast the second shadow. (R6) Io has an extended atmosphere of sodium and sulphur. (WRC)

X5. March 22, 1966. At three different times during Io's transit, its shadow appeared momentarily doubled. It was not as dark or as large as the true shadow. (R6)

References

R1. "A Double Shadow of Jupiter's First Satellite," English Mechanic, 46:252, 1887. (X1)

R2. Hoffman, H. O.; "Double Shadow of Jupiter's Satellite I," Sidereal Messenger, 10:517, 1891. (X3)

R3. Hurlbert, Henry S.; "Unusual Phenomenon Observed on Jupiter, Sept. 29, 1891," Astronomical Society of the Pacific, Publications, 3:380, 1891. (X2)

R4. Hulbert, H. S.; "Double Shadow of Jupiter's Satellite I," Astronomy and Astrophysics, 11:87, 1892. (X2)

R5. Wend, Richard E.; "The Doubling of Jovian Satellite Shadows," Strolling Astronomer, 19:79, 1965. (X4)

R6. "Another Doubled Jovian Satellite Shadow," Strolling Astronomer, 19:143, 1965. (X4, X5)

AJX5 Limb Phenomena during Occultations and Transits

Description. (1) The apparent projection of the object being occulted onto Jupiter's limb for several minutes; (2) The "hanging" of the occulted object of the edge of Jupiter's disk for several minutes; (3) The temporary reappearance of a transiting satellite outside Jupiter's disk. The occulted objects may be satellites or stars.

Data Evaluation. Many instances of anomalous occultations have been acquired, although they are all almost a century or more old. Except for the bizarre transit anomaly (X1, X17), the data are good. Rating: 1.

Anomaly Evaluation. One is tempted to write off this phenomenon as a minor problem of optics and possibly human perception. But seemingly ideal opportunities for the phenomena to occur are available frequently, yet the phenomena do not oblige. As with the analogous lunar occultations (ALX8), the phenomena are not well understood. Rating: 2.

Possible Explanations. Jupiter possesses a dense atmosphere, so abnormal refraction seems a likely explanation, except for the fact that the same phenomena occur when the airless moon occults objects. Diffraction effects, which seem instrumental in producing the "black drop" effect (AHO0), may also be useful here.

Similar and Related Phenomena. The projection of stars on the moon's limb (ALX8); the black drop effect (AHO0).

Examples

X1. June 26, 1828. Transit phenomenon. "On the evening of June 26th, in the year 1828, the celebrated Piazzi Smyth sat down to the telescope, applied his practiced eye to the ocular, and began a survey of Jupiter. At about the same time a certain Mr. Maclear, 12 miles from Bedford, was similarly engaged, while yet another observer, a Mr. Pearson, 35 miles from Bedford, was also examining the planet. And this is what all three of them independently observed: A transit of J II (Europa) was in progress. Whether the beginning of the transit was observed we are not told by Webb, from whom the above particulars were taken, but after the satellite had 'fairly entered on Jupiter' it was subsequently found outside the limb 12 or 13 minutes later 'where it remained visible for at least 4m, and then suddenly vanished.'" (R5) The Webb mentioned above is T. W. Webb, the author of the celebrated Celestial Objects for Common Telescopes. (WRC)

X2. January 3, 1857. Occultation phenomenon. Hodgson saw Io projected on the disc of Jupiter for nearly one minute, with a clear space around it. (R4)

X3. April 26, 1863. Occultation phenomenon. Wray saw Europa projected distinctly within Jupiter's limb for nearly 20 seconds. (R4) S. Gorton, reporting the same occultation stated, 'the occultation occupied nearly seven minutes, during which time, owing apparently to the movement of the atmosphere, the satellite seemed to disappear and appear again several times.' (R3)

X4. June 9, 1877. Occultation phenomenon. Ganymede seen through Jupiter's limb at reappearance. (R2)

X5. June 19, 1877. Occultation phenomenon. Europa seen within Jupiter's limb at disappearance. (R2)

X6. July 2, 1877. Occultation phenomenon. Io observed through the southern dark belt for one minute. (R2)

X7. August 29, 1877. Occultation phenomenon. Europa seen within Jupiter's limb at disappearance. (R2)

X8. September 20, 1878. Occultation phenomenon. Carlisle saw Europa within the planet's limb for 45 seconds after last contact. (R4)

X9. October 5, 1878. Occultation phenomenon. Callisto temporarily disappeared several times during occultation commencement. (R3)

X10. September 14, 1879. Occultation of the star 64 Aquarii. "At the moment of contact the star did not instantly disappear, but seemed to possess a visible disc, the limb of Jupiter seeming to advance gradually upon it, the star, by-and-by, appearing to be bisected and then gradually disappearing altogether. The time of final disappearance was 10h. 7m. 47. 6s. M. M. T., at which instant the circle of Jupiter's limb appeared perfect; previous to this the star appeared as a small protuberance upon the limb, which gradually got smaller until final disappearance... For about 10 sec. after disappearance the star could be seen through Jupiter's atmosphere as a speck of light seen through ground glass." the protuberance also preceded the star's reappearance. (R3)

X11. April 14, 1883. Occultation phenomenon. Callisto disappeared slowly, flashing up again at intervals. (R3)

X12. February 25, 1893. Occultation phenomenon. Ganymede hung on Jupiter's limb for about 2 minutes. (R4)

X13. March 22, 1895. Occultation phenomenon. Io appeared on the limb for fully 2 minutes as a faint white projection. (R4)

X14. February 23, 1896. Occultation phenomenon. Io seen through Jupiter's limb for 4 minutes. (R4)

X15. April 12, 1898. Occultation phenomenon. Europa seen as a bright point of light just inside Jupiter's limb by Molesworth. (R4)

X16. October 3, 1903. Occultation phenomenon. "On Friday, October 3, the third satellite (Ganymede) was due to be occulted at 10h 25m. I watched it carefully, but without thought of this question. After occultation began, it appeared to pass entirely within the limb, and took exactly the appearance of a transit before it finally vanished. It was visible with 12-1/4 in. and power of 226 for at least one full minute after it was clean within the apparent outline of the planet. It was considerably brighter than the edge, upon which it appeared as if projected, and looked merely as if slightly obscured." (R4)

X17. No date given. Transit phenomenon. Europa approached Jupiter and seemed to remain in contact for some minutes on the edge of the disk before moving onto the planet itself. Exactly 12 minutes later, the same satellite was again seen outside the disk. It remained there for precisely 4 minutes and then vanished. (R1) This phenomenon was attributed to Schroeter, who was not mentioned in X1, which is almost identical in content to the above account. (WRC)

References

R1. "Signs in the Heavens," English Mechanic, 23:37, 1876. (X17)

R2. Todd, C.; "Observations of Jupiter at Adelaide," Observatory, 2:226, 1878. (X4-X7)

R3. Ranyard, A. C.; "Note with Respect to the Limb of the Planet Jupiter," English Mechanic, 37:522, 1883. (X2, X9-X11)

R4. Holmes, Edwin; "An Occultation Phenomenon," British Astronomical Association, Journal, 14:25, 1903. (X2, X3, X8, X12-X16)

R5. Bartlett, James C., Jr.; "Peculiar Optical Phenomena Associated with the Transits and Occultations of the Four Large Satellites of Jupiter," Strolling Astronomer, 7:5, 1953. (X1)

AJX6 Post-Eclipse Brightening of Io

Description. The occasional, temporary brightening of Jupiter's first satellite, Io, as it emerges from Jupiter's shadow after an eclipse. The effect lasts for about 20 minutes and amounts to a 10% brightness increase over Io's normal brightness.

Data Evaluation. The phenomenon has been noted many times with photometers and other scientific instrumentation. The effect is erratic, however, and many astronomers have doubted its reality. Rating: 2.

Anomaly Evaluation. This is a marvelous anomaly, though perhaps not one of great importance. Very likely an explanation exists in terms of a thermally controlled precipitant. Rating: 2.

Possible Explanations. Albedo increases caused by the precipitation of a substance, such as SO_2, on Io's surface during eclipse-cooling. The brightening is reduced to normal levels as the sun vaporizes the "snow". Solar particulate radiation may contribute a warming effect. The brightening might possibly be an emission effect, as opposed to an albedo change. In this case, emission might be stimulated by solar-flare particles.

Similar and Related Phenomena. Terrestrial cooling during solar eclipses.

Examples

X1. General observations. Occasional erratic phenomenon. "The post-eclipse brightening, first noted in 1962, is an unpredictable anomaly.... Brian O'Leary of the University of Massachusetts and Ellis Miner of JPL, Pasadena, add another observation to the pot. The graph shows brightness measurements after an Io eclipse, taken quite independently at the Kitt Peak and Table Mountain observatories. A bump in the curve, when levels off after about 20 minutes is evident. These are ultraviolet measurements, and O'Leary and Miner suggest that all future monitoring be done in this waveband where the effect shows greatest contrast. In a companion paper, University of Hawaii astronomers D. P. Cruickshank and Robert Murphy describe how they saw Io brighten on some occasions in 1971-72, but not on others. Satellites Europa and Ganymede never showed the anomaly." (R3, R2)

In 1983, Lane et al monitored three Io eclipses; two showed no brightening, but the third found Io 30% brighter 3 minutes after the end of the eclipse than it was 6 minutes after. (R9)

During 1970, Fallon and Murphy studied four Io eclipses, three of Europa, and one of Ganymede. No post-eclipse brightenings were seen for any of the satellites. (R1)

Three eclipse reappearances of Io were observed by Voyager instruments during the two encounters with Jupiter. No post-eclipse brightenings occurred. (R7)

X2. Spectrophotometric observations. In 1978, Witteborn et al monitored Io's brightness between 1.2 and 5.4 micrometers. At one phase angle, Io's brightness at 4.7 to 5.4 micrometers was found to be three to five times greater than it was 5.5 hours before and 20 hours after. They concluded that the observed transient must have been the consequence of an emission mechanism. The interaction of Io with Jupiter's magnetosphere was considered a possible explanation. (R6)

X3. Correlation with major solar flares. In 6 out of 7 reported brightenings of Io, a major solar flare had occurred within 10° of the sub-Jovian longitude in the 100-day period prior to the observation. In none of the 18 cases of no posteclipse brightening was there such a flare. Thermoluminescence of Io's surface material was considered a possible mechanism. Also suggested: the warming of Io's surface by trapped solar particles around Jupiter, which in turn released gases, which precipitated during eclipse cooling, and then vaporized after the eclipse. The temporary atmosphere would change Io's brightness. (R5)

X4. Additional explanations. Ammonia or some other heavy gas falls as snow during eclipse cooling, brightening Io after it emerges from Jupiter's shadow for the few min-

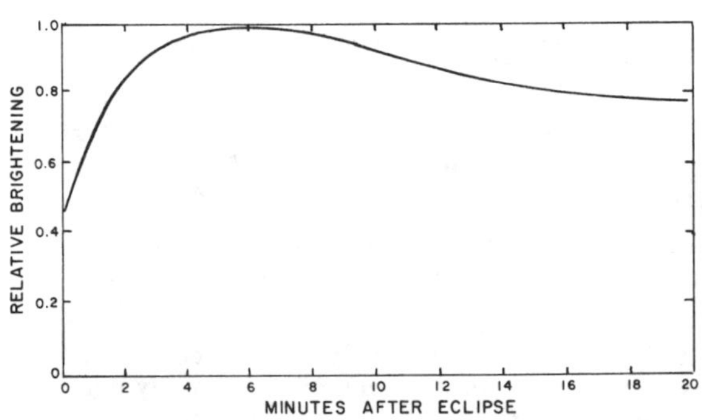

Typical brightening curve of Io after emerging from Jupiter's shadow. (X1)

utes it takes to vaporize the snow in the sunlight. (R3, R4) More recently, after Io was found to host sulphur volcanos, the condensation of sulphur dioxide during eclipse cooling has been considered the most likely precipitation involved in Io's anomalous brightening. (R8)

References

R1. Fallon, Frederick W., and Murphy, Robert E.; "Absence of Post-Eclipse Brightening of Io and Europa in 1970," Icarus, 15:492, 1971. (X1)

R2. Cruikshank, D.P., and Murphy, Robert E.; "The Post-Eclipse Brightening of Io," Icarus, 20:7, 1973. (X1)

R3. "A Satellite Looks on the Brighter Side after Eclipses," New Scientist, 60:239, 1973. (X1, X4)

R4. Metz, William D.; "Moons of Jupiter: Io Seems to Play an Important Role," Science, 183:293, 1974. (X4)

R5. Nelson, Robert M., and Hapke, Bruce W.; "Possible Correlation of Io's Post-eclipse Brightening with Major Solar Flares," Icarus, 33:203, 1978. (X3)

R6. Witteborn, F.C., et al; "Io: An Intense Brightening near 5 Micrometers," Science, 203:643, 1979. (X2, X4)

R7. Veverka, J., et al; "Voyager Search for Posteclipse Brightening of Io," Icarus, 47:60, 1981. (X1)

R8. Morrison, Nancy D., and Morrison, David; "Io: Post-Eclipse Brightening Still Mysterious, Mercury, 11:27, 1982. (X4)

R9. Nelson, R.M., et al; "Report of a Post-Eclipse Brightening of Jupiter's Satellite Io," Eos, 64:746, 1983. (X1)

AJX7 Discrepancies in Predictions of Eclipses and Transits

Data Evaluation. Relatively large discrepancies between observed and predicted times of transit and eclipse of the Galilean satellites.

Data Evaluation. Few specifics have been found so far, although general comments in the century-old literature imply that the discrepancies are frequent. A 1910 study confirmed the existence of large residuals. Rating: 2.

Anomaly Evaluation. It is not known if modern celestial mechanics has resolved this residual problem, although it seems to be too large to be merely the consequence of poor calculations or theoretical deficiencies. More likely, the problem is with the precise determination of the beginnings of eclipses and transits due to atmospheric effects. Rating: 3.

Possible Explanations. The atmosphere of Jupiter makes the precise timing of eclipses and transits difficult.

Similar and Related Phenomena. Discrepancies in the prediction of solar eclipses (ASX); also transits of Mercury (AHB3); the projection or hanging of satellites on Jupiter's limb (AJX5); the physical distortion of Jupiter's disk (AJO1).

Examples

X1. March 25, 1874. Transit phenomenon. "It will be seen that the III. satellite was a few minutes 'behind time.' I believe this is nearly always the case with III. and IV. (R1) III is Ganymede; IV, Callisto.

X2. 1884. Transit phenomenon. "Last night I observed the transit of Jupiter's fourth satellite (Callisto); the first contact was about two minutes later than the Almanac time." (R2)

X3. Theoretical analysis. Eclipses. Using refined photometric observations of the eclipses of the Galilean satellites, Professor Sampson, of Harvard Observatory, predicted eclipses between 1850 and 2000. He discovered, however, residuals so large that he believed some physical cause was required to explain them. His conclusions follow: "(1) The photometric method of observation of the eclipses gives the instant of half brightness with an accuracy distinctly greater than the residuals from gravitational theory would indicate. (2) The anomalies noticed in the observations are real features due to the departure of Jupiter's figure from a perfect spheroid. (3) These departures are probably of an irregular and transient character; on the average they may amount to 1/500 of the radius, but exceptionally to about three times as much. We see therefore, that there is a certain amount of indirect evidence in favour of the 'humpiness' of Jupiter." (R3, R4)

References

R1. Durrad, J. W.; "Jupiter," English Mechanic, 19:72, 1874. (X1)

R2. Sidereal Messenger, 3:126, 1884. (X2)

R3. Downing, A. M. W.; "Is Jupiter 'Humpy?" British Astronomical Association, Journal, 21:150, 1910. (X3)

R4. Downing, A. M. W.; "Is Jupiter 'Humpy?" English Mechanic, 93:51, 1911. (X3)

AJZ JUPITER'S MAGNETIC FIELD

Key to Phenomena

AJZ0 Introduction

Jupiter possesses a powerful magnetic field estimated at 0.16 gauss at a distance of three radii, according to Pioneer spacecraft measurements. It rises to several gauss near the planet's surface. This field results in an extensive magnetosphere with intense fluxes of captured charged particles, which generate the Jovian radio noise. At this time, the only anomaly that seems worth recording here is the offset axis of Jupiter's magnetic field---it does not coincide with the axis of rotation.

AJZ1 Offset Magnetic Field

Description. The failure of Jupiter's magnetic axis to coincide with the planet's axis of rotation. The magnetic axis is offset by about 30,000 miles.

Data Evaluation. The unusual position of the magnetic axis was determined by terrestrial radio astronomical observations. So far, no reports from the Pioneer and Voyager spacecraft programs have been found to confirm this observation. Rating: 3.

Anomaly Evaluation. The implication is that the supposed dynamo generating Jupiter's intrinsic magnetic field is not centered on the planet itself. The convection cells constituting the dynamo must then be asymmetrical---a situation hard to account for. Rating: 2.

Possible Explanations. Some of all of the field distortion may be induced by charged particles in the solar wind.

Similar and Related Phenomena. Mercury's offset magnetic field (AHZ2).

Examples

X1. General observations. "The magnetic poles of Jupiter are in a strange position, far to one side of the axis about which the planet rotates every ten hours, radio astronomers have found. Dr. David Morris, a research fellow in radio astronomy, and Glenn Berge, a graduate student, both at the California Institute of Technology's Owens Valley Radio Observatory, located the magnetic axis some 30,000 miles from the center of Jupiter, which is 89,000 miles in diameter. The center of the axis, they found, is 40,000 miles above the equatorial plane." (R1)

References

R1. "Jupiter's Magnetic Poles Found in Unusual Position," Science News Letter, 86:331, 1964. (X1)

AL INTRODUCTION TO THE MOON

Key to Categories

ALB THE MOON'S ORBITAL ANOMALIES
ALE LUNAR GEOLOGY PROBLEMS
ALF LUNAR LUMINOUS PHENOMENA
ALL THE MOTION OF LUNAR SATELLITES
ALO ANOMALOUS TELESCOPIC AND VISUAL OBSERVATIONS
ALW LUNAR "WEATHER"
ALX LUNAR ECLIPSE AND OCCULTATION PHENOMENA
ALZ THE ENIGMA OF LUNAR MAGNETISM

Down the centuries, astronomers have lavished more effort of the moon than any other astronomical object. This is natural, for the moon is close-by, intriguing, and a spectacular object on a clear night. We know a lot about the moon by virtue of all this attention; but this same scrutiny has uncovered a wide array of lunar anomalies and curiosities.

One class of anomalous phenomena apparently involves the play of light on the moon, transitory venting of lunar gases, the earth's atmosphere, and the eye-brain combination of the observer. Included in this class are such phenomena as variable spots and streaks on the lunar surface, which may be either dark or bright. And, as in the cases of Mercury and Venus, curious lineaments can be detected by some (not all) human observers, although they seem to disappear under telescopic scrutiny. The moon also exhibits an "annular" phase like Venus in which a thin ring of light seems to encircle the dark limb of the new moon. This kind of phenomenon may arise in the observer's eye-brain complex, for the moon does not possess enough atmosphere to refract sunlight all the way around the sphere.

The moon also plays host to a wide range of transient luminous phenomena (TLPs). These may be starlike points of light or gentle reddish glows covering several square miles. They may also be mists or obscurations that veil lunar detail on occasion. Almost a thousand of these ephemeral lunar phenomena have been recorded. They occur all of the time.

Lunar eclipses seem to multiply the anomalies. Many of the eclipse phenomena suggest the presence of a lunar atmosphere. Chief among these is the frequent hanging or projection of a star or planet on the edge of the moon during an occultation. Even the radio signals of spacecraft well behind the moon have been picked up on earth. What medium bends these electromagnetic waves?

One of the major Space Age lunar enigmas came with the discovery of a patchy magnetic field on the lunar surface and remanent magnetism in the lunar samples returned to earth. It is unclear how lunar magnetism originated, for the moon is too small and cold (at least it seems to be) to conceal a hot core of molten metal; and it doesn't spin rapidly enough to set up dynamo action. But the presence of lunar magnetism is undeniable. Lunar magnetism in just one of over 50 lunar anomalies described in this section.

ALB THE MOON'S ORBITAL ANOMALIES

Key to Phenomena

ALB0 Introduction

The moon's orbit exhibits several ostensibly small deviations from theory. One class of discrepancies involve differences between the moon's measured position and the computed ephemeris. These deviations seem to indicate the presence of unidentified, long-period gravitational perturbations. The most serious anomalies center on the often-discussed retreat of the moon from the earth, which is classically attributed to tidal forces. Not only does this earth-moon acceleration show unexplained short term variations but, when projected far back into solar system history, suggests that the prevailing theory of the moon's formation is in error. Perhaps not surprisingly, some recent theoretical studies predict long-term solar system instability, thus permitting controversial scenarios involving earth-moon fission and the capture of the moon from a nearby solar system orbit.

ALB1 Earth-Moon Instability

Description. The mathematical/theoretical determination that the earth-moon system is now unstable and, by inference, that the moon may have originated somewhere besides its present orbit and may eventually either escape or impact earth.

Background. The stability of a multi-body astronomical system is most difficult ascertain. In the case of the solar-system-as-a-whole, the assessment of stability is almost guesswork (ABB), for the variables are many. The earth-moon system is much simpler, but even here the sun's influence must be included for realism. The gravitational influences of the other planets are usually neglected.

Data Evaluation. The classical astronomical claim is that the earth-moon system is stable---meaning stable over a long, but unspecified period of time. Only one modern study has been found so far that disputes this claim. Rating: 2.

Anomaly Evaluation. Since other astronomical and geological research has led scientists to suspect that the moon was either captured or fissioned from the earth, it is not surprising to find that the projection of the moon's orbit backward and forward in time shows the possibilities of escape or collision with earth; i.e., the earth-moon system is still evolving. Rating: 3.

Possible Explanations. Previous stability assessments have involved simplifying assumptions.

Similar and Related Phenomena. Earth-moon acceleration due to nongravitational forces (ALB3); solar system instability (ABB).

Examples

X1. Theoretical analysis. "Earlier studies had concluded that the earth-sun-moon system is stable, but they had ignored certain of the smaller dynamic variables (the eccentricity of the earth's orbit and the effect of the moon's mass). Including these variables leads (Victor) Szebehely and (R.) McKenzie to conclude that the system is in fact slightly unstable. Someday in the future the moon may come loose and become a planet in its own right. Alternately, the calculation may support a history in which the moon was originally part of the earth, then became detached and went into orbit and will eventually fall back." (R1, R2)

References

R1. "Moon, June: Gone Balloon," Science News, 111:361, 1977. (X1)

R2. Szebehely, V., and McKenzie, R.; "Stability of the Sun-Earth-Moon System," Astronomical Journal, 82:303, 1977. (X1)

ALB2 Discrepancies in the Moon's Ephemeris

Description. Discrepancies between the measured position of the moon and its theoretically computed position, taking into account all known gravitational perturbations. Earth-moon acceleration due to nongravitational forces is treated in ALB3 and ALB6.

Data Evaluation. As in the case of earth-moon instability (ALB1), the preparation of the moon's ephemeris requires various assumptions and simplifications. However, the discrepancies discussed here are the perturbations that suggest the existence of real, unidentified physical forces and are presumably not the consequence of inaccurate mathematical modelling. Several rather old articles refer to some significant, long-period discrepancies. Despite their age, these articles are given considerable weight, because precision observations of the moon and mathematical descriptions of its orbit predate them by a century or more. Still, it is strange that no recent discussions of this subject have been found in the large volume of literature examined to date. Rating: 3.

Anomaly Evaluation. Solar system perturbing forces of the proper periods are not known, but the discrepancies could be artifacts of the computational methods used. Rating: 3.

Similar and Related Phenomena. Discrepancies in the earth-moon acceleration (ALB3).

Examples

X1. General observations. "Referring to a lunar problem, a note from Professor Newcomb's 'Investigations of Inequalities in the Motion of the Moon,' 1907, published in the Carnegie Institute, Washington, reads as follows: 'The representation of the moon's mean longitude during the period from 1650-1875 showed a discrepancy between existing theory and observation which might be indicated by a term having a period of two or three centuries, and a co-efficient of about 15". There must be some term of long period still undiscovered in the actual mean motion of the Moon.' The question then occurs, might not the 'term of long period still undiscovered' be looked for in the motion of the Earth's magnetic poles." (R1)

"...the Moon exhibits two oscillations from unknown sources, one with a period between 250 and 300 years and a coefficient of 10" to 15", the other with a period of 60 to 70 years and a coefficient of some 3". Other differences are more doubtful: the secular acceleration of the Moon's mean motion appears to be greater than its theoretical value by about 2" per century in a century, while there is evidence that its node or the mean motion of the Earth shows a similar change, besides these greater changes, small terms of much shorter period in the Moon's motion which are not quite satisfactorily accounted for, but some of which have lately yielded to a more careful comparison between theory and observation." (R2)

References

R1. Mumford, N.W.; "Mercurial and Lunar Perturbations," Popular Astronomy, 18: 319, 1910. (X1)

R2. Brown, Ernest W.; "The Problems of the Moon's Motion," Observatory, 37: 206, 1914. (X1)

ALB3 Nongravitational Forces and Earth-Moon Acceleration Discrepancies

Description. Discrepancies between recently observed earth-moon accelerations and those calculated from known gravitational and nongravitational forces. "Recent" here means within historical times.

Background. George Darwin, son of Charles, was one of the first scientists to estimate the effect of tidal friction on the moon's orbit. He projected the acceleration back in time and concluded that the moon had been created by fission from earth.

Data Evaluation. Radar can accurately measure the instantaneous rate of the moon's recession from earth. To estimate the recession rate throughout historical times, one relies upon eclipse data, which becomes less accurate in antiquity. Rating: 2.

Anomaly Evaluation. Both radar and the analysis of eclipse data reveal unexplained variations in the moon's recession rate. Some of these variations are secular but others are sudden and temporary. Scientists can only guess at the reasons for these variations. Rating: 2.

Possible Explanations. Unappreciated dissipative forces inside the earth, perhaps involving core-mantle interactions; the general expansion of earth due to the progressive weakening of gravity; the shrinking of earth due to phase changes in the core.

Similar and Related Phenomena. The long-term consequences of earth-moon recession. (ALB4)

Examples

X1. Radar-derived discrepancies. "Abstract. Precise measurements of the Doppler shift of radar waves reflected from Moon disclose unexpectedly large discrepancies---averaging about 0.6 centimeter per second---between the radial velocities and the predictions based on the Eckert-Brown lunar ephemeris. These residuals have a rapidly changing component corresponding to a relatively large, variable, and unexplained discrepancy in radial acceleration of about 10^{-4} centimeter per second, per second, in magnitude and about 1 day in period." The authors could not identify any specific cause for these discrepancies. (R1)

X2. "Sudden" perturbations in nongravitational forces. "The parameter [D''], which is a linear combination of the accelerations of the Earth and Moon, can be followed as a function of time with high confidence from about 700 BCE to the present. From its behavior, we are apparently forced to conclude that there was something like a 'square wave' in the non-gravitational forces that began about 700 CE and that lasted until about 1300 CE. During the time of this square wave, the accelerations apparently changed by factors of around 5.

We are seriously lacking in mechanisms to explain the non-gravitational forces. The only mechanism of tidal friction (the 'shallow seas' model) that has been evaluated quantitatively provides only about one fourth of the necessary amount of friction, and it does not provide for much change with time within a period as short as historic times. Forces of non-tidal origin, which are of the same order as the tidal forces, may be due largely to core-mantle interactions. There are no quantitative theories of these interactions; there are only models whose parameters are uncertain within many orders of magnitude." (R2) CE equals Christian Era.

X3. Discrepancy between the observed rate of the moon's recession and that attributable to tidal action. "Now Thomas Van Flandern of the U.S. Naval Observatory reports that he has discovered such evidence for a weakening of gravity. He told the meeting of the American Geophysical Union in Washington this week that the evidence comes from a study of the motions of the moon.

If gravity is weakening, the orbits of planets around the sun or of satellites around planets will expand, and the orbital period of these bodies will correspondingly increase. Some such expansion is provided by tidal forces in these systems, and the trick is to subtract out the tidal part and see if there is any left over.

Working with the calculations of two other observers, Van Flandern reports he has found there is an increase of four centimeters a year in the radius of the moon's orbit that is not accounted for by tidal action. 'This is the first numerical result which appears to have as its most probable explanation that gravity is decreasing." (R3)

X4. "Nongravitational" drift of the moon's longitude. "(Ray) Lyttleton's view cannot be dismissed out of hand. The argument starts out from observational evidence that the Moon, as viewed from Earth, seems to be subject to an 'extra' acceleration apart from the expected influence in line with Newton's laws of motion. The result is a drift of the longitude of the Moon compared with Newtonian predictions, amounting to 0.5 degrees of arc per thousand years, and 3 or 4 degrees for the dates of the oldest eclipses on record (Quarterly Journal of the Royal Astronomical Society, vol. 20, p. 243).

Lyttleton says that calculations based on Newtonian mechanics suggest that a certain eclipse two or three millenia ago would have been visible only in America. But contemporary records report sightings from Europe." Lyttleton takes these figures as evidence that the earth is slowly shrinking and that its moment of inertia is likewise changing. (R4) In contrast to Lyttleton's view, some astronomers believe the earth is expanding. (WRC)

References

R1. Smith, Carl R., et al; "Discrepancies between Radar Data and the Lunar Ephemeris," Science, 160:876, 1968. (X1)

R2. Newton, R.R.; "Astronomical Evidence Concerning Non-Gravitational Forces in the Earth-Moon System," Astrophysics and Space Science, 16:179, 1972. (X2)

R3. "Evidence for Weakening Gravity," Science News, 105:237, 1974. (X3)

R4. "Mountains Come from Earth Shrinkage," New Scientist, 84:110, 1979. (X4)

R5. Lyttleton, R.A., and Fitch, J.P.; "On the Apparent Secular Accelerations of the Moon and the Sun," The Moon and the Planets, 22:99, 1980. (X4)

ALB4 Earth-Moon Acceleration Incompatible with Moon's Origin in Earth Orbit

Description. The incompatibility of the prevailing theory of the moon's accretion in earth orbit with the projection of the earth-moon acceleration backward in time. Instead, the following scenarios are suggested: earth-moon fission, geologically recent close approach of earth and moon, the earth's capture of another solar system planet, and the outright impossibility of the moon's formation in terrestrial orbit. Some of these alternate moon histories are not supported by terrestrial geological evidence. Very succinctly, something is wrong some place.

Data Evaluation. The present earth-moon acceleration is observable (ALB3), but its accurate projection backward and forward over geologically long time spans depends upon good estimates of terrestrial ocean basin configuration, sea level, the earth's spin rate, the influence of other planets, etc., as a function of time. This sort of information is not well known. Rating: 3.

Anomaly Evaluation. The mode, place, and time of the moon's origin are in serious doubt, as underscored by the debate over several alternative scenarios. Most competing scenarios deny the current hypothesis that the moon accreted while in terrestrial orbit. Considering the fact that the moon is our closest celestial neighbor of appreciable size, this anomaly is considered very important. Rating: 2.

Possible Explanations. Earth-moon fission and the gravitational capture of the moon, perhaps from a nearby planetary orbit, are possible alternatives to "in situ" accretion. Both of these hypotheses possess theoretical and geological problems.

Similar and Related Phenomena. Mercury's large eccentricity and inclination (AHB2); earth-moon instability (ALB1).

Examples

X1. Moon a captured former planet. "About a century ago George Darwin showed that the tides made the moon's distance from earth increase and that, consequently, the moon earlier must have been much closer to the earth. He calculated the orbit of the moon back to a period when it was only a few times the earth's radius away. Darwin had developed a theory according to which the moon was formed by tidal friction from an elongated, rapidly rotating earth. However the coordination of this theory with his calculations, just mentioned, led to difficulties concerning the inclination of the lunar orbit relative to the plane of rotation, at least in the case of low viscosity. In the popular literature Darwin's hypothesis has been 'developed,' and it is often claimed that when the moon was thrown out it left a hole which is now the Pacific Ocean. This idea, of course, does not stand a serious scientific criticism..... H. Gerstenkorn, a high school teacher in Hannover, Germany, has taken up the problem again and has repeated Darwin's calculations, using modern values for some of the parameters. Especially important is the fact that he continued the calculation beyond the point where Darwin interrupted it. He thus found that, when the moon was close to the earth, the inclination of the moon's orbit was much greater than it is now, indeed so great that the moon passed over the poles of the earth. Still earlier the moon moved in a retrograde orbit. Gerstenkorn has traced the moon's orbit back still farther and has found that once the orbit was so eccentric that it was a parabola. In this way he has demonstrated that the moon may have been an independent planet which was captured by the earth in a parabolic retrograde orbit. The tidal action explains the development of the earth-moon system to its present state." (R1)

J. Martyn Bailey explores possible consequences of Mercury's high eccentricity and inclination. (See also AHB2.) "The capture of Planet Luna would explain a number of anomalies in the Earth-Moon system. The closeness of the two bodies and the relatively large mass of the satellite are unlike those of other planetary satellites. In the Jupiter system, for example, the mass ratio is of the order 1:10,000. The Earth-Moon system is best described as a double planet. But the present rates of tidal friction, if extended into the past, predict a maximum age for the system which is less than half that of the Earth itself. This has been interpreted by many, therefore, to mean that the Moon was derived relatively recently, either by fission from the Earth or by capture. The much greater inclination (23 1/2°) of the earth's spin axis to the plane of the ecliptic than that of the orbit of the Moon (5°) supports origin by capture rather than by fission. It has also been noted that origin by fission would require that the primitive Earth had an unusually high rotation rate compared with the other planets.

The capture of Planet Luna in a direct orbit by the process described can be compatible with both the observed rotational angular momentum of the Earth-Moon system, and the time scale proposed on the basis of tidal friction rates. The orbit of the former Planet Luna and the pre-encounter orbit for Mercury are in accord with the Bode-Titius law. The hypothesis accounts for the deviation in the present orbital velocity of Mercury, and predicts an eccentricity for the orbit of Mercury which is close to the observed value. It is concluded therefore that the Moon may be the former Planet Luna which once occupied an orbit between those of Mercury and Venus at a mean distance of 51,000,000 miles from the Sun." (R2)

X2. Moon created by fission from earth. "Several major criticisms commonly claimed to invalidate theories of lunar origin by fission from earth are shared by other lunar theories or are answerable by mechanisms associated with fission. Arguments based on the present rate of retreat, which require the moon to be a newcomer to the earth about 1.8 b.y. ago, are in direct conflict with the geologic record; MacDonald's calculations of the amount of tidal energy to be dissipated by lunar capture would require, in middle Precambrian time, surface temperatures of earth approaching those of the sun. One interpretation of his method of many-moon capture to extend the time scale fails by at least two orders of magnitude. The simplest conclusion is that something is wrong with present rate-of-retreat determinations or projections. Discrepancies of present angular momentum of the earth-moon system versus pre-moon values for earth are almost as great for the retrograde capture theory as for the fission theory. The angular momentum at fission can be halved from earlier estimates by using a gravity-stratified ellipsoid plus Chandrasekhar's calculation that all Jacobian ellipsoids are in unstable rotation. The entire remaining angular momentum discrepancy can be accounted for by escape of an incandescent silicate atmosphere volatized by the huge tidal frictions immediately following fission. By the use of a reasonable set of assumptions, escape of 4% of the mass of the earth using 3/4 of the available tidal energy can account for the entire angular momentum discrepancy. Goldreich's calculation of the history of the lunar orbit, with which he argues strongly against fission and against all other lunar theories as well, omits one significant possibility, that the earth's equator had been tilted about 10° to the ecliptic prior to the time of fission. If this is true, the inclination of earth's equator and of the lunar orbit to the ecliptic fit nicely into Goldreich's curves. It is concluded that there are fewer outstanding criticisms of the hypothesis of lunar orbit by friction than of other hypotheses. The average lunar density and the results of the Surveyor chemical analyses lead us to argue that fission from the earth triggered by the earth core formation is the most probable origin of the moon, considering the data available now." (R3)

Evaluation of the fission theory by T. A. Van Flandern. "Previously, this breakaway theory was condemned because of dynamical difficulties. If the moon (about 1 percent of the earth's mass) broke off, figured scientists, its angular momentum would have been so great that it would have flown away. (The angular momentum of an orbiting object is the product of its mass, speed and distance from the center of rotation.) NASA geophysicist John O'Keefe has shown that more of the earth's mass---perhaps 10 percent---may have split off. The bulk of this mass did escape into space. Only the moon remains.

Another criticism deals with the moon's orbit throughout its past. Since scientists know the moon's current orbit and the rate at which the orbit is slowing down and increasing in diameter, they can trace the evolution of its orbit backward in time. By thus method, it was once calculated that the moon's initial orbit could not have been in the plane of the earth's Equator if the earth were semi-molten. But a chunk of the earth could have broken off only from the Equator. Yet, recently, along with Robert Harrington at the U.S. Naval Observatory, I traced the moon's orbit backward and discovered an instability point. Before that time, the orbit could have been either in or out of the equatorial plane. We then traced the orbit forward, assuming the moon broke off from the Equator, and we arrived at the moon's current orbit." (R7)

X3. Impossibility of the moon's formation in orbit around the earth. "Abstract. A new method for determining the early history of the Earth-Moon system is described. Called the study of lunar paleotides, it describes a method for explaining features of the remnant lunar gravity field, and the generation of the lunar mascons. A method for the determination of Earth-Moon distances compared with the radiometric ages of the maria is developed. It is shown that the Moon underwent strong anomalous gravitational tidal forces, for a duration $t \cong 10^6$ yr. prior to the formation of the mascon surfaces. As these tidal forces had not been present at the time of the formation of the Moon, this shows that the Moon could not have been formed in orbit about the Earth." (R4)

X4. Moon's close approach to earth was too recent. The usual extrapolations of the moon's orbit back in time lead to close approaches of 1.5 billion years ago (R5) and 1.8 billion years ago (R3) and similar estimates, all of which are about one-third the earth's actual age according to present thinking. There seem to be no indications in the geological record of such a devastating encounter. However, the calculation of tidal effects is very sensitive to assumptions about the character of the earth's ocean basins during geological times. Some ocean-basin models, such as that of D. J. Webb, do not lead to such anomalous consequences. (R8, R9)

X5. Possible conflict between contributions to earth-moon acceleration from tidal effects and the general expansion of the universe. "A remarkably close agreement between the rate of increase of the Earth's radius and that of the universe, according to Hubble's law, was recently pointed out by MacDougall et al. Extending this line of investigation to the recession of the Moon from the Earth, assuming the same expansion rate, Hubble's constant H = 100 km/s megaparsec. We then arrive at a recession rate for the Moon of 3.9 cm per annum, whereas the observed recession, found as far back as 1878 by G. H. Darwin, amounts to 13 cm per annum. This recession is usually ascribed to a tidal effect. Does perhaps part of this recession stem from the general expansion of the universe?" (R6) If this "local" application of universal expansion is valid, it changes all the ground rules used in the above discussions of earth-moon history, as moderated by tidal effects. (WRC)

X6. Modification of prevailing earth-moon history by the changing rate of the earth's rotation. "Kirk Hansen of the University of Chicago (now at Shell Development Company, Houston) suggested recently that it is not so much the changing ocean basins but the changing rotation rate of Earth that determines the rate of tidal dissipation over geologic time." As Hansen's model is projected backward in time, the earth spins faster as it regains energy lost to the moon, the rate of tidal dissipation lessens, and the moon's inward march slows. In this model, the moon's closest approach is 225,000 kilometers. Hansen says that, "The apparent dominance of rotation rate over basin configuration argues against a fission origin for the moon, a recently revived theory, or capture by Earth during a near miss... The alternative is simultaneous accretion of Earth and the moon from the solar nebula as a 'double planet.'" (R10)

References

R1. Alfven, H.; "Origin of the Moon," Science, 148:476, 1965. (X1)

R2. Bailey, J. Martyn; "The Moon May Be a Former Planet," Nature, 223:251, 1969. (X1)

R3. Wise, Donald U.; "Origin of the Moon from the Earth: Some New Mechanisms and Comparisons," Journal of Geophysical Research, 74:6034, 1969. (X2)

R4. Anderson, Allen Joel; "Lunar Paleotides and the Origin of the Earth-Moon System," The Moon and the Planets, 19:409, 1978. (X3)

R5. Hughes, David W.; "Why Is the Moon Slowing Down?" Nature, 290:190, 1981. (X4)

R6. Klepp, H. B.; "Terrestrial, Interplanetary and Universal Expansion," Nature, 201:693, 1964. (X5)

R7. Van Flandern, Thomas; "Exploding Planets," Science Digest, 90:80, April 1982. (X2)

R8. Lambeck, Kurt; "Where Has That Moon Been?" Nature, 298:704, 1982. (X4)

R9. "Long Ago, When the Moon Was Closer," New Scientist, 95:301, 1982. (X4)

R10. Kerr, Richard A.; "Where Was the Moon Eons Ago?" Science, 221:1166, 1983. (X6)

ALE LUNAR GEOLOGY PROBLEMS

Key to Phenomena

ALE0 Introduction

Lunar geological anomalies are mainly diverse, but here and there one espies interrelatedness. Lunar asymmetry, for example, is exhibited by topography, chemical differences, and the possible relation of the mascons to the farside highlands. The centuries-old controversy between the proponents of volcanism and meteor impact as the primary shaper of the moon's surface can be seen in the discussions of the lunar rays, the rilles, and lunar crater distribution. Although the moon is certainly a very dry place right now, some scientists believe they see the hallmarks of water and/or ice action on lunar surface features. Perhaps the most significant anomalies of all are those that bear on the origin of the moon. It seems that the three favorite hypotheses of lunar origin: earth-moon fission, capture, accretion in orbit; are all flawed seriously in some way in the eyes of many astronomers.

ALE1 Asymmetrical Distribution of Maria and Large Basins

Description. The much larger number of dark maria and light-toned large basins on the moon's nearside compared to its farside. Even the few dark maria that can be found on the farside are filled only sparingly with dark material.

Data Evaluation. Large numbers of photos from the Lunar Orbiter and Apollo programs attest to the strong asymmetry of dark and light basins. Rating: 1.

Anomaly Evaluation. Since the dark and light basins taken together show strong asymmetry, the first anomaly appears to be one of differential meteor bombardment, as in the cases of Mercury and Mars. Since the dark maria are also distributed nonuniformly, the second anomaly may be one of thinner crust on the nearside, which would allow for easier penetration to the dark basalt reservoirs. This possibility is supported by the approximate 2 kilometer bulge on the farside compared with a 2 kilometer depression on the nearside. Both anomalies are hard to explain in terms of the moon's accretion in terrestrial orbit. Rating: 2.

Possible Explanations. Differential meteor bombardment could be the consequence either of the gravitational shielding of the moon by nearby earth or a highly directional meteor flux. The asymmetry of dark maria could the the result of thinner nearside crust, which, in turn, might be the consequence of differential accretion, the differential accumulation of crater ejecta, or the cataclysm of earth-moon fission.

Similar and Related Phenomena. The asymmetry of craters on Mercury (AHE1) and Mars (AME8).

Examples

X1. Permanent feature. "A major surprise in the early days of lunar exploration was the discovery that the soft maria visible from earth were far more rare on the moon's farside, presumably because of some one-sided influence of the earth." (R2, R1, R4)

The nearside-farside asymmetry is evident from the dark mare material which fills the circular basins on the nearside. Although there are similar basins on the farside, they are apparently not filled with mare basalt. Laser altimeter data suggest that this may be a result of the greater thickness of low density crust on the farside. The surface of the farside is largely elevated from a sphere of radius 1,738 km, centred on the Moon's centre of gravity, and the nearside is correspondingly depressed. Although this may explain why maria occur only on the nearside, the differences in crustal thickness are still unaccounted for. Either the Moon accreted heterogeneously, or else low density crust was, at some early stage of lunar evolution, transferred from the nearside to the farside. A possible mechanism for the transfer of vast quantities of crustal material could involve a very major impact on the nearside." (R3)

"The Lunar farside is markedly different from the frontside hemisphere in that it has fewer very large basins and virtually no maria comparable in the extent of fill to those on the frontside. Mare Orientale, the largest of all circular features on the farside, is only partially filled with mare material. Mare Moscoviense is comparable to Mare Nectaris, one of the smaller frontside circular maria. Although the floors of a few individual craters, most notably Tsiolkovsky, are substantially covered with smooth, dark mare material, the predominant impression is of the scarcity of such material. In this respect much of the farside could be compared in gross morphology to the frontside's southern highlands. The latter region's large basins with light-toned floors, exemplified by Bailly and Clavius, have about a score of farside counterparts, most of which are larger." (R5)

References

R1. Wood, J. A.; "Geographic, Geophysical, and Chemical Asymmetry of the Moon: Why?" Meteoritics, 8:82, 1973. (X1)

R2. "The Mystery of the Hemispheres," Science News, 105:241, 1974. (X1)

R3. Cadogan, P. H.; "Oldest and Largest Lunar Basin?" Nature, 250:315, 1974. (X1)

R4. Barricelli, Nils Aall, and Metcalfe, Ralph; "A Note on the Asymmetric Distribution of the Impacts Which Created the Lunar Mare Basins," The Moon, 12: 193, 1975. (X1)

R5. Kosofsky, L. J., and El-Baz, Farouk; "The Moon's Farside," The Moon as Viewed by Lunar Orbiter, NASA SP-200, 1970. (X1)

ALE2 Sinuous Rilles and Formations Resembling Terrestrial Water-Formed Features

Description. The existence on the moon of geological features resembling terrestrial river beds, arroyos, wave-cut beaches, benches, and terraces.

Background. When astronomers first studied the moon telescopically, they tended to interpret lunar topography in terms of what they saw around them on earth. The lunar maria were thought to be real seas or perhaps dried-up seabeds and thus received the Latin name for 'seas'. Today, however, the moon is known to be exceedingly dry; and the tendency is to explain topography employing lava flows, meteor impacts, dust flows, and gas venting.

Data Evaluation. That many lunar topographical features superficially look like terrestrial water-formed features is not in question. The Lunar Orbiter and Apollo photos are very suggestive. Detailed scrutiny, however, reveals that the sinuous rilles are actually quite different from terrestrial river beds in some respects. The data are excellent, it is their interpretation that is at issue. Rating: 1.

Anomaly Evaluation. Strong evidence for the existence of considerable water on the lunar surface, past or present, would be highly anomalous. However, the detailed study of sinuous rilles and other supposed water-formed features do not support origins by water erosion or deposition. Nevertheless, the sinuous rilles in particular remain mysterious, with no generally accepted explanation at hand. Rating: 2.

Possible Explanations. Water erosion is still a remote possibility, perhaps when capped with a permafrost cover. Collapsed lava tubes are today's preferred explanations of the sinuous rilles, although one wonders at the great lengths of the rilles and the seeming disappearance of the lava. High velocity ash flows and crustal faulting have been proposed, but they seem inconsistent with the long, meandering character of the sinuous rilles.

Similar and Related Phenomena. Lunar features resembling terrestrial glacier effects (ALE4); the lunar mascons as mare sediments (ALE21); the Martian channels (AME1); terrestrial riverbeds and arroyos.

Examples

X1. Sinuous rilles. These river-like lunar features have long been considered the best evidence that water once flowed on the moon's surface or perhaps below covering layers of permafrost.

"The highly favourable atmospheric conditions at Arequipa have enabled Prof. W. H. Pickering to make numerous observations which have a special bearing on the question of the existence of water on the moon. In addition to the ordinary rills (sic), Prof. Pickering has catalogued thirty-five narrower ones, which, from their resemblance to terrestrial watercourses, he does not hesitate to regard as 'river beds.' These are wider at one end than at the other, and the wide end always terminates in a pear-shaped craterlet. Most of them are only a few miles in length, and a few hundred feet in width at the widest part, and, except when very deep, they are very difficult objects. The largest and most readily observed has its origin in the Mount Hadley range in the Apennines; its course lies a little north of west, and its total length is about sixty-five miles. There does not appear to be any reason to suppose that these formations actually contain water at the present time, but Prof. Pickering brings forward other evidence in favour of the presence of a small amount of moisture on the lunar surface." (R1) See the final entry under the heading (X1) for the Apollo astronauts' exploration of this rille. (WRC)

"Schroeter's Valley, to the west of Aristarchus, has been known since the late eighteenth century, but terrestrial observations could not establish its origin. Orbiter 4 and 5 pictures show this valley in great detail as well as many neighboring valleys. Schroeter's Valley begins in a mountainous region in an enlarged area called the 'Cobra Head', and extends in a generally westerly direction into a smooth maria area. Its greatest width is some 8 km and its depth some 600 or 700 m. Other smaller valleys with 'cobra heads' are found north of Aristarchus and Prinz. Branching occurs in a few cases but mostly there are no tributaries and no deltas. Possibly there are small deposits at the maria ends, but they are much too small to account for the materials eroded from the valleys. South of Aristarchus and northerly from Marius is a valley extending to more than 160 km in a very smooth maria area. It decreases gradually and uniformly in width from the south-easterly end to the north-westerly end where it runs off the picture available to me (Orbiter 4, frame 150, high resolution). Near the larger end it is about 900 m wide and apparently goes to near zero at the other end. It is a very crooked rille and cannot possibly be a physical fracture. It has no tributaries. Similar rilles are found in many places, for example, within and near the Alpine Valley and many other areas. They seem to be a general lunar feature. But why are there no sediment deposits at either end of these valleys and no tributaries? (There are very few examples of tributaries.) The walls seem to be formed by slumping, and north of Krieger there is a row of craters which seem to form an initial stage of such slumping. Water must have run below the surface and formed a tunnel which then caved in. But where did the excavated material go? If the maria are underlaid with ice, it is only necessary to melt the ice and let the water drain into the desert sands. (The explanation offered here was suggested by Gold.) But could it have been lava? Would lava flow in a narrow stream for hundreds of kilometers and disappear without a trace? I believe that the answer to both these questions is 'No'." (R2, R3). Above, "Orbiter" refers to the Lunar Orbiter spacecraft.

The permafrost suggestion. "Abstract. Mature meanders in lunar sinuous rills strongly suggest that the rills are features of surface erosion by water. Such erosion could occur under a pressurizing ice cover in the absence of a lunar atmosphere. Water, outgassed from the lunar interior and trapped beneath a layer of permafrost, could be released by a meteoric impact and overflow the crater to form an ice-covered river. A sinuous rill could be eroded in about 100 years." (R4) Further discussions of this theory may be found in R5-R7. The permafrost theory if also employed to explain the apparent watercourses observed on Mars. (AME1)

Additional rille features. "....a sinuous rille may issue from a small crater on relatively high ground then wind its way for many miles to end on the plains below. Such a rille is very meandering, usually rather uniform in width (some gradually taper), and perfectly continuous. The secondary rille on the floor of Schroeter's Valley even shows oxbows. These properties hardly suggest tensile cracks or dust flows. The sinuous rilles cannot be collapsed lava drainage ditches, for no residual bridges can be seen. Moreover, the great length of these rilles seems to require some liquid much less viscous than lava." (R10) Despite the difficulties just mentioned, the rilles are now generally thought to be collapsed lava tubes. (WRC)

More rille details. "Sinuous rilles are winding, meandering channels or valleys in the lunar surface, unlike normal rilles that are either straight or very gently curved (large radius of curvature). The latter are generally considered to be graben-type faults. Crater chains (linear arrays of partially overlapping craters) are excluded from this definition, as are the highly irregular fractures (rilles) confined entirely within some crater floors. Sinuous rilles may occasionally grade into normal rilles or crater chains. The majority of sinuous rilles occur in the maria on the lunar nearside, at mare-highland boundaries, but a few begin in the highlands. Most are concentrated around the margins of Mare Imbrium and Oceanus Procellarum. Somewhat over half of the observed rilles appear to originate in craters or irregular depressions; about a fifth taper---becoming narrower and shallower with increasing distance from their presumed source. Sinuous rilles can be over 200 km long and 4-5 km wide, but more typical dimensions are lengths of 30 to 40 km and widths of less than one km.

Various theories have been advanced for the origin of sinuous rilles, but none has been entirely satisfactory. Some scientists have proposed that the rilles represent water-eroded streams, or valleys carved by ice-melted water under a rubble-covered permafrost. The basic argument in favor of water erosion lies in the alleged morphological similarity between sinuous rilles on the Moon and terrestrial rivers, particularly with respect to meandering oxbow bends or goosenecks. However, evidence to be presented below will show that lunar rille geometry differs considerably from terrestrial rivers." (R10)

Hadley Rille data from Apollo 15. "Hadley Rille lies at the base of the Apennine Mountains, which form the southeast boundary of the large, multiringed Imbrium Basin. The mountains have prominent fault patterns trending northeast and northwest, respectively concentric and radial to Imbrium. For much of its course, Hadley Rille follows a mare-filled graben valley, trending northeast between two high mountain massifs. Most of the rille is incised in mare material, but locally the rille cuts into the premare mountains. To the northwest, the mare plain extends through a gap in the mountains to join the main part of Palus Putredinus (the Marsh of Decay), and the rille reaches the wider mare plain through this gap. Continuing north, the rille becomes shallower and indistinct, ultimately intersecting a segment of the linear Fresnel II Rille. The south end of the rille adjoins an elongate cleftlike depression that cuts into old highlands covered by a dark mantle. This cleft is commonly compared to the source craters of terrestrial lava channels. ..... The V-shaped profile that characterizes Hadley Rille appears to be a consequence of recession of the walls by mass wasting, so that the talus aprons of the two sides coalesce. Between the landing site and Fresnel Ridge, the rille is discontinuous and consists of a series of coalescing bowl-shaped depressions that are clearly the result of collapse..... Analysis of detailed topography of the part of the rille shown in figure 25-46 indicates an intriguing relationship between depth and width in that the rille is deepest at the widest point. This relationship differs from that shown by river channels, in which depth and width vary inversely, so that the cross-sectional area remains approximately constant. Possibly Hadley Rille was formed by incomplete collapse of a buried lava tube, with more ex-

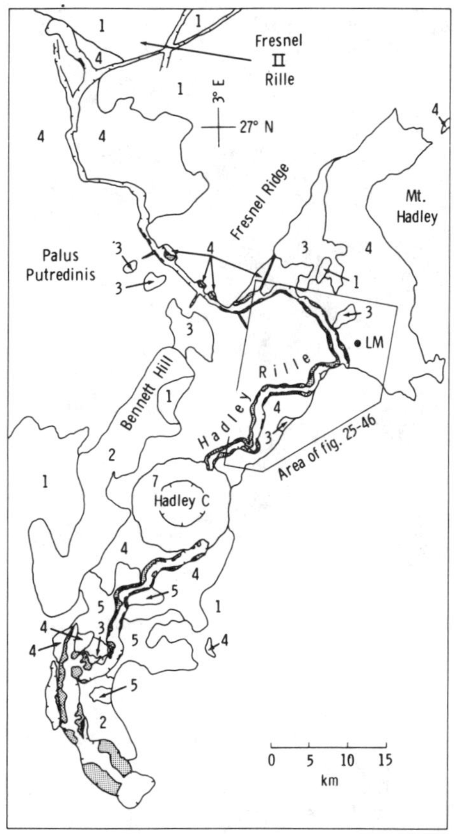

Geologic map of Hadley Rille. The numbers indicate different surface materials; i.e. 4 = mare basalt. LM = position of Lunar Module, Apollo 15 flight. (X1)

tensive foundering at the widest points." (R11)

X2. Miscellaneous lunar features similar to terrestrial water-cut landforms.

H. C. Urey suggested that the lake-like areas near and on the walls of the crater Alphonsus were once actually lakes. T. Gold has wondered if the central masses in some lunar craters, such as Alpetragius, were once pingos (ice-formed cones of debris). (R2)

"Krieger, north of Aristarchus, is an irregular crater, with a recent smaller collision crater on the southern wall, a curious depression in the plain south of the crater, and an irregular mass covering the north wall and an area north of the crater. There is a break in the western wall, and from this break an irregular gully or stream bed

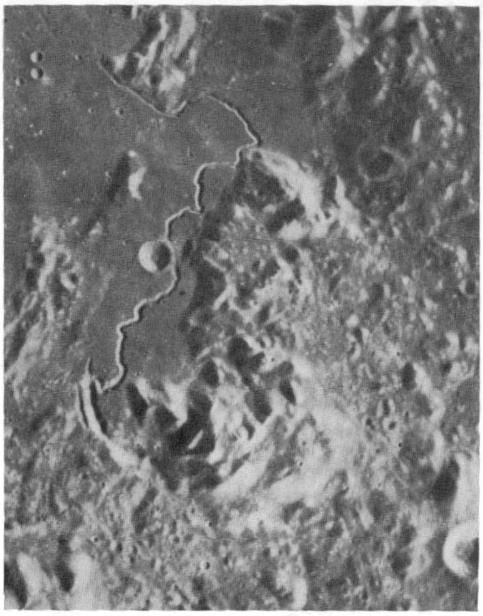

Photograph of Hadley Rille. (X1) (Courtesy NASA)

extends a short distance to the west. This seems to have been a surface stream, and a suggestion of a sediment deposit can be seen at its western terminus. This appears to be a surface stream bed quite different from the other rilles discussed. North of Prinz one of the small rilles of the Schroeter valley type crosses an elongated sunken area, traverse to this valley. It has a flat floor, and a small stream bed, similar to that coming from Krieger, crosses it. These two stream beds indicate that streams flowed on the surface and thus that an atmosphere had a pressure equal to and probably much greater than the vapour pressure of water at its melting point, that is, 4.6 mm of mercury. Could Krieger be a pingo that has melted and collapsed? If so, it is a very large example as compared with terrestrial pingos." (R2, R3)

Survey of purported water-associated lunar features. Abstract. "Since discovery of the mascons (mass concentrations) in the ringed lunar seas, a model explaining their cause has been sought. One of the competing theories is that of J. J. Gilvarry, which explains the mascons as sedimentary deposits, the result of erosion from ancient lunar seas and rivers into the ringed sea basins. An objection frequently raised to this theory has been the apparent absence of water-produced features such as beaches, benches, arroyos and rivers. Examination and interpretation of selected Lunar Orbiter and Apollo 8 pictures has resulted in identification of these features to the authors' satisfaction. The paper presents this photographic evidence with appropriate interpretation. Water cut beaches and benches are common in the areas surrounding the lunar seas. Arroyos, and less commonly, rivers (specific type of sinuous rille) can also be identified. These results constitute support for the Gilvarry hypothesis." (R8, R13)

Dark bands along the bases of lunar mountains. The Apollo-15 astronauts saw and photographed a dark, horizontal band near the base of Mt. Hadley. Similar bands occur near the bases of other lunar mountains where they contact the maria. The astronauts had the impression these bands were high-lava marks associated with the lava-filled maria. (R12) Could these bands be high-water marks or terraces? Probably not, but they are thought-provoking. (WRC)

References

R1. "Lunar River Beds and Variable Spots," Nature, 51:589, 1895. (X1)
R2. Urey, H. C.; "Water on the Moon," Nature, 216:1094, 1967. (X1, X2)
R3. Stubbs, Peter; "Where Moon Rivers Ran," New Scientist, 36:708, 1967. (X1, X2)
R4. Lingenfelter, Richard E., et al; "Lunar Rivers," Science, 161:266, 1968. (X1)
R5. "Through Caverns Measureless to Man," New Scientist, 39:251, 1968. (X1)
R6. "A Link between Sinuous Rilles and Lunar Maria," New Scientist, 41:30, 1969. (X1)
R7. "Thick Permafrost beneath Moon's Maria Impossible," New Scientist, 41: 465, 1969. (X1)
R8. Muller, P M., and Sjogren, W. L.; "Photographic Evidence for the Presence of Water Cut Beaches, Benches, Arroyos, and Rivers on the Moon," Eos, 50:457, 1969. (X2)
R9. "Lunar Riverbeds," Sky and Telescope, 36:91, 1969. (X1)
R10. Gornitz, Vivien; "The Origin of Sinuous Rilles," The Moon, 6:337, 1972. (X1)
R11. Howard, Keith A., and Head, James W.; "Regional Geology of Hadley Rille," Apollo 15 Preliminary Science Report, NASA SP-289, 1972. (X1)
R12. Swann, G. A., et al; "Preliminary Geologic Investigation of the Apollo 15 Landing Site," Apollo 15 Preliminary Science Report, NASA SP- 289, 1972. (X2)

R13. Gilvarry, J. J.; "Evidence for the Pristine Presence of a Lunar Hydrosphere," Astronomical Society of the Pacific, Publications, 76:245, 1964. (X2)

ALE3 The Lunar Rays

Description. Long, quite straight, streaks of white material that radiate outward from large, fresh, bright lunar craters. Dazzling when the sun is high, the lunar rays may stretch hundreds of miles with occasional gaps. They seem independent of the local geology, but they are definitely related to the rays extending from nearby craters. The rays of large craters do not radiate from the crater center, but are more often tangent to the crater rim.

Data Evaluation. Lunar rays are prominent in even small telescopes. Rating: 1.

Anomaly Evaluation. None of the several explanations noted below require significant readjustment of prevailing astronomical and geological theories. The central problem associated with the lunar rays is selecting the correct mechanism of formation. Rating: 3.

Possible Explanations. (1) Dust sprays from a system of cracks in the lunar crust caused by the impact of the crater-forming body; (2) Dust and rock flour splashed out during the cratering process; and (3) splashes of fine material accompanying the impacting body, for example, cometary material guided, perhaps, by magnetohydrodynamic forces.

Similar and Related Phenomena. Many large lunar craters possess surrounding aprons of debris splashed out during their formation; but these splashes are usually formless and very close to the crater.

Examples

X1. Permanent lunar features. The following summary, written by the compiler, is based mainly on R4, R7, and R8.

Many of the large, fresh-looking lunar craters---Copernicus, Tycho, Kepler, and others---are the hubs of systems of white streaks that seem to emanate radially. Through the telescope, these "rays", as they are called, present an imposing picture. Visualize Copernicus, a gaping round wound in the moon's crust, some 50 miles in diameter. This crater is ringed with high, rugged walls. Outside the crater walls a complex, splash-like array of white streaks extend ten crater-diameters and more in all directions. When the sun is low, these rays are inconspicuous. Under a high sun they become so brilliantly white that the terrain around the crater is lost in the glare. The splash-like configuration of the rays make it easy to believe them to be the pattern made by rock flour ejected from Copernicus during a cataclysmic impact.

The lunar rays, however, are not what they superficially seem to be. In fact, no one is quite sure what they really are. No one contests that the rays are definitely associated with large, bright craters and with some small craters as well; the "big splash" theory, however, does not seem to explain all of the facts.

Three general properties of the rays must be explained by any ray theory: (1) The rays are predominantly associated with bright, recent craters. We know the rayed craters are young because their topography is sharp and has not been smoothed by eons of erosion by the perpetual rain of cosmic dust. (2) Single rays may strike out for hundreds of miles over all sorts of geological formations. In other words, the rays are basically independent of local geology. (3) Virtually invisible at lunar sunrise and sunset, the rays become very bright at high sun angles. This fact suggests that their prominence is due to the optical properties of the ray material rather than the topography. The upshot is that the rays are pretty certainly streaks of material of no great thickness deposited on the surface around young craters. The older, darker craters possess no ray systems; or, if they once had them, they have become darkened with age.

Detailed examination of single rays provide additional clues about the nature of this phenomenon: (1) Single rays often display obvious gaps. (2) Most rays, where studied closely, are seen to be peppered with small craters and craterlets. (3) Rays traced back to the suspected parent crater may miss it completely, or, very frequently, be tangent to the crater's rim. Only the small craters seem to have truly radial streaks issuing from them. The gaps in the rays and their frequent tangency suggest that they were probably not deposited from the dust cloud raised after the formation of the crater. Why, for that matter, would a dust cloud settle back to the surface in immensely long, very narrow streaks? All this does not mean that the formation of the crater was not involved in the creation of the rays. After all, just about all rays are intimately associated with craters.

The ray problem becomes even more fascinating when one follows the rays of a large crater towards a second, nearby rayed crater. Sometimes a ray from one crater simply blends in, end-to-end, with a ray from the adjacent crater. If, however, the ray approaches the ray of another crater at a large angle, it ends abruptly at the point of intersection. Apparently, the rays of adjacent craters are not necessarily independent of one another, even though the craters were created at different times. (R4, R7, R8)

References

R1. "Lunar Streaks," English Mechanic, 54:13, 1891. (X1)

R2. Holden, Edward S.; "The Systems of Bright Streaks on the Moon," Astronomical Society of the Pacific, Publications, 4:81, 1892. (X1)

R3. Schaeberle, J. M.; "The Bright Streaks on the Moon," Astronomical Society of the Pacific, Publications, 5:198, 1893. (X1)

R4. Alter, Dinsmore; "Nature of the Lunar Rays," Astronomical Society of the Pacific, Publications, 67:237, 1955. (X1)

R5. O'Keefe, John A.; "Lunar Rays," Astrophysical Journal, 126:466, 1957. (X1)

R6. Giamboni, Louis A.; "Lunar Rays, Their Formation and Age," Astrophysical Journal, 130:324, 1959. (X1)

R7. Devadas, P.; "The Origin of Lunar Rays," British Astronomical Association, Journal, 72:380, 1962. (X1)

R8. Oberbeck, Verne R.; "A Mechanism for the Production of Lunar Crater Rays," The Moon, 2:263, 1971. (X1)

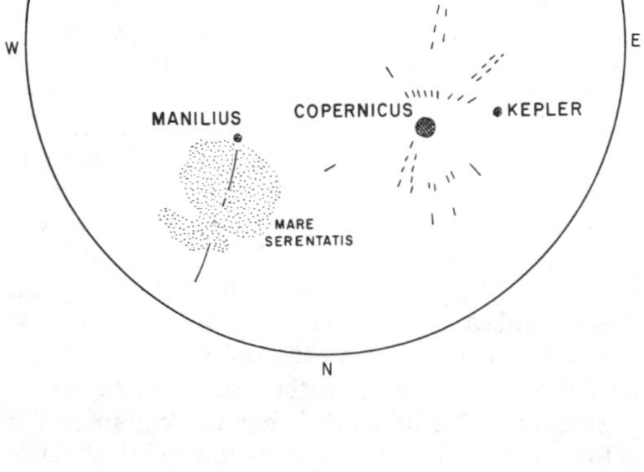

Selected rays from Tycho and Copernicus, illustrating how some do not converge on the craters per se. (X1)

ALE4 Lunar Features Seemingly Shaped by Ice

Description. Lunar features resembling terrestrial features that are believed to have been shaped by ice action, such as moraines, pingos, shrinkage features, and terrain with a pronounced linear texture.

Data Evaluation. The lunar features involved only "look like" terrestrial glacier and ice-worked features. Terrestrial analogies are dangerous to use on an object as alien as the moon. Rating: 3.

Anomaly Evaluation. The past presence of large accumulations of ice on the moon is a rather startling thought, for the moon is now an extremely arid body. Rating: 2.

Possible Explanations. Ice meteors or comets (normally icy) impacting at rather low velocities have been suggested as both the cratering objects and the sources of large quantities of ice.

Similar and Related Phenomena. Sinuous rilles and other apparently water-worked lunar features (ALE2); terrestrial ice falls (GWF1).

Examples

X1. General observations. "Some of the Moon's visual features are faintly suggestive of glacial till in various forms. For example, striations in the northwest wall of Copernicus look a little bit like a medial moraine. Vast areas of the Moon have scattered low domes, of which the largest is Reumker in Sinus Roris, and which have prompted several authors to suggest that they are like terrestrial pingos. The more popular explanation, of shield volcanos, suffers from the lack of more common and conspicuous volcanic features which always accompany shield volcanos on Earth. Finally, there are in the floors of certain craters (always of the mare type) low mesalike domes, flatter than those mentioned above and sometimes having a faint suggestion of layers or terraces. It should be mentioned at this point, that the ice collision hypothesis does not deny the possibility of volcanism, but only its importance in shaping the major features.

The most important indication of the former presence of ice is in the extensive shrinkage at various points on the Lunar surface. These shrinkage features always fit the notion of granular material loosely suspended in ice, settling downward as the ice is lost through sublimation or melting. Contour estimates of the maria, for example, indicate that their centers are depressed to a surprising degree. The center of Mare Imbrium is 6 km lower than its outer regions. It is striking that the total volume of this depression, some 8×10^5 km^3, is in the range of size estimated for the colliding object which created the mare. Many smaller features, such as craters having reticulate rilles or clustered domes, may be readily explained by the loss of volume in the form of ice. Explanation of the same features by the notion of volcanic calderas suffers from the fact that no blocks of hard material are visible." (R1)

X2. The crater King. A camera on the orbiting command module of Apollo 16 took a photo of a section of the farside crater King, which showed a marked linear texture. The surface looks strikingly like the result of glacier action. Several "ponds" are also revealed. (R2)

References

R1. Bruman, J.R.; "Ice on the Moon," Icarus, 8:198, 1968. (X1)

R2. "Puzzling Lunar Terrain Snapped by Apollo 16," New Scientist, 55:127, 1972. (X2)

ALE5 Swirl Markings

Description. Bright/dark ribbon- and loop-like patterns, 10-50 kilometers in size, draping craters and other features and crossing both mare and highland terrains. The swirls are associated with strong magnetic anomalies and are usually located antipodal to large impact basins.

Data Evaluation. The photographs and magnetic measurements from the Lunar Orbiter and Apollo programs are of very high quality. Rating: 1.

Anomaly Evaluation. The swirl patterns cannot be explained as simple crater ejecta. Their association with magnetic anomalies seems to indicate some poorly understood magnetohydrodynamic process created by impacting objects. Rating: 2.

Possible Explanations. Magnetization and patterned distribution caused by cometary impact (R4); lunar basin ejecta that is more strongly magnetized than normal (R5)

Similar and Related Phenomena. Similar swirl markings on Mercury (AHE3); the lunar rays (ALE3); lunar magnetic anomalies (ALZ1).

Examples

X1. Visual discovery of swirls. "Groups of light colored markings with discontinuous sinuous patterns occur in three maria: (1) The most typical example is in Mare Inginii where the swirls occupy much of its SW quadrant; (2) A similar group lies north of Mare Marginus especially west of Goddard and NE of Ibn Yunus; (3) A somewhat atypical occurrence includes the Reiner γ Structure and associated bright markings in Oceanus Procellarum. These swirls are best seen at small phase angles, however, because of their high albedo, they are also detectable at low sun. The swirls vary considerably in shape and size. In all cases there are no topographic expressions associated with them. The following relationships are noted: (1) The swirls of Mare Inginii lie diametrically opposed to the center of Mare Imbrium; (2) The swirls of Mare Marginus lie on the opposite side of Mare Orientale; and (3) The markings in Oceanus Procellarum are opposite to a point within one crater diameter of Tsiolkovsky. It is proposed that these markings are produced as a result of disturbances caused by major impacts; the disturbances may have been caused by seismic wave penetrations at the antipodal areas of impact points." (R1) At this point in time (1971), there was no correlation of the swirls with magnetic anomalies. (WRC)

X2. Swirls associated with lunar magnetic anomalies. "Bright/dark swirl patterns occur in three regions of the Moon. One of the best known examples is Reiner γ, which is near the western limb of the nearside (5°S, 60°W). The concentrations of patterns near Mare Marginis (15°N, 90°E) and Mare Ingenii (35°S, 180°E) are more impressive. Bright/dark swirls have not previously been studied in detail, but have generated several possible interpretations including nué ardente deposits, antipodal effects of major basins, volcanically derived sublimates, secondary impact effects, unusual secondary cratering phenomena associated with a cometary impact, and selective preservation of albedo (crater rays) controlled by local enhancement of magnetic fields. More detailed examination, however, reveals features that are suggestive of remnants of the impacting nuclear region of a comet.

Swirl patterns range in size from nearly 10 km to < 50 m across and form a variety of characteristic geometries: ribbon-like patterns, open loops and closed loops.... these patterns are commonly crossed by dark lanes. Both dark and bright patterns drape relief such as crater walls and rims and cross both the highlands and mare terrains. At the highest resolutions available (~ 10 m), alteration of the surface (scouring) is not visible. Rather, the patterns represent diffuse brightening/darkening of unmodified terrains and commonly exhibit sharply defined boundaries over 50-100 m scales. In several examples, higher albedo regions correspond to sloped surfaces (for example, walls of degraded craters), whereas lower albedo regions correspond to low-lying regions (for example, crater floor). Under high illumination, such patterns form a bright ring that strongly contrasts with other adjacent degraded craters. Ring patterns occur, however, in plains regions, not in association with changes in relief. Consequently there may be several processes contributing to the formation of the swirls."

The swirls are also associated with strong magnetic anomalies (ALZ1) and seem to be very young lunar features, perhaps less than 10^8 years old. (R4) The authors of R4 contend that the swirls and magnetic anomalies were both created by cometary impacts. Abbreviated discussions of the swirls and magnetic anomalies may be found in R2, R3. See also ALZ1.

In a letter to Nature, L. L. Hood questions the cometary impact theory of Schultz and Srnka (R4), and suggests that "meteorite impacts occurring over a span of time comparable to the age of the Solar System can explain their (the swirl's) existence at least as well as can recent comet impacts." An important point emphasized by Hood is the nonrandom distribution of the swirls; they seem to be antipodal to large impact basins. A letter from Schultz and Srnka follows, defending their position. (R5)

References

R1. El-Baz, Farouk; "Light Colored Swirls in the Lunar Maria," Eos, 52:856, 1971 (X1)

R2. "Magnetic Deposit on the Moon," Eos, 60:28, 1979. (X2)

R3. "Magnetism Is Patchy on the Moon," New Scientist, 8 :925, 1980. (X2)

R4. Schultz, Peter H., and Srnka, Leonard J.; "Cometary Collisions on the Moon and Mercury," Nature, 284:22, 1980.(X2)

R5. Hood, L. L., et al; "Cometary Collisions on the Moon and Mercury," Nature, 287: 86, 1980. (X2)

ALE6 Anomalous Red Formations

Description. Red areas observed on the lunar surface which seem to be associated with high levels of radioactivity.

Data Evaluation. Lunar Orbiter and Apollo data. Rating: 1.

Anomaly Evaluation. The anomaly here does not concern the red color or the radioactivity of the features, but rather our inability to identify these features in the context of lunar geology and history. Rating: 3.

Possible Explanations. Exposures of pre-mare basalt containing unusually high quantities of radioactivity.

Similar and Related Phenomena. Lunar radioactivity concentrations (ALE13).

Examples

X1. Permanent features. "Several anomalous red features observed by Whitaker have been examined in detail. Crater counts and regional stratigraphy suggest ages comparable to that of highland material. Initial gamma ray spectrometry indicates increases in native radioactivity are associated with certain red areas. A possible explanation for these observations is that the red objects are the surface manifestation of high radioactivity, old, possibly pre-mare basalts (KREEP/Norite Material). Conclusive proof of the nature of the objects would depend on the study of returned material, making them ideal candidates for future space missions ....." (R1)

References

R1. Malin, Michael C.; "Lunar Red Spots: Possible Pre-Mare Basalts," Eos, 54: 359, 1973. (X1)

ALE7 Layered Structures

Description. Visual, seismic, and radio-sounder detection of layered structures and lineaments on the moon's surface and beneath it. The separation planes vary from a few inches to thousands of feet.

Data Evaluation. Many convincing photographs and radio soundings from the Apollo Program. As the Apollo scientists discovered, visual and photographic observations of lineaments have to be checked carefully to eliminate artifacts due to lighting and viewing angles. Rating: 1.

Anomaly Evaluation. Layered structures created by natural fracturing and jointing are to be expected, as are superimposed layers of crater ejecta (volcanic and impact types). However, layering due to compositional changes are harder to explain in terms of the accepted theory of lunar evolution. Some of the lunar "strata" are highly inclined to the horizontal---a condition indicating considerable crustal deformation. These suggestions of extensive, active crustal evolution are more surprising than layering per se. Rating: 3.

Possible Explanations. Compositional changes, the natural fracturing and jointing of bedrock; episodic deposition from volcanic and impact craters. Fluid- or ice-carried sediments would imply the past existence of much more water on the moon than now considered possible.

Similar and Related Phenomena. The possible past existence of water and ice on the moon (ALE2 and ALE4).

Examples

X1. West face of Mt. Hadley, Apollo 15 site. "Area 5 is the entire west face of Mt. Hadley, partly shown in figures 5-17 and 5-19. Lineaments measured in an intermediate-illumination panoramic camera photograph are grouped around three major trends that correlate with the lineaments photographed from the lunar surface. The predominant set trends north 55°E and corresponds with the well-defined set of linears dipping steeply left in figures 5-17 and 5-19. Although no single linear can be traced across the entire outcrop face, the parallelism and crisp definition of some bright linears through distances of tens or hundreds of meters give the impression that they are the surface traces of compositional layers or of a system of well-defined, uniform fractures. The impression is heightened in a few places where the same linear apparently emerges both upslope and downslope from beneath the cover of one of the many subhorizontal, smooth, regolith-covered benches on the mountain face. In other places, faint traces of the linears extend across these benches." (R1)

West face of Mt. Hadley, showing northeast-trending lineaments. Apollo 15 photograph. (X1) (Courtesy NASA)

X2. Silver Spur, Apollo 15 site. "The striking 500-mm photograph taken from the LM (fig. 5-21) shows massive ledges apparently dipping gently to the left and crossed by finer, more nearly horizontal lineaments that give the impression of crossbedding. The finer lineaments are probably caused by the foreshortened view across an undulating cratered surface, an effect similar to that on the south wall of St. George crater." (R1)

X3. Outcrop near Apollo 15 site. "At least 12 layers occur in the outcrop (fig. 5-37). Several of the more massive of these layers (1 to 3 m thick) contain less well defined internal layering or parallel banding. Thinner

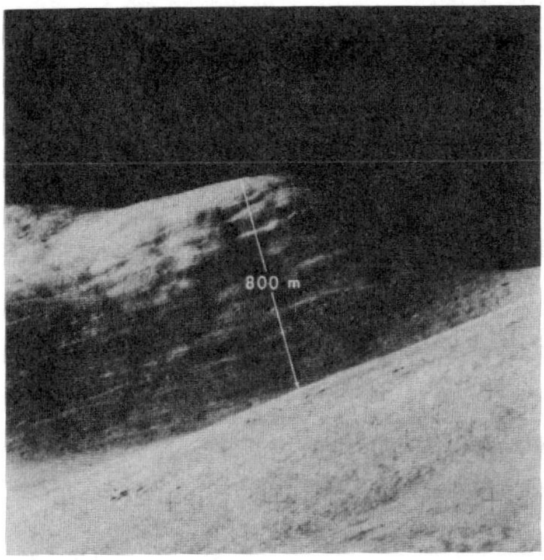

Apollo-15 photograph of Silver Spur, 20 kilometers distant, showing massive ledges. (X2) (Courtesy NASA)

layers less than a meter thick occur together or separate the more massive layers from each other. These thinner layers weather out distinctly." (R1)

X4. Mare Serenitatis, Apollo 17 flight. "The first clear evidence of subsurface layering in the flooded basins which form the lunar maria has come from the 5 MHz lunar sounding experiment on the Apollo 17 flight. An eight-man team from the Department of Geology and Geophysics, University of Utah and the Jet Propulsion Laboratory, Pasadena, California, has found two clear interfaces separating different layers at mean depths of 900 and 1600 metres in Mare Serenitatis and one interface at 1400 metres depth in Mare Crisium.... The layering becomes indistinct close to major features such as craters and shows changes in form underneath wrinkled ridges. It also shows evidence of faulting systems. When the interfaces reach the surface they correlate well with visible surface units." (R2, R3)

References

R1. Swann, G. A., et al; "Preliminary Geologic Investigation of the Apollo 15 Landing Site," Apollo 15 Preliminary Science Report, NASA SP-289, 1972. (X1-X3)

R2. "Apollo Gets under the Moon's Skin," New Scientist, 80:437, 1978. (X2)

R3. Peeples, Wayne J., et al; "Orbital Radar Evidence for Lunar Subsurface Layering in Maria Serenitatis and Crisium," Journal of Geophysical Research, 83:3459, 1978. (X4)

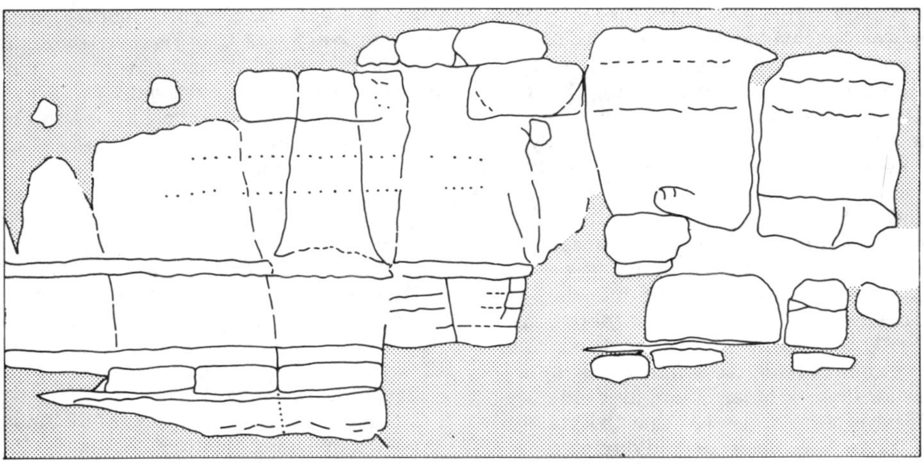

Drawing of layered outcrop near Apollo-15 landing site. Outcrop is about 8 meters high. (X3)

ALE8 Lunar Glasses

Description. The appearance of glassy veneers on lunar soil and rocks.

Data Evaluation. Photographs and samples from the Apollo Program. Rating: 1.

Anomaly Evaluation. The glassy clumps discovered in some shallow lunar craters by the Apollo 11 astronauts were initially very puzzling, but on later missions glassy veneers were found in diverse situations. These findings led to a consensus that impact melting was the common cause of almost all lunar glass as well as the ubiquitous glass beads in the lunar soil. Lunar glass, therefore, although once anomalous is now considered fairly well explained---more curious than enigmatic. Rating: 4.

Possible Explanations. The glassy veneers are splashes of lunar material that was melted by high velocity impacting meteors. The glassy craters are thought to be secondary craters excavated by molten or partially molten ejecta from primary craters.

Similar and Related Phenomena. On the earth: the Libyan desert glass, the Australian Darwin glass, and tektites (ER).

Examples

X1. General observations. Summarized by the compiler from the references listed below.

The lunar surface is liberally pockmarked with small craters in the 1-to-5-foot range. As the Apollo 11 astronauts walked among these depressions, they noted clumps of glass-crusted soil near the centers of some of these small craters. On this lunar landing this was the only type of glazing observed. With a close-up stereo camera, the astronauts photographed the glassy clumps for later analysis, noting that the clumps appeared to have been deposited accidently in the craters.

The Apollo 11 photos created a stir, for the glazing was entirely unexpected. Was the glass coating applied to the clumps when the craters were excavated by meteors? The kinetic energy of the meteor's impact could have been converted to heat, raising the soil above its melting point, and thus coating the clumps with a sheen of glass. The glassy veneer seen in the photos looked much like that acquired by pottery in a glazing oven. On the other hand, the clumps could have originated when neighboring craters were blasted out by meteors. Then, glass-encrusted debris might have fallen into the nearby craters and rolled toward their centers. Still, the Apollo 11 astronauts saw no glassy clumps outside the small craters on the surrounding surface. Thomas Gold, a Cornell physicist, suggested that the glassy clumps might have been formed during an intense burst of solar radiation which, when focussed by the crater walls, was powerful enough to fuse the surfaces of clumps of soil in the crater centers.

The astronauts of Apollo flights 15, 16, and 17 ranged farther around their landing sites, both on foot and while riding in the Lunar Rover. Alert for glassy objects, they found many. The small, shallow craters with glass-coated clumps of soil turned out to be rather common. Glassy patches of soil were also seen inside these craters in addition to the glazed clumps. The Apollo 15 crew asserted that they could always predict from afar which craters would contain glass. Glassy craters were always fresh-looking and not rounded by the erosion of the steady rain of micrometeorites. Bits of ejecta were also scattered around these craters. These later astronauts also discovered isolated lunar rocks partially coated with veneers of glass. Such rocks did not seem to be associated with any specific craters. Another pertinent discovery was that glass-coated rocks were incorporated in lunar breccias (small angular rocks cemented together). Finally, the lunar soil proved to be full of tiny glass beads. Lunar glass was almost everywhere and in many forms. (R1-R8)

References

R1. Gold, T.; "Apollo 11 Observations of a Remarkable Glazing Phenomenon on the Lunar Surface," Science, 165:1345, 1969. (X1)

R2. "Gold and the Glassy Craters," Nature, 226:598, 1970. (X1)

R3. Green, Jack, et al; "Origin of Glass Deposits in Lunar Craters, Science, 168: 608, 1970. (X1)

R4. Morgan, John W., et al; "Glazed Lunar Rocks: Origin by Impact," Science, 172:

556, 1971. (X1)
R5. "Not All That Glisters Is Moonstuff," New Scientist, 50:438, 1971. (X1)
R6. Schaber, Gerald G., et al; "Glass in the Bottom of Small Lunar Craters: An Observation from Apollo 15," Geological Society of America, Bulletin, 83:1573, 1972. (X1)
R7. "The Orange Glass and the Lunar Highlands," Science News, 103:7, 1973. (X1)
R8. Wilshire, H. G., and Moore, H. J.; "Glass-Coated Lunar Rock Fragments," Journal of Geology, 82:403, 1974. (X1)

ALE9 Nonrandom Distribution of Lunar Craters

Description. The occurrence of lunar craters in recognizable patterns or in statistical departures from randomness. Of most interest here are lines and clumps of craters.

Data Evaluation. This type of anomaly requires two kinds of information: (1) Observational data on crater sizes, locations, ages, etc.; and (2) The determination of nonrandomness by either subjective evaluation or statistical analysis. Although subjective decisions as to nonrandomness are risky (viz., some lunar lineaments in ALE7), this does not seem to be a serious problem here. The data and analyses seem very reliable. Rating: 1.

Anomaly Evaluation. The location of large lunar craters along great circles as consequences of the impacts of ancient lunar satellites, while a fascinating possibility, is not really incompatible with any of the three favored hypotheses of lunar origin (accretion, capture, fission). More serious,however, are the implications of nonrandomness of crater origin, especially the favoring of volcanism over meteor impact. Rating: 2.

Possible Explanations. Internal lunar forces (volcanism, gas emissions) occurring along crustal cracks or in areas of crustal weakness; the nonrandom incidence of meteors, possibly because of their concentration in favored lunar orbits.

Similar and Related Phenomena. Lunar satellites (ALL1); Martian crater belts (AME13); linear patterns of terrestrial volcanos (ET).

Examples

X1. Linear arrays of small craters. "In discussions of geological processes on the Moon, frequent reference is made to various linear arrays of small craters. Such arrays occur, for example, in the vicinity of Copernicus, as well as at several locations in the highlands, such as the edge of the crater Mueller. It is often pointed out that such linear formations are difficult to explain as resulting from random impacts. On the basis of this difficulty, the linear arrays are commonly held up as evidence of geologic activity on the Moon. That is, some subsurface phenomenon gave rise to a linear crack on the surface, along which volcanic activity occurred, which resulted in craters also arrayed along a line." The author then ventures that a large body approaching the moon might disintegrate inside the Roche limit in such a way as to produce linear strings of objects. (R1)

X2. Crater clumpiness. "Summary. A fresh attack on the vital problem of the origin of the lunar craters has been made by analyzing the surface distribution of craters of given diameter. The distribution shows a general clumpiness in both the lunarite (bright regions) and the lunabase (dark regions). Craters between 30 and 40 km in diameter, situated in the lunarite, are nonrandomly distributed at the 2 per cent level of significance. This argues against the impact theory.

It is found that the number-density of differently-sized craters is slightly greater in the following half of the Moon than in the preceding half. This result is shown to apply equally to the lunarite and, taken separately, to the lunabase, and again argues against the theory that the craters were produced exclusively by impact.

In assessing the origin of the craters on the basis of the observed frequencies and distribution of craters alone, it is concluded that the ratio of the number of impact craters to the number of endogenic craters is not very large. If only one theory is allowed, it must be that the craters are of internal origin." (R2)

"Summary. In a recent study of the distribution of centres of lunar craters, Fielder found two apparent anomalies which he used as arguments against the meteoroidal impact hypothesis for the origin of lunar craters. First of all, the Poisson distribution gave a very bad fit to the number of crater centres in equiareal sectors of the continents and the maria, especially among craters smaller than 40 kilometres in diameter. The second difficulty was a systematic excess in the number density of small craters in the western (trailing) half of the Moon. We will show that these observations could also have been expected even under the impact hypothesis, since the numbers of small and of large craters in a finite region are negatively correlated. Available crater statistics therefore neither preclude nor establish the impact or volcanic hypothesis for the origin of craters." (R3)

X3. Arcuate patterns of very large craters. "Abstract. The circular maria---Orientale, Imbrium, Serenitatis, Crisium, Smythii, and Tsiolkovsky---lie nearly on a lunar great circle. This pattern can be considered the result of a very close, non-capture encounter between Moon and Earth early in solar-system history. Of critical importance in analyzing the effects of such an encounter is the position of the weightlessness limit of the Earth-Moon System which is located at about 1.63 R_e, measured from the center of Earth to center of Moon. Within this weightlessness limit, material can be pulled from the lunar surface and interior by Earth's gravity and either escape from the Moon or be redistributed onto the lunar surface. In the case of an encounter with a non-spinning Moon, backfalling materials would be distributed along a lunar great circle. However, if the Moon is rotating during the encounter, the backfall pattern will deviate from the great circle, the amount depending on the rate and direction of spin. Such a close encounter model may be related to the pattern of circular maria if materials departing from the source region are visualized as spheroids of molten lunar upper mantle basalt. These spheroids, then, would impact onto the lunar surface to form a pattern of lava lakes. Radiometric dates from mare rocks are consistent with this model of mare formation if the older mare rock dates are considered to date the encounter and younger dates are considered to date subsequent volcanic eruptions on a structurally weakened Moon." (R4)

"Abstract. An analysis has been made of the tendency of large lunar craters to lie along circles. A catalog of the craters >50 km in diameter was prepared first, noting position, diameter, rim sharpness and completion, nature of underlying surface, and geological age. The subset of those craters 50-400 km in diameter was then used as input to a computer program which identified each 'family' of four or more craters, of selected geological age, lying on a circular arc. For comparison, families were also identified for randomized crater models in which crater spatial density was matched to that on the Moon, either overall or, separately, for mare and highland areas. The observed frequency of lunar arcuate families was statistically highly significantly greater than for the randomized models, for craters classified as either late pre-Imbrian (Nectarian), middle pre-Imbrian, or early pre-Imbrian, as well as for a number of larger age-classes. The lunar families tend to center in specific areas of the Moon: these lie in highlands rather than maria and are different for families of Nectarian craters than for pre-Nectarian. The origin of the arcuate crater groupings is not understood." (R5)

References

R1. Hibbs, A.R.; "Linear Arrays of Small Craters," Planetary and Space Science, 8:121, 1961. (X1)

R2. Fielder, Gilbert; "Distribution of Craters on the Lunar Surface," Royal Astronomical Society, Monthly Notices, 129: 351, 1965. (X2)

R3. Marcus, Allan; "Comments on 'Distribution of Craters on the Lunar Surface,'" Royal Astronomical Society, Monthly Notices, 134:269, 1966. (X2)

R4. Malcuit, R.J., et al; "The Great Circle Pattern of Large Circular Maria: Product of an Earth-Moon Encounter," The Moon, 12:55, 1975. (X3)

R5. Jaffe, L.D., and Bulkley, E.O.; "Lunar Crater Arcs," The Moon, 16:71, 1976. (X3)

ALE10 Unexplained Minor Surface Features

Description. Minor, local features of the lunar surface which are different from the surrounding terrain in structural detail, reflectivity, color, etc.

Data Evaluation. For the single example found so far, Apollo photography provides excellent evidence. Rating: 1.

Anomaly Evaluation. Although the photography is good, it is impossible to come to any firm conclusion about the nature of the example reported below. We can only guess that it is only a minor eccentricity of the lunar surface. Rating: 3.

Possible Explanations. None.

Similar and Related Phenomena. None.

Examples

X1. Permanent feature near the Apennine Mountains. " While the Apollo 15 panoramic camera photography was being rapidly scanned, a most peculiar feature was noted that presented a totally different appearance from anything seen in all the rest of the Lunar Orbiter and Apollo photography. The feature was missed on Lunar Orbiter IV frame H-102 because it is situated in a group of bimat marks. The feature is located at latitude 18°40' N, longitude 5°20' E in a small patch of mare material lying between the Haemus and the Apennine Mountains. This patch is abnormal in that it is an unbordered plateau; the surface appears to lie several hundred meters above adjacent mare patches.

The feature is D-shaped with a 3-km-long straight edge. Viewed stereoscopically, it is seen to lie perhaps a few tens of meters below the level of the surrounding mare, the latter presenting a convex meniscus at the line of contact. About half the floor is covered with blobs of marelike material, reminiscent of dirty mercury. Contacts between the floor and both the mare and blobs are frequently outlined with highly reflective material, perhaps sublimates. The floor also displays some darker areas that have noticeably different photometric properties from the mare surface and the blobs. The whole feature is seen to be almost devoid of small impact craters, thus differing from the surrounding mare." (R1)

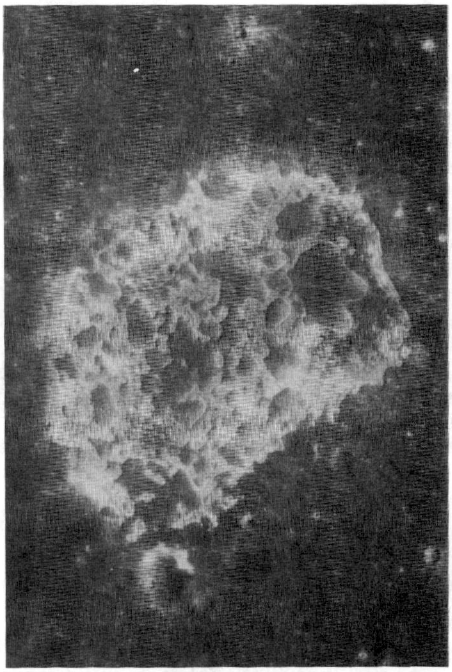

Peculiar 3-kilometer-wide feature near the Apennine Mountains. (X1) (Courtesy NASA)

References

R1. Whitaker, Ewen A.; "An Unusual Mare Feature," Apollo 15 Preliminary Science Report, NASA SP-289, 1972. (X1)

ALE11 Large-Scale Asymmetries in Composition

Description. Large regions of the lunar surface which differ in composition from adjacent or analogous regions. These areas are generally mare-sized or larger.

Data Evaluation. All of the data sources (photographs, γ-ray spectrometry, actual samples) are of high quality. Rating: 1.

Anomaly Evaluation. Since very large areas of the moon's surface differ significantly in composition from pre-Apollo expectations, some important revisions in lunar history may be indicated. Rating: 2.

Possible Explanations. The unexpected concentrations of KREEP material and its surprising presence in the Appenines may be the consequences of a very large impact predating the formation of the present mares, and exceeding them in size. This catastrophe could have released KREEP material over a wide region. (R2) The general asymmetry of the lunar mares (impact basins) is covered in ALE1.

Similar and Related Phenomena. Lunar topographical asymmetry (ALE1); small-scale concentrations of radioactivity (ALE13); volatile-rich areas of small scale (ALE16)

Examples

X1. Concentrations of KREEP (basalt enriched with trace elements) in Oceanus Procellarum and Mare Imbrium. "Measurements with γ-ray spectrometers have revealed that the levels of natural radioactivity (which results from U, Th and K) are considerably higher in Oceanus Procellarum and Mare Imbrium than anywhere else on the lunar surface. These regions are therefore presumed to be richer in KREEP material. It is usually assumed that this distribution of KREEP arose as a result of the Imbrium impact. It has, however, been pointed out that if KREEP was ejected by the Imbrium event then the Haemus Mountains to the east of the Imbrium basin should be as radioactive as the Fra Mauro formation to the south. This has not been found to be the case." (R2, R1)

X2. Lack of anorthosite highland crust in the Appenines. "Before the Apollo 15 mission it had been expected that the lunar Appenines would consist of anorthositic crust, supposedly lifted up by the Imbrium impact. When samples were returned from the Appenine front, however, they were found to have predominantly a KREEP composition, and there were few truly anorthositic rocks. This lack of anorthositic crust in the region has since been confirmed by the results of the X-ray fluorescence experiments. The measured Al:Si ratio from the Appenines is about half the value obtained from the Descartes highlands. These results suggest that the original crust in this region was removed and replaced with KREEP basalts, before the formation of the Imbrium Basin." (R2)

X3. Concentration of mare material on the lunar nearside. A consequence of this important asymmetry is the concentration of KREEP on the nearside. (R3) This particular asymmetry in composition is, of course, initimately related to the topographical asymmetry of the lunar farside and nearside, as presented in ALE1. (WRC)

References

R1. Wood, J. A.; "Geographic, Geophysical, and Chemical Asymmetry of the Moon: Why?" Meteoritics, 8:82, 1973. (X1)

R2. Cadogan, P. H.; "Oldest and Largest Lunar Basin," Nature, 250:315, 1974. (X1, X2)

R3. Stevenson, D. J.; "Lunar Asymmetry and Palaeomagnetism," Nature, 287: 520, 1980. (X3)

ALE12 Dark-Haloed Lunar Craters

Description. A rare type of lunar crater that is surrounded by a halo of material that is significantly darker than the surrounding terrain. Three varieties are recognized: (1) Non-impact craters with dark halos and, frequently, dark rays. This variety is not randomly distributed and is concentrated along rilles, fractures, etc. (2) Impact craters with dark halos that are composed of impact melt, either in the form of ejecta or solidified flow. (3) Impact craters with dark halos that do not seem to be composed of impact melt. All types of dark-haloed craters apparently have bright centers.

Data Evaluation. High-quality photographs from spacecraft. Rating: 1.

Anomaly Evaluation. Convincing explanations of most dark-haloed craters are at hand, although these explanations have not yet been verified by actual samples. Rating: 3.

Possible Explanations. The first type of dark-haloed crater is generally agreed to have a volcanic origin. Type two seems to be the consequence of either ejecta of impact melt or, perhaps, dark projectile material, or dark impact melt that flowed outward from the crater. The genesis of the third type is more controversial but may be due to the excavation of dark, subsurface mare material from the impact.

Similar and Related Phenomena. Lunar rays (ALE3).

Examples

X1. Rare, permanent features. "A dark-haloed crater, unlike a typical recent crater on the lunar surface, is generally surrounded by material of a lower albedo than that of its surroundings. This low-albedo material may be either in the form of a roughly circular aureole, or in the form of rays. Whatever the form of its halo, this sort of crater is relatively rare on the visible disk. Identification is made difficult because dark-haloed craters are typically small (less than 5 km in diameter) although a few such craters have been identified which are greater than 19 km in diameter. In addition, the halo varies both in degree of darkness and of contrast with the surrounding surface......

(1) The most prominent dark-haloed craters typically displayed a bright central crater, with a well-defined, very dark halo, which is roughly one crater diameter wide. Some of these craters have dark rays that extend several crater diameters beyond the halo. The less prominent dark-haloed craters displayed the same morphological characteristics, but the haloes were less dark and the craters less bright.

(2) The larger dark-haloed craters typically have haloes less than one crater diameter wide---e.g. Picard. One large crater, Dionysius, has an extensive system of very dark rays, but this is an unusual feature.

(3) Intense dark spots occur on these photographs, which are typically quite small (less than 1 km in diameter) and possess no discernable crater. Often many such dark spots are found together in fields, usually on a darker than usual mare surface. Presumably, these are also dark-haloed craters, with the crater itself too small to be resolved. ......

The distribution of dark-haloed craters ....is obviously not random. These features are concentrated on the maria, and commonly clustered around recent light-rayed craters, such as Copernicus and Aristillus, or along the margins of the maria. This distribution appears to be real, rather than the result of observational bias." (R1) The authors believe that the evidence supports a volcanic origin for the dark-haloed craters, noting also a possible correlation between their locations and transient lunar phenomena.

X2. General observations and description of a different type of dark-haloed crater. "Early observers of the lunar surface noted the presence of a small number of atypical craters surrounded by deposits of low-albedo material. Before spacecraft images of the moon were available, these features were usually interpreted as ash deposits surrounding volcanic vents. Later studies have confirmed this interpretation for some dark-haloed craters, most notably, those on the floor of Alphonsus crater. Other examples of volcanic dark-haloed craters can be found on the floors of the J. Herschel, Atlas, and Franklin craters and in many other regions. However, many dark-haloed craters exhibit morphologies characteristic of impact craters. Head and Wilson proposed the following morphological criteria for distinguishing dark-haloed craters of volcanic origin: noncircular shape, alignment with rilles, fractures, or lineaments; absence of raised rim; smooth, untextured exterior deposits; and a depth/diameter ratio which is generally less than that of impact structures." The halos around impact craters seem to be either deposits of impact melt or, as in the case of the well-known dark ring around Tycho, rings of impact melt resulting from fluid flow as opposed to ejecta.....

Neither volcanism nor impact melt appears adequate to explain the origin of another class of dark-haloed craters that differs considerably from the types already described. These craters typically resemble normal impact craters with near-circular rims and ejecta blankets. The entire ejecta blankets are darker than their backgrounds, while the crater interiors are typically bright. In high resolution photographs the dark ejecta exhibit no evidence for extensive impact melt deposits." The authors suggest that the dark material surrounding these craters may be subsurface basalt excavated by the impacts. (R2)

References

R1. Salisbury, John W., et al; "Dark-Haloed Craters on the Moon," Royal Astronomical Society, Monthly Notices, 138:245, 1968. (X1)

R2. Bell, Jeffrey F., and Hawke, B. Ray; "Lunar Dark-Haloed Impact Craters: Origin and Implications for Early Mare Volcanism," Journal of Geophysical Research, 89:6899, 1984. (X1, X2)

ALE13 Local Concentrations of Radioactivity

Description. Enhanced fluxes of alpha and gamma radiation attributable to surface radioactivity in specific, localized areas.

Data Evaluation. Alpha particle and gamma ray spectrometer data from the Apollo spacecraft. Rating: 1.

Anomaly Evaluation. Radioactive radon-222 is released frequently from the earth's crust in association with earthquakes and other crustal activity. Similar events are to be expected on the moon, especially in areas where transient lunar events are common, such as the region around Aristarchus. Thus, it is not surprising to detect the alpha particles characteristic of radon-222 in some lunar localities. However, the concentration of radioactive elements, as indicated by gamma ray emission in the Imbrium and Procellarum areas is much more difficult-to-explain. Rating: 2.

Possible Explanations. The emission of radon-222 from the lunar crust during quakes and other crustal activity; the concentration of radioactive minerals during past lunar igneous activity.

Similar and Related Phenomena. Lunar transient phenomena (ALF3, ALF4).

Examples

X1. Probable transient event. Abstract. "The alpha particle spectrometer aboard the Apollo 15 command/service module was designed to detect alpha particles from radon decay and to locate regions with unusual activity on the moon. A significant increase in radon-222 activity was detected from a region containing the crater Aristarchus. The result is interpreted as probably indicating internal activity at the site. By analogy with terrestrial processes, increased radon emanation may be associated with the emission of other volatiles." (R1) The spacecraft made its measurements at 110 km altitude. The radon-222 activity in the region of Aristarchus and Schroeter's Valley was about four times that for the rest of the moon surveyed. (R2) The release of radon in the Aristarchus region is significant because this crater is notorious for transient lunar events. Radon releases may well be associated with such events (R3).

X2. General observations. Apollo 15 gamma ray spectrometer data. "The region around the Imbrium basin, however, appears to have some unusual features. Observations of the gamma rays given off by radioactive materials showed much higher concentrations of uranium, thorium, and potassium in Mare Imbrium and in the neighboring lava flows of Oceanus Procellarum than elsewhere on the moon. The observations were made from lunar orbit, and the flight path of the Apollo 15 spacecraft allowed about 15 per cent of the moon's surface to be mapped. The results of the experiment, conducted by a team headed by James Arnold, of the University of California at San Diego, indicate concentrations of about 10 parts per million of thorium in the Mare Imbrium soil, compared with 1 ppm in the eastern part of the moon. The high concentrations of radioactive elements in the Imbrium region are similar to the concentrations found in KREEP basalts, which are found in large quantities at the Apollo 14 landing site. This material has also been found, in smaller quantities, at all the landing sites, but why the Imbrium and Procellarum regions should be the overwhelming source of the radioactive elements on the moon's surface, as they appear to be, has been difficult for geochemists to explain." (R4)

References

R1. Gorenstein, Paul, and Bjorkholm, Paul; "Detection of Radon Emanation from the Crater Aristarchus by the Apollo 15 Alpha Particle Spectrometer," Science, 184:792, 1973. (X1)

R2. "Radioactivity of Aristarchus," Sky and Telescope, 45:277, 1973. (X1)

R3. Srnka, L. J.; "On the Detection of Lunar Volatile Emissions," Nature, 278:152, 1979. (X1)

R4. Hammond, Allen L.; "Lunar Research: No Agreement on Evolutionary Models," Science, 175:868, 1972. (X2)

ALE14 Scarcity of Dust and Meteoric Material

Description. The obvious lack of dust and meteoric material in comparison with the amounts expected from the measured influx of meteoric material on the earth, the radiometric age of the moon, and the dust-creating potentials of solar X-rays and ultraviolet light.

Background. Prior to the descent of lunar spacecraft, scientists expressed concern that these craft would founder in a sea of dust. They didn't. The subject seems to have been forgotten except for the creationists, who claim that the lack of lunar dust indicates a young moon.

Data Evaluation. The depth of the lunar dust and the general scarcity of meteoric material of any kind are derived from astronaut observations. The radiometric age of the moon is based on analysis of returned samples. The degradation of lunar rocks by solar radiation is mainly theoretical. Overall, the data are very good, although only a very tiny portion of the lunar surface has been explored by astronauts. Rating: 2.

Anomaly Evaluation. Although barely addressed in the scientific literature, we have here an important anomaly, regardless of whether the dust-influx estimates are wrong, or the dust has somehow been concealed, or the moon is a recent addition to the earth's orbital space. Rating: 2.

Possible Explanations. The moon may be a newcomer to our region of the solar system. The dust may have been buried by crater ejecta and lava flows, or possibly removed electrostatically. (R1)

Similar and Related Phenomena. The apparent incompatibility of present earth-moon acceleration and the origin of the moon in earth orbit. (ALB4)

Examples

X1. Permanent feature. "From the reports of the first lunar landing, the accumulation of dust on the surface of the moon in the vicinity of the touch-down was very small (a layer not much more than 1/8 inch in thickness). The later landings were in 'seas' that have larger dust accumulations, but these layers were still very small in thickness compared to the earlier predicted amount. The dust layer is 1/8" to 3" in thickness. The moon moves through the same region of space that the earth does and, consequently, should have about the same influx of cosmic dust on its surface as on the earth. Astronomers had been concerned that a lunar spaceship upon landing would sink into the supposed huge amount of dust that should have accumulated on the surface of the moon in about 4.5 billion years of assumed time. The rocket would be stuck in this layer of 'mud' and not be able to leave the moon. Also, in the 'sea' areas, where lunar ships landed, there should have accumulated more dust than elsewhere on the moon. Yet, the amount of dust is amazingly small. What could have happened to all the dust, assuming a 4.5-billion-year-old moon?

Strong ultraviolet light and solar X-rays should also produce dust, perhaps at the rate of a few ten-thousandths of an inch per year, as it destroys the surfaces of exposed lunar rocks. At this rate 6.3 miles of dust would accumulate in a billion years. Such vast dust thicknesses are not observed. (R1)

Similar sentiments are expressed in R2-R3.

X2. General observations. "On the other hand, meteoric material does seem to be in strangely short supply, and one of the world's leading authorities on meteorites, G. J. H. McCall, has even asked plaintively, 'Where have all the meteorites gone?'" (R4)

References

R1. Slusher, Harold S.; "Cosmic Dust Influx," Age of the Cosmos, San Diego, 1980, p. 41. (X1)

R2. Barnes, Thomas G.; "Young Age for the Moon and Earth," ICR Impact Series No. 110, 1982. (ICR is the Institute for Creation Research) (X1)

R3. "Moon Dust," Ex Nihilo, 1:13, 1983. (X1)

R4. Moore, Patrick; "Focus on the Moon, Star & Sky, 1:10, April 1979. (X2)

ALE15 Young Lunar Surface Ages

Description. Radiometric measurements of surface soils and rocks consistent with very young surface ages; specifically, surface exposure ages of a million years or less.

Data Evaluation. Although the radiometric measurements were made directly upon returned samples, the technique requires that assumptions be made about lunar composition and the space radiation environment. Nevertheless, several different samples display surprisingly recent surface dates. Rating: 2.

Anomaly Evaluation. The soil and rocks exposed by very young crater-producing meteorites could account for some recent surface ages. However, several different samples show the same young ages, implying that a fairly large area of the lunar surface was recently reworked. This is contrary to the expectations of most scientists. Rating: 2.

Possible Explanations. The measurements may be in error, perhaps due to faulty assumptions. Some regions of the lunar surface have been altered recently.

Similar and Related Phenomena. Erroneously young radiometric ages have been measured for some terrestrial rocks, notably some lavas; lack of lunar dust (ALE14).

Examples

X1. Frequent occurrence. Abstract. "A unique suite of Apollo 15 samples collected near the St. George crater at the Apennine-Hadley site has been analyzed for its cosmogenic and primordial radionuclide content by multidimensional gamma-ray spectroscopy. Two chips (15205 and 15206) from the top surface of a meter size boulder together with soil samples from beneath the boulder (15231), adjacent to the boulder (15211), and about 0.5 (15231) and 10 meters (15091) from the boulder were studied. The ^{22}Na concentrations in the boulder chips were at saturation while the ^{26}Al concentrations are at approximately one-half to two-thirds of their saturation value, indicating a short lunar surface life of about 1 million years. The soil (15231, 1) directly beneath the boulder has a higher ^{26}Al concentration than its saturation value for its rather highly shielded location, indicating it had been covered for about 1 million years. Two very interesting lunar specimens were collected at Stations 9 and 9a of the Hadley-Apennine site. Sample 15501, 2, a soil clod from the ejecta blanket of an apparently young crater, showed saturation with respect to ^{22}Na, however the ^{26}Al concentrations indicate a surface age of only 0.7 to 1 million years. A breccia, 15505, was also found to have a low ^{26}Al content compatible with a surface age of about 0.75 million years." (R1)

References

R1. Perkins, R. W., et al; "Apollo 15 Samples with Short Lunar Surface Ages," Eos, 53:725, 1972. (X1)

ALE16 Local Concentrations of Volatiles

Description. Localized areas of the lunar surface where the concentrations of volatiles are significantly higher than the average level for the moon. Typical volatiles are potassium, rubidium, water, etc.

Background. The normal, average concentration of volatiles on the moon is considerably lower than it is on the earth (ALE19). The enhanced concentrations discussed here are measured relative to the normal lunar concentrations.

Data Evaluation. The data from from instruments left behind at the Apollo landing sites and from the analysis of samples returned to earth. The information radioed back to earth from the unattended instruments is of the transient type and sometimes subject to various interpretations. The analysis of the carefully collected and preserved lunar samples is considered highly reliable. Overall rating: 2.

Anomaly Evaluation. At least two reasonable explanations of localized concentrations of volatiles are at hand (see below). This phenomenon, therefore, is not especially anomalous. Rating: 3.

Possible Explanations. The impacts of comets bearing ices and other volatiles may have "salted" localized portions of the moon with volatiles. Releases of internal gases might have triggered the Apollo instruments.

Similar and Related Phenomena. Transient lunar phenomena (ALF); concentrations of radon (ALE13); the lunar swirl markings, which are thought by some to be comet impact sites. (ALE5).

Examples

X1. Water-rich deposits. "In the face of voluminous chemical evidence that the moon does not now, and probably never had much if any water, scientists continue to search for its traces. Their findings always stir up lunar scientists.

One report by John W. Freeman Jr. and H. Kent Hills of Rice University was the detection of water vapor by the Suprathermal Ion Detectors left at the Apollo 12 and 14 sites. The Rice group still cannot account for the vapor for the source other than the moon.

Now, S.O. Agrell, J.H. Scoon, J.V.P. Long and J.N. Coles of the University of Cambridge have reported finding a hydrous mineral, geothite, in a sample from Fra Mauro. The goethite occurs as a surface corrosive layer on iron grains---like rust on earth."

The water vapor and the water in the hydrous mineral, goethite, may have come from external sources, such as comets or carbonaceous chondrites. (R1)

X2. Deposits enriched in a variety of volatiles.

"Abstract. A subsurface Apollo 16 soil, 61221, is much richer in volatile compounds than soils from any other locations or sites as shown by thermal analysis-gas release measurements. A weight loss of 0.03 percent during the interval 175° to 350° was associated with the release of water, carbon dioxide, methane, hydrogen cyanide, hydrogen, and minor amounts of hydrocarbons and other species. These voltaile components may have been brought to this site by a comet, which may have formed the North Ray crater." (R2)

"Microscopic study of large rusty rock returned by Apollo 16 astronauts done at the Max Planck Institute in Heidelberg and reported by Ahmed El Goresy has identified a set of minerals and chemical compounds that are new to lunar samples, adding further support to the hypothesis of a cometary impact there. One mineral never before reported in a lunar sample is sphalerite, a zinc sulfide that is a common ore for zinc on earth. Goethite, a hydrous iron oxide that was reported before also was found. Others that included a set of volatiles unique for the moon were zinc iron sulphates, associated with chlorine, zinc iron chlorine, phosphates and a lead-rich mineral with lead on the order of 4,000 parts per million, unusually high for the moon." The enhancement of these volatiles supports the hypothesis that a comet formed North Ray crater. (R3)

References

R1. "Another Vote for Moon Water," Science News, 101:73, 1972. (X1)

R2. Gibson, Everett K., Jr., and Moore, Gary W.; "Volatile-Rich Lunar Soil: Evidence of Possible Cometary Impact," Science, 179:69, 1973. (X2)

R3. "Study Supports Comet Impact," Aviation Week, 55, May 7, 1973. (X2)

ALE17 Lunar Soils Older Than Associated Rocks

Description. All methods for dating lunar rocks and soils agree that the soils are considerably older than the rocks from which they were presumably derived.

Data Evaluation. The radiometric analyses were made on samples returned from the moon. All of the radioactive dating techniques, however, involve assumptions relative to the purity of the samples. Some of the problems involving adulteration are discussed in X1. Rating: 2.

Anomaly Evaluation. If the data are correct, as seems likely, a first class anomaly would seem to exist, for it seems impossible to have soils older than their source rocks. But as explained in the quotation from R3 below, the lunar surface rocks are probably derived from lavas that erupted long after the formation of the rocks from which the soils are derived. In other words, the soils are not derived from the surface rocks in the vicinity but rather from older basement rocks. If this is true, the anomaly is a minor one. Rating: 3.

Possible Explanations. The measured ages could be in error due to the admixture of mare materials, meteoric material, or even cometary material to the soils. Another possibility is that the surface rocks may have somehow been depleted in the crucial elements used for radiometric dating. Perhaps most convincing is the explanation mentioned above, in which the surface rocks are truly younger than the soils surrounding them because they are the products of later lava flows.

Similar and Related Phenomena. Short lunar surface ages of rocks and soils (ALE15).

Examples

X1. Widespread lunar phenomenon. "But the most puzzling data so far from the age dating by all three methods have been the apparent ages of the soil from all of the sites---ranging from 4.2 to 4.9 billion years---considerably older on the average than the ages for the rocks. How could the rocks be younger than the soil? Lunar scientists have sought several explanations. One is that some magic ingredient such as KREEP (material with high concentrations of potassium, uranium, thorium, rare earth elements, and phosphorous) from the highlands has been mixed in with the soil. Another is that perhaps the rocks were depleted in radioactive isotopes before the rocks crystallized. And a third is that perhaps the soil has incorporated meteoritic material from impacting bodies." (R2, R1)

"After Apollo 11, the first manned lunar landing mission in 1969, there was much confusion about lunar ages; it was reported that the soil was much older than the rocks from which it was presumably derived! The oldest Apollo 11 rocks dated back 3.9 billion years, whereas the soil was several hundred million years older. What the age of the soil defines, however, is not the average age of all the particles of rock in the soil, but rather the time since the various chemical rock types that make up the soil were separated from one another, deep inside the Moon. This separation event may have occurred long before the lavas derived from these rock types erupted at the surface of the Moon." (R3)

References

R1. Hammond, Allen L.; "Lunar Science: Analyzing the Apollo Legacy," Science, 179:1313, 1973. (X1)

R2. Driscoll, Everly; "Dating of Moon Samples: Pittfalls and Paradoxes," Science News, 101:12, 1972. (X1)

R3. Cadogan, Peter; "The Moon's Origin," Mercury, 12:34, 1983. (X1)

ALE18 Problems in Dating Lunar Rocks and Soils

Description. Discordances in the ages of lunar rocks and soils when determined by different radiometric methods and conflicts between measured ages and geological expectations.

Data Evaluation. Radiometric dating is subject to many uncertainties. The uranium-lead method, for example, is complicated by the inability to tell how much of each lead isotope was originally in the sample under analysis and just how much lead was lost or added to the sample during its lunar history. Rating: 3.

Anomaly Evaluation. The seriousness of the dating conflicts and discordances is lessened by the realization that many sources of error and misinterpretation exist. It is likely that the anomalies mentioned below will eventually be resolved, as they have in analogous terrestrial situations. Rating: 3.

Possible Explanations. The loss or addition of various lead isotopes to the lunar samples during the moon's long history.

Similar and Related Phenomena. Short surface ages of some lunar samples (ALE15); lunar rocks younger than lunar soils (ALE17).

Examples

X1. General observation. The uranium-lead method of measuring the ages of lunar rocks and soils consistently yields older ages than the rubidium-strontium method. (R1)

X2. Apollo 11 samples. "Assuming a basaltic rock from Mare Tranquillitatis was formed 4.6 billion years ago, the uranium in it would produce daughters until something happened to that rock. The rubidium-strontium ages say that something happened at that site 3.6 billion years ago. What then could have happened is that some of the lead daughters in the rock volatilized at that time, but not all of them. But after the event, the uranium began producing other daughters. Thus the older daughters (those produced from 4. 6 to 3. 6) became mixed in with the younger daughters (produced from 3. 6 to the present). When the rocks from Apollo 11 were dated, they all gave an apparent uranium-lead age of 4. 1. The unexplainable fact is that not one, but all of the rocks from the site had a 4. 1 age, which means that the lead had to be boiled off in all of the rocks at a fixed rate. This, says Silver, should not have happened, especially when one assumes that all the rocks would not have had the same ratios to start with. What was expected was that the ratios of lead would be spread out---say from 3. 6 to 4. 6 billion years, but they weren't. Why is not yet understood." (R1)

X3. Analysis of lunar sample 14163. The ratios of lead-207 to lead-206 were 1. 2 to 1. 3, giving apparent ages of up to 5. 5 billion years. "This isotopic composition has never been observed anywhere in the material of the solar system." (R1)

References

R1. Driscoll, Everly; "Dating of Moon Samples," Science News, 101:12, 1972. (X1-X3)

ALE19 Compositional Differences between Earth and Moon

Description. Relative to the earth's mantle, the moon is depleted in volatiles (e. g., C, Br) and siderophiles (iron-loving elements like Pt and W) and enhanced in refractory elements (U and the rare earths).

Background. The prevailing view of solar system history has all of the planets accreting as the solar nebula cooled over 4 billion years ago. The proximity of a planet to the sun and in consequence its temperature should have strongly affected its retention of various classes of elements. The volatiles, for example, should be more depleted in the planets closer to the sun.

Data Evaluation. The relative abundances of the elements are based on the analyses of lunar and terrestrial samples. Rating: 1.

Anomaly Evaluation. The differences in lunar and terrestrial element abundances do not strongly support any of the three major hypotheses of lunar origin: fission from earth, capture, and accretion in terrestrial orbit. Rating: 2.

Possible Explanations. Other scenarios of lunar origin may be more in accord with the facts. The so-called "big splash" theory, for example, claims to solve some of the compositional problems. Here, the moon is formed from the ejecta of a massive collision between the earth and a Mars-sized object.

Similar and Related Phenomena. Noble gas anomalies on Venus (AVW4) and Mars (AMW5).

Examples

X1. General observations. "Whatever the cause of the terrestrial siderophile element abundance patterns, the concentrations of refractory siderophile elements Os, Ir, Ni, Pd and Re are enriched by large factors in the Earth compared with the Moon. Au, Sb and Ge are depleted by factors of about 100 in the Moon while the very volatile elements (Bi, Tl, In) show rather uniform depletions of about 50 in the Moon. The alkali elements K, Rb and Cs similarly show depletions by factors of about 2 between the Moon and the mantle of the Earth. All these comparisons are complicated by the differing evolutionary histories of the Earth's mantle and the Moon. Simple comparisons, for example, of basalts from both bodies, encounter this problem. Elements such as K and U whose bulk composition can be estimated from a variety of techniques (K-Ar systematics, K/Rb/Sr, Cs/Al/U ratios and heat flow values) are useful in this context. Terrestrial bulk U values of 15-20 p.p.b. are a factor of 2 lower than the lunar abundances. Similar constraints apply to the refractory elements Ca, Al and Ti. The Moon also contains more Fe and less Mg than the Earth's mantle. Thus, the weight of the chemical evidence seems to indicate that the terrestrial mantle and the Moon differ substantially in composition for many elements, refractory, volatile and siderophile." These facts make the fission hypothesis of lunar formation improbable and support the theory that the moon accreted in orbit about the earth. (R4)

Another summary of compositional differences. "Perhaps the most significant factor in this respect, relative to the chondritic meteorites, is the overall dearth in the moon of those elements that dissolve easily in molten iron or iron sulfide (such as platinum and tungsten)(the siderophile elements) and of those which are relatively volatile (such as carbon and bromine). There may also be a slight overall enrichment in refractory elements such as uranium and the rare earths. (Refractory elements can only remain as a gas at very high temperatures.) Some investigators believe that, rather than there being an overall enrichment in refractory elements, the Moon could have accumulated them later ---separately from the other elements--- and thus enriched only the surface. But is is difficult to see why these elements should have been collected last, even though they presumably condensed first out of the cooling material which formed the Moon." This author finds no theory of lunar origin completely satisfactory, but thinks the capture hypothesis looks best. (R5)

Another author summarizes the situation in a somewhat different but equivalent way. "Models of the bulk composition of the Moon show various differences compared to estimates of terrestrial mantle composition. They contain higher Fe/Mg ratios and display a general enrichment of refractory elements. Most lunar models contain more than 30 ppb uranium, in contrast to values of less than 20 ppb for the terrestrial mantle. The Moon is probably depleted in strongly siderophile elements, even when the lower total iron content of the Moon is taken into account. The presence of a small lunar core could account for some of this depletion, and early crystallization and sinking of olivine could deplete the source regions of mare basalts in nickel. The comparison is also complicated by the apparently Cl patterns of the Pt group elements in the terrestrial mantle. There is general agreement that the Moon is depleted in volatile elements (e.g. K, Rb) relative both to the Earth and to the Cl abundances taken as representative of primordial solar nebula values. This depletion is shown by Rb/Sr and K/U ratios. Rb/Sr$\sim$ 0.30 for Cl, 0.031 for the terrestrial mantle and 0.009 for the Moon. K/U$\sim 6 \times 10^4$ for Cl, 10^4 for the Earth and 2500 for the Moon. These differences have to be established, in the fission hypothesis, not only following accretion and core separation in the Earth, but also in a separate element fractionation event during or following fission, but before the accumulation of the Moon." The author concludes that the fission hypothesis of lunar formation is ruled out. (R6)

X2. General observations concerning the implications of the composition of the moon relative to that of earth.

The earth-moon fission hypothesis is supported by the data. (R1)

Volatile-rich material may have been acculated preferentially by the earth in the later stages of the evolution of the earth-moon system. (R2)

Since the earth is not depleted in sodium relative to the carbonaceous chondrites but the moon is, the moon may have fissioned from Mercury instead of the earth. (R3)

References

R1. O'Keefe, John A.; "Apollo 11: Implications for the Early History of the Solar System," Eos, 51:632, 1970. (X2)

R2. "Accretion of Volatiles?" Nature (Physical Science), 232:178, 1971. (X2)

R3. Seyfert, C. K.; "The Origin of the Earth-Moon System," Eos, 58:989, 1977. (X2)

R4. Taylor, Stuart Ross; "Structure and Evolution of the Moon,: Nature, 281:105, 1979. (X1)

R5. Cadogan, Peter; "The Moon's Origin," Mercury, 12:34, 1983. (X1, X2)

R6. Taylor, Stuart Ross; "The Lunar Fission Hypothesis and the Volatile/Refractory Element Fractionation," Meteoritics, 18:405, 1983. (X1, X2)

ALE20 Apparently Anomalous Long-Term Persistence of Craters

Description. The equations and data of rheology seem to indicate that the lunar craters should have disappeared long ago under the influence of gravity; but they have not!

Background. The only discussion of this seeming paradox was found in a creationist journal.

Data Evaluation. The equations of rheology are well-verified; but the viscosities of lunar rocks at lunar temperatures are inferred from measurements of terrestrial rocks, although this is not unreasonable. Rating: 3.

Anomaly Evaluation. Either the lunar craters are very, very young or the viscosities of lunar rocks are about 10^5 higher than those of similar terrestrial rocks. Rating: 1.

Possible Explanations. The lunar craters are youthful, belieing radiometric dating; there is a flaw in the whole argument somewhere.

Similar and Related Phenomena. Simulations of the Martian erosion environment (wind-blown dust) suggests that the Martian craters should have been worn down long ago (AME18).

Examples

X1. An apparently unresolved paradox. Compiler's summary. The science of rheology deals with the flow of solids under stress. Equations describing the deformation of solids under the influence of prolonged forces, such as gravity, are well-developed. That terrestrial rocks can deform in relatively short periods of time can be seen in the movements of salt and ice glaciers and the sagging of old tombstones under their own weights. The thrust of this article (R1) is that the moon's crater walls must flow like any other solid. In fact, after 3 or more billion years, the lunar craters should all but have disappeared!

The unit of viscosity is the poise. Terrestrial rocks have viscosities well under 10^{23} poises; e.g., granite is 10^{20} poises. No direct measurements of the viscosities of lunar rocks at lunar temperatures have been found in the literature but, by analogy to terrestrial rocks, they should be under 10^{23} poises also. Lunar basalts, for example, are very much like terrestrial basalts.

Using the equations of rheology, the lunar craters should be flattened by gravity in less than a million years if the viscosities of similar terrestrial rocks (basalt) are used. Actually, some scientists have estimated the viscosity of the lunar crust at 10^{27} poises based on the assumption that the lunar crust solidified over 3 billion years ago. In contrast, the viscosity of the terrestrial mantle is usually taken as 10^{22} poises, a factor of 10^5 lower!

The paradox, of course, is that the lunar craters are still very much in evidence despite the rheological analysis. (R1)

References

R1. Morton, Glenn R., et al; "The Age of Lunar Craters," Creation Research Society Quarterly, 20:105, 1983. (X1)

ALE21 Alignment of Mascons and Lunar Moments of Inertia

Description. The principal moments of inertia of the lunar mascons (buried concentrations of mass taken collectively) and the moon-as-a-whole are approximately parallel.

Data Evaluation. The moments of the entire moon are well-known from astronomical and spacecraft tracking measurements. The properties of the mascons are less certain since they must be inferred from the perturbations of spacecraft in orbit around the moon. Rating: 2.

Anomaly Evaluation. The data indicate that the emplacement of the mascons was generically related to the formation of the lunar farside highlands; i.e., the thicker crust on the other side of the moon. The nature of this event or series of events is unknown. Rating: 2.

Possible Explanations. The scenario that produced the mascons also created the farside highlands. A less probable explanation is that the moon's orientation is controlled by the mascons and nearside basalt flows.

Similar and Related Phenomena. Perturbations of orbiting lunar spacecraft (ALL1); the asymmetry of lunar topography (ALE1), Mercury's topography (AHE1), and Martian topography (AME8).

Examples

X1. Permanent feature of lunar structure. Abstract. "This letter reports the discovery of a relation between the moments of inertia of the mascons (taken about the moon's center) and the moon's moments of inertia. It is found that the principal axes of the mascons alone are nearly parallel to those of the moon. Possible explanations of this parallelism are discussed. If the mascons are associated with a layer of uncompensated basalt on the moon's nearside, then the parallelism can be adequately explained on the grounds that the mascons and basalts together determined the moon's orientation. On the other hand, the third-order harmonics of the moon's gravity field indicate that the excess mass controlling the moon's orientation is on the farside. It thus appears that the mascons have been emplaced in special sites whose position was controlled by the processes which produced the farside highlands." (R1)

References

R1. Melosh; H. J.; "Mascons and the Moon's Orientation," Earth and Planetary Science Letters, 25:322, 1975. (X1)

ALE22 Geological Changes within Historical Times

Description. Lunar geological features that have apparently disappeared, appeared, or altered noticeably during historical times. Such changes are detected by comparing old and new drawings and photographs. Generally speaking, the time frame is restricted to "telescopic times" or the last four centuries.

Data Evaluation. The old lunar maps are not completely reliable. Two centuries ago, telescopes were small and imperfect. In addition, lunar observers must always combat subtle changes in the lunar landscape as illumination conditions change. However, when two or more of the old maps agree on a feature, we may have some confidence that it was real. Overall the data used to establish lunar changes are only fair. Rating: 2.

Anomaly Evaluation. When the moon was considered a dead world, any demonstrable change in its surface would have high anomaly value. Today, lunar changes are widely accepted; viz., the moonquakes and TLPs are symptomatic of some low level of lunar activity. The subject is no longer controversial, although the appearance of an active lunar volcano would be a great surprise. Rating: 3.

Possible Explanations. Moonquakes; meteorite impacts.

Similar and Related Phenomena. Lunar changes resulting from changes in lighting conditions (ALO2); mist and obscurations (ALW1); large-scale lunar catastrophism (ALF2); transient lunar phenomena (TLPs) (ALF3, ALF4).

Examples

X1. Cichus, near the southern extremity of Mare Nubium. "Here, many years ago, in comparing Schroeter's drawings with the moon, I was struck with the apparent enlargement of the small crater which has defaced one side of the ring. On procuring the map of Beer and Maedler, I found that they had also seen it enlarged. Could we in this instance depend upon the older drawings, we might presumably infer the probability of a change since the year 1792." (R1) Most of the purported permanent changes on the moon are based upon comparing old maps and records with more recent observations. In many cases the evidence for change is vague and weak. (WRC)

X2. Linne. The case of Linne is by far the most famous, perhaps because the pronouncements by astronomers have been as conflicting as the data.

"'So far as can be ascertained by the most careful examination, not the slightest change ever takes place on its surface.' Thus wrote one of the world's greatest astronomers, S mon Newcomb, in 'Astronomy for Everybody,' page 128, published in 1910. This appears to be the consensus of opinion among astronomers, notwithstanding the evidence that the crater Linne was prior to the year 1866, a six-mile crater, very deep, visible under all illuminations, and that in that year it became invisible for some months, seemingly concealed by a fog or haze that hung over it, and upon clearing up, a much smaller crater was in its place, and surrounding it a white spot like a drop of whitewash on the dark sea floor, occupying the space of the original crater." (R22)

In 1960, Joseph Ashbrook, editor of Sky and Telescope reviewed the evidence for a change in Linne. He found the evidence to be weak. The drawings of J. F. J. Schmidt made between 1841 and 1843, thought to be persuasive that Linne was once a well-marked, deep crater, are inconclusive. Of the eleven drawings Schmidt made of this region, only five showed Linne as a crater. Ashbrook notes that Schmidt was only in his teens when he made this set of drawings! Other early astronomers saw Linne as a crater, too, but a few recorded only a white spot. (R27)

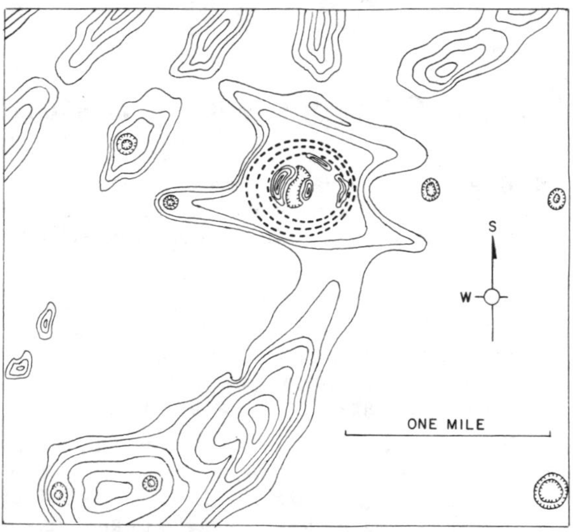

Topography in the vicinity of Linne as drawn by P. Fauth in 1906. The dashed ellipses indicate the present white spot, within which are a craterlet and some small ridges. (X2)

In 1977, Patrick Moore again went over the facts about Linne and concluded as follows: "In my view, at least, we can at last reject the idea that any variation in Linne has occurred in telescopic times. No TLP events have been recorded in the area; it is not an active region. And perhaps the early results are not so surprising as has often been thought---and as I once thought myself. Schroeter can tell us nothing; Beer and Maedler used a telescope which was, by any standards, tiny; and when Schmidt drew Linne as a crater he was still young and inexperienced. Linne is a small object, and it is not hard to believe that observers of the years following 1866 were deceived into reporting changes which did not really occur. Unconscious prejudice may well have played a major role." Moore considers that other reported changes of the lunar surface are even less convincing. (R28) Of course, the remark Moore made about prejudice works both ways and bedevils all scientific observation. (WRC)

The above reviews by Ashbrook and Moore both question the accuracy of the pre-1866 drawings of Linne, which showed it as well-defined crater. In this light, consider the following: "The crater Linne was discernible on the early successful lunar photographs taken in America from 1850 by Bond, H. Draper, and Rutherfurd." (R26) These photos are not mentioned in the reviews of the Linne problem. Indeed, the photos may not be convincing either. (WRC)

The Linne controversy generated a considerable literature. We have collected, in addition to the items mentioned above, the following: R2-R6, R8, R11, R13, R18-R20, R22, R24. There are many more that have not yet been assimilated. (WRC)

X3. Linnemann. This crater appeared on the maps of Riccioli and Mayer at longitude 55°E on the equator. It is no longer there. (R7)

X4. Hyginus N. "In May, of last year (1877), however, Dr. H. J. Klein, of Koln, while examining the moon, noticed a great black crater on the Mare Vaporum, and a little to the Northwest of the well-known crater Hyginus. He describes it as being nearly as large as Hyginus, or about three miles in diameter, as being deep and full of shadow, and as forming a conspicuous object on the dark gray Mare Vaporum. Having frequently observed this region during the last twelve years, Dr. Klein felt certain that no such crater existed there at the time of his previous examinations. He communicated his observations to Dr. Schmidt, of Athens, the veteran selenographer, who assured him that this crater was absent from all his numerous drawings of this part of the lunar surface; neither is it shown by Schroeter, Lohrmann, nor Maedler. Last April Dr. Klein made his discoveries known generally, and they seem to have been confirmed by other observers." (R9, R10, R11, R13, R16, R15, R14, R19, R20) For some unknown reason this lunar change did not cause the furor that Linne's supposed disappearance did, although it was widely reported. No debunking of this crater has been found to date. (WRC)

X5. Plato. "The eastern part of the wall of Plato is very suitable for a strict scrutiny of the kind suggested: it attains an elevation of about 3000 ft. above the interior plain, and just at the eastern extremity it is surmounted by a needle-pointed rock, the summit of which rises to about 4000 ft. above the wall. Near this rock, which is marked Z by B. and M. (Beer and Maedler), is a triangular formation, which in 1860, May 28, was seen by the writer to be inclosed on all sides by a low wall: but in 1866, April 25, the Rev. T. W. Webb noticed that the western wall was absent, and under this date he speaks of the formation exhibiting the appearance of a landslip." (R11)

X6. Plinius. In 1888, J. T. Stevenson, of Auckland, examined the terraced ring called Plinius in response to a cablegram that changes were taking place there. Instead of the hillocks recorded in Webb's Celestial Objects, he found four craters. (R17) No other reference to the cablegram or changes in Plinius have been found. (WRC)

X7. Eimmart. "The crater is situated on the northwestern border of Mare Crisium, in longitude 295°.... and is about 40 kilometers (25 miles) in diameter. The general nature of the change observed is that while formerly at each lunation, the crater apparently filled up and overflowed with a white material, whose source was at a point at the foot of the nothern interior slope, that this change now no longer occurs." (R21, R12) In many respects the changes formerly observed in Eimmart each lunation resemble those in ALO2. (WRC)

X8. Vendelinus. An obvious cleft in the northern interior of Vendelinus was discovered by T. G. Elger in 1891. A careful search with the Mount Wilson 60-inch reflector in 1954 could not find this cleft. (R25)

X9. Messier. In the Mare Foecunditatis,

two small craters lie side by side at the end of two parallel white streaks. Beer and Maedler examined this region more than 300 times between 1829 and 1837 and recorded the two craters as identical in size, shape, and all other respects. In 1859, T. W. Webb found that the craters were no longer alike, and that the difference was so great that some permanent surface disturbance must have occurred between 1837 and 1859. (R1)

References

R1. Webb, T. W.; "Notice of Traces of Eruptive Action on the Moon," Philosophical Magazine, 4:18:80, 1859. (X1, X9)

R2. "Disappearance of a Lunar Volcano," Eclectic Magazine, 5:775, 1867. (X2)

R3. Birt, W. R.; "On the Obscuration of the Lunar Crater 'Linne'," American Journal of Science, 2:43:411, 1867. (X2)

R4. Birt, W. R.; "On the Present State of the Question Relative to Lunar Activity or Quiescence," English Mechanic, 12:10, 1870. (X2)

R5. Birt, W. R.; "Evidence of Recent Changes in the Moon's Surface," English Mechanic, 13:55, 1871. (X2)

R6. Birt, W. R.; "Linne," English Mechanic, 17:376, 1873. (X2)

R7. Birt, W. R.; "Missing Lunar Crater," English Mechanic, 22:503, 1876. (X3)

R8. Backhouse, Thos. Wm.; "The Ring of Linne," Astronomical Register, 15:125, 1877. (X2)

R9. "Volcanoes in the Moon," Eclectic Magazine, 28:383, 1878. (X4)

R10. "An Active Volcano in the Moon," Scientific American, 39:54, 1878. (X4)

R11. "Changes on the Moon's Surface," English Mechanic, 29:25, 1879. (X2, X4, X5)

R12. Birt, W. R.; "White Spot West of Picard," English Mechanic, 29:518, 1879. (X7)

R13. Flammarion, Camille; "Is the Moon Inhabited?" Scientific American Supplement, 7:2696 and 2711, 1879. (X2, X4)

R14. Klein, Hermann J.; "The Moon Not a Dead Star," English Mechanic, 31:152, 1880. (X4)

R15. Klein, Hermann J.; "The Moon Not a Dead Star," Scientific American Supplement, 9:3552, 1880. (X4)

R16. "Supposed Changes on the Moon," Scientific American Supplement, 9:3700, 1880. (X4)

R17. Stevenson, John T.; "Changes in the Moon," English Mechanic, 50:242, 1889. (X6)

R18. Pickering, William H.; "Is the Moon a Dead Planet?" Century Magazine, 64:90, May 1902. (X2) (New series vol. 42)

R19. Saunder, S. A.; "Changes on the Surface of the Moon," English Mechanic, 78:304, 1903. (X2, X4)

R20. Pickering, William H.; "Changes upon the Moon's Surface," Nature, 71:226, 1905. (X2, X4)

R21. Pickering, William H.; "Recent Change in the Lunar Crater Eimmart," Astronomische Nachrichten, 196:415, 1914. (X7, X9)

R22. Cook, John A.; "Is the Moon a Dead World?" Scientific American, 115:549, 1916. (X2)

R23. Haas, Walter H.; "The Problem of Lunar Changes," Royal Astronomical society of Canada, Journal, 32:347, 1938. (X2)

R24. Wilkins, H. Percy; "A Vanished Lunar Crater," Popular Astronomy, 55:197, 1947. (X2)

R25. "A Reported Lunar Change," Sky and Telescope, 14:254, 1955. (X8)

R26. Kwaterniak, W.; "Lunar Craters," New Scientist, 3:35, 1958. (X2)

R27. Ashbrook, Joseph; "Linne in Fact and Legend," Sky and Telescope, 20:87, 1960. (X2)

R28. Moore, Patrick; "The Linne Controversy: A Look into the Past," British Astronomical Association, Journal, 87:363, 1977. (X2)

ALF LUNAR LUMINOUS PHENOMENA

Key to Phenomena

ALF0 Introduction

For a celestial body long considered to be a "dead world," the moon exhibits a surprising variety of luminous phenomena. While the Dead Moon Dictum held sway (roughly the first half of this century), lunar luminous phenomena were seldom reported in the scientific literature. After all, they "couldn't exist"! However, the arrival of the Space Age brought the moon under detailed scrutiny; and both professional and amateur astronomers began reporting flashes of light, transient color phenomena around some craters, and strange obscurations of lunar detail. These kinds of phenomena were labelled TLPs, for Transient Lunar Phenomena. But there were also other light-emitting events, such as episodes of enhanced luminescence over large regions of the moon and hot spots that could be seen with infrared imaging equipment. Manifestly, the moon was emitting energy that had to come from somewhere. It was certainly not a "dead world," but it wasn't very lively either.

The modern spate of TLPs encouraged researchers to look at the older literature. There they found hundreds of sightings of lights on the moon. Before circa 1900, such points of light were thought by many to be erupting volcanos, for in those days the whole of the moon's surface was believed to be sculpted by volcanic action.

Some luminous anomalies, especially the infrared hot spots, may indeed be due to internal lunar heat sources. Other luminous phenomena, though, may be the result of such exotic mechanisms as large-scale electrostatic discharges in dust clouds raised by vented gases. High energy particles from solar flares may induce lunar materials to fluoresce. Finally, the simple reflection of sunlight from lunar peaks has high explanatory value when points of light are observed on the darkside of the moon near the terminator.

Note that the observations of the washing out and obscuration of lunar detail, which are usually considered a variety of TLP, are to be found in this Catalog in section ALW.

ALF1 Infrared Anomalies

Description. Regions of the moon that are brighter than their surroundings in the infrared portion of the spectrum and, therefore, warmer. Such infrared anomalies are most easily detected during lunar eclipses and on the dark side of the moon.

Data Evaluation. All of the studies found so far have been carried out in recent years with state-of-the-art infrared instrumentation. Rating: 1.

Anomaly Evaluation. Since most lunar thermal anomalies are associated with young, bright craters, they are probably the consequence of the meteoric removal of insulating soil and rock, exposing hotter rock underneath. Such anomalies are not "anomalous" in the sense used in this Catalog; i.e., they are easy-to-explain in terms of current scientific knowledge. However, those thermal anomalies not correlated with bright craters may reveal places where hot gases are being expelled or where localized heat sources exist. For these few lunar thermal anomalies only: Rating: 3.

Possible Explanations. Lunar thermal anomalies may be the consequence of: (1) the meteoric removal of insulating rock and soil; (2) the expulsion of hot gases; and (3) the presence of local, possibly radioactive, heat sources near the surface.

Similar and Related Phenomena. Lunar color phenomena (ALF4).

Examples

X1. Thermal anomalies on the eclipsed moon. "Abstract. Infrared scanning during a total eclipse has revealed hundreds of hot spots, many identified with craters smaller than the detector resolution. Areal corrections show that some of these features may have the thermal properties of bare rock. Correlation of thermal response with albedo and radar reflectivity shows discrepancies. There is a concentration of hot spots in Mare Tranquillitatis." Some highlights from the paper follow: The lunar surface does not cool uniformly; a number of ray craters cool more slowly than their surroundings. Of the 400 thermal anomalies cataloged, over 90% are craters which are visually bright in some respect at full moon. However, there are bright areas which are not thermally anomalous. There is no simple correlation between radar reflectivity and infrared measurements. (R1)

X2. A thermal anomaly that may be of internal origin. "Abstract. Infrared images of the lunar eclipse of April 13, 1968, were obtained and compared with infrared images of the December 19, 1964, eclipse. A similarity of apparent strength and distribution of most thermal anomalies on the maria is evident from inspection of these images, indicating that these features are not ephemeral. One new linear thermal anomaly was discovered, which is thermally enhanced during the lunar afternoon. Its close relation to a lunar crustal fracture line and other features of probable internal origin suggests that this anomaly may be of internal origin." (R2)

"The interesting hot streak coincides with a previously recognized step fault running southwards from the crater Gassendi A. While the whole of Mare Humorum is thermally anomalous, the odd thing about this fault line is that its extra heat is still detectable during the lunar afternoon when the surrounding mare has cooled down. Such behavior is atypical of other lunar hot spots and cannot be caused by the presence of a rough surface which would be cooler than its surroundings at this time. There is no evidence of unusual surface features on Lunar Orbiter pictures, and hence the AFCL researchers favour the idea that extra internal heat may be escaping here, aided by rocks of greater bulk thermal conductivity along the broken fault zone. Hot gases leaking to the surface along this feature could provide an alternative mechanism. The hypothesis is supported by reports of reddish flashes or patches in near-by Gassendi; and by the occurrence of four, possibly gas-emitting, craters along the actual fault line. This strange feature obviously merits further study." (R3)

X3. Thermal anomalies on the dark part of the moon. "Abstract. A program of lunar infrared radiometry which uses large area scanning is described. Procedures for atmospheric attenuation correction and data reduction to temperature by relative radiometry are outlined. The scan data of the waning moon taken during ten evenings in the 10- to 12-μm window are presented as isothermal contour maps of the lunar disc. More than 160 areas of anomalous thermal emission have been found in the lunar darkside data......" (R4)

References

R1. Shorthill, R. W., and Saari, J. M.; "Nonuniform Cooling of the Eclipsed Moon: A Listing of Thirty Prominent Anomalies," Science, 150:210, 1965. (X1)

R2. Hunt, Graham R., et al; "Lunar Eclipse: Infrared Images and an Anomaly of Possible Internal Origin," Science, 162:252, 1968. (X2)

R3. "A Lunar Gash That May Be Heated from Below," New Scientist, 40:206, 1968. (X2)

R4. Raine, William L., et al; "Thermal Study of the Unilluminated Surface of the Waning Moon," The Moon, 12:407, 1975. (X3)

ALF2 Lunar Catastrophism within Historical Times

Description. Visible, apparently catastrophic phenomena engulfing large areas of the moon with fire, smoke, ashes, lava, etc.

Data Evaluation. A single, second-hand, medieval account. Rating: 3.

Anomaly Evaluation. Two theories purport to explain the single, rather questionable observation on file. Neither stretches our credulity or any scientific laws. Superficially, then, the event of 1178 would seem to possess only curiosity value. However, the account of the event mentions a twelve-times repetition of the phenomenon; and the two theories at hand do not seem to do justice to this aspect of the observation. Rating: 3.

Possible Explanations. The impact on the moon of a large meteorite and the consequent excavation of a large crater. A large meteor in the terrestrial atmosphere traversing the moon. Both theories have their pros and cons. (R1, R3)

Similar and Related Phenomena. None.

Examples

X1. July 18, 1178. The following observation was taken from the medieval chronicles of Gervase of Canterbury, as translated from the Latin by R. Y. Hathorn.

"In this year, on the Sunday before the Feast of St. John the Baptist, after sunset when the moon had first become visible a marvelous phenomenon was witnessed by some five or more men who were sitting there facing the moon. Now there was a bright new moon, and as usual in that phase its horns were tilted toward the east; and suddenly the upper horn split in two. From the midpoint of this division a flaming torch sprang up, spewing out, over a considerable distance, fire, hot coals, and sparks. Meanwhile the body of the moon which was below writhed, as it were, in anxiety, to put it in the words of those who reported it to me and saw it with their own eyes, the moon throbbed like a wounded snake. Afterwards it resumed its proper state. This phenomenon was repeated a dozen times or more, the flame assuming various twisting shapes at random and then returning to normal. Then after these transformations the moon from horn to horn, that is along its whole length, took on a blackish appearance. The present writer was given this report by men who saw it with their own eyes, and are prepared to stake their honor on an oath that they have made no addition or falsification in the above narrative." J. B. Hartung has interpreted this description as the consequence of a meteoric impact on the moon, specifically the impact that excavated the fresh crater Giordano Bruno. (R1, R2) H. H. Nininger, to the contrary, believes that the men saw only a meteor in the earth's atmosphere that happened to cross the moon's face. (R3)

References

R1. Hartung, Jack B.; "Was the Formation

of a 20-km-diameter Impact Crater on the Moon Observed on June 18, 1178?" Meteoritics, 11:187, 1976. (X1)

R2. Hughes, David W.; "Giordano Bruno; the Moon's Latest Large Crater," Nature, 264:212, 1976. (X1)

R3. Nininger, H. H., and Huss, Glenn I.; "Was the Formation of Lunar Crater Giordano Bruno Witnessed in 1178? Look Again," Meteoritics, 12:21, 1977. (X1)

ALF3 Transient Points of Light

Description. Bright, temporary, starlike points of light observed on the moon's surface. This type of TLP (transient lunar phenomenon) may occur as a simple flash, several successive flashes, or it may persist as a steady light for anywhere from a few seconds to an hour or two. Some even last a couple days or more. The light's color is usually white, although yellowish, reddish, and bluish lights have been reported. Most occur on the dark side of the moon, frequently near the terminator. A few dayside reports exist.

Background. In the earlier days of astronomy, when the lunar craters were considered volcanic in origin, many TLPs were reported, and were supposed to be volcanos in eruption. Around the turn of the century, the meteoric hypothesis became dominant, and the moon was looked upon as a dead world. As if in response to this outlook, TLPs do not appear in any quantity in the scientific literature for the first half of this century.

Data Evaluation. The reality of transient points of light on the moon is confirmed by hundreds of reports in the scientific literature. In fact, only a small fraction of the total has been processed so far for this Catalog. Those interested in large numbers of additional observations should consult one of the modern TLP catalogs, such as R55. Rating: 1.

Anomaly Evaluation. Many TLPs are probably only reflections of sunlight and, therefore, not anomalous. Even those sightings due to the presence of incandescence of lunar material, such as lava and gases, would be only surprising---hardly very anomalous. Lunar lightning and the equivalent of earthquake lights, however, would definitely be of great interest, for we do not understand completely all terrestrial manifestations of these phenomena. Overall rating: 2.

Possible Explanations. Reflection of sunlight from lunar features (usually near the terminator); incandescent lunar materials (lava and gases); triboelectric phenomena; piezoelectric phenomena (perhaps akin to earthquake lights); meteors in the earth's atmosphere (i.e., spurious TLPs).

Similar and Related Phenomena. Lunar color phenomena (ALF4); lunar obscurations (ALW1); earthquake lights (GLD8); flashes of light from Mars (AMF1).

Examples

X1. About 577. Gregory of Tours observed a light on the moon. (R50)

X2. November 26, 1540. Several individuals reported a starlike appearance on the dark side of the moon. (R50, R51)

X3. November 1668. Cotton Mather, in Boston, recorded that "a star appear'd below the body of the moon within the horns of it." This observation was apparently the basis of Coleridge's famous lines:

"Till clomb above the eastern bar
The horned moon with one bright star
Within the nether tip"

(R27, R29, R41)

X4. May 4, 1783. William Herchel saw what he believed was a lunar volcano in eruption. (R36, R39)

X5. April 19-20, 1787. William Herschel's famous description of lunar volcanos. "I perceive three volcanoes in different places of the dark part of the new moon. Two of them are either already nearly extinct, or otherwise in a state of approaching eruption;

which, perhaps, may be decided next lunation. The third shows an actual eruption of fire, or luminous matter.

April 20. 1787, the volcano burns with greater violence than last night. I believe its diameter cannot be less than 3", by comparing it with that of the Georgian planet; as Jupiter was near at hand, I turned the telescope to his third satellite, and estimated the diameter of the burning part of the volcano to be equal to at least twice that of the satellite. Hence we may compute that the shining or burning matter must be above three miles in diameter. It is of an irregular round figure, and very sharply defined on the edges. The other two volcanoes are much farther towards the centre of the moon, and resemble large, pretty faint nebulae, that are gradually much brighter in the middle; but no well defined luminous spot can be discerned in them. These three spots are plainly to be distinguished from the rest of the marks on the moon; for the reflection of the sun's rays from the earth is, in its present situation, sufficiently bright, with a ten-feet reflector, to show the moon's spots, even the darkest of them; nor did I perceive any similar phenomena last lunation, though I then viewed the same places with the same instrument.

The appearance of what I have called the actual fire or eruption of a volcano exactly resembled a small piece of burning charcoal, when it is covered by a very thin coat of white ashes, which frequently adhere to it when it has been some time ignited; and it had a degree of brightness, about as strong as that with which such a coal would be seen to glow in faint daylight. All the adjacent parts of the volcanic mountain seemed to be faintly illuminated by the eruption, and were gradually more obscure as they lay at a greater distance from the crater." (R11, R25, R27, R36, R39, R42) This observation could be called a color phenomenon (ALF4) with equal justification. (WRC)

The interior of the crater Aristarchus often seems luminous shortly after new moon due to reflected light---not volcanic action! (R36) Whether Herschel mistook this bright reflection for volcanic action will never be known for certain. He claims he never saw such phenomena during previous lunations. (WRC)

X6. December 1787. A luminous point was seen on the moon by the Maltese observer d'Angos. (R47)

X7. April 9-11, 1788. The German selenographer, Schroeter, reported seeing a bright spot 26" north of the rim of Aristarchus. (R21, R26)

X8. September 26, 1788. "At 4.30 a.m. on this date Schroeter remarked 'a whitish bright spot, shining somewhat hazily and 4" to 5" in diameter' and as bright as a star of the fifth magnitude appears to the naked eye. Nothing similar could be seen elsewhere although detail on the earth-lit moon was very distinct and many familiar objects were recognized. The spot lay about 1'18" southwest of Plato and in the bright mountainous region bounding Mare Imbrium. 'But now my bright spot became inconspicuous at times, finally uncertain, and soon thereafter---it disappeared entirely.' It had been visible for fully 15 minutes." (R25)

X9. October 15, 1789. Another Schroeter observation. "As I was observing the night side of the Moon on the 15th October, 1789, in the morning with the marking Plato, together with the Mare Imbrium in view, but absolutely nothing of the illuminated lunar hemisphere in the field, there arose, soon after 5 o'clock, in, or rather in front of the dark lunar disk, and just, in fact, in the centre of the Mare Imbrium, as far as I was able to report, after my sudden astonishment, and almost right in the middle of the field of the telescope, quite suddenly and quickly, a bright burst of light, which was composed of many single, separate small sparks which had just such a white light as the illuminated day side of the Moon, and which all moved away together in a straight line northwards over the northern part of the Mare Imbrium and the other parts of the Moon's surface bordering this to the north, but from there onwards through the remaining empty field of the telescope.

When this light shower had travelled half its distance, a further, similar burst of light arose to the south of it over just the same place or perhaps somewhat more to the east of where the first originated: the second was exactly the same in appearance as the first, composed of like, small, white sparks of light which flashed away in the same direction and in an exact parallel direction towards the north, as far as the edge of the telescope field.

Startling as this impression was also, which this distant light phenomenon made on me, nevertheless, I quickly recovered my wits, re-imagined this phenomenon repeatedly, and vividly after it had disappeared and estimated, using a pocket watch, the time it took each shower of light to travel from the position of origin to the edge of the telescope field as about 2 seconds, and thus the complete duration of this phenomenon as 4 seconds, in which time it was all over without leaving the faintest trace behind." (R45)

X10. October 22, 1790. Herschel saw about 150 bright red, luminous points when the moon was totally eclipsed. The points were small and round. (R1) This sighting seems superficially like the lunar thermal anomalies detected on the eclipsed moon in the infrared portion of the spectrum. (WRC)

X11. March 7, 1794. A bright, starlike point of light appeared on the darkside of the moon. Observed for at least 5 minutes. The brightness increased just before it disappeared. (R2) Confirmed by other observers. (R3, R27)

X12. February 4-6, 1821. A "volcano" was reported on the moon for three successive nights. (R5, R8) The sighting was confirmed by Olbers, who located it in Aristarchus. Olbers, however, believed the point of light to be merely a reflection from some smooth surface. (R6, R7) Such bright spots are frequently seen in Aristarchus when the moon is only a few days old, as it was in this 1821 case, supporting the theory that these particular spots of light are true reflections. (WRC)

X13. November 28-29, 1820. One large and three smaller luminous spots were seen on the dark side of the moon. The large spot seemed like a star of the sixth magnitude. On the 29th, the large spot was still brilliant but the smaller spots had faded, one completely. Weather prevented observations on the 30th. (R4, R9)

X14. December 22, 1835. Another bright spot in Aristarchus, when the moon was young. The observer, a Captain Smythe, linked his sighting to earlier, similar observations (viz., X12) and recognized that reflection was the probable cause. Yet, he wondered why the spot's intensity varied so much---from the 6th. to the 10th. magnitudes, in his experience. (R10)

X15. 1843. A Dr. Gerling, of Marberg, saw a bright spot on the moon, apparently in the same area of Herschel's famous 1787 "volcano" sighting. (R21)

X16. March 18, 1847. Large luminous spots on the shaded portion of the moon. No further details. (R12)

X17. December 27, 1854. "She was 8^{d} 4^{h} old at the time, and just on the edge of the light.....there were two luminous spots, one on either side of a small ridge, which ridge was in the light, and of the same colour as the moon; but these spots were of a yellow flame colour, while all the rest of the enlightened part was of a snowy white, and the mountain-tops that were coming into the light, and just on the shadow side of these spots, were of the same colour as the as the moon. The lights of these spots were like the light of the setting sun reflected from a window a mile or two off. I observed it for five hours. I thought them rather less bright than as first seen, but very little less; so bright were they, when the instrument was the least thing out of focus, they showed rays around them as a star would do." (R13-R15)

X18. January 1, 1865. "In a very clear sky Grover observed a bright spot at, or very close to, the position of Schroeter's second spot. Instantly reminding him of Herschel's idea of a glowing coal covered by ashes, the spot resembled a star of the fourth magnitude slightly out of focus. This bright speck remained changeless for fully 30 minutes, and its light was steady."(R25, R16, R21)

X19. April 9, 1867. A bright spot, about equal to a 7th. magnitude star, was observed near Aristarchus. (R17)

X20. April 18, 1874. "Whilst glancing at the N.E. quadrant a bright spot suddenly caught my eye, which turned out to be Aristarchus. It was, at the time of my first observing it, about as bright as a 9.5 magnitude star, but faded at 9h.30m. to the 10th magnitude; very faint streaks of light issued from it in a N.E. and W. direction. Three other spots, fainter, and shining with a more diffused light, were seen; their positions I thought coincided with the craters Menelaus, Manilius, and Sosigenes; the latter was much the faintest." (R18) Other observers have noted additional spots of lesser brightness. The streaks of light, though, are new. (WRC)

X21. November 12, 1878. More volcano-like phenomena. "I take the liberty to offer you a sketch of an observation of the moon, taken November 12, hour 8:30 evening, seen by me, my son, and several guests who were present, at the town of Oskaloosa, Iowa, about latitude 41°30'---what I supposed to be an eruption of a volcano. It was only seen for one half hour through my 6 1/2 inch telescope, as plain as any other mountain scenery in the moon is seen, and of the same color. I would like to hear what you think of it." This letter was sent to the Superintendent of the U.S. Naval Observatory, who (understandably) asked that the observer, a John Hammes, be vouched for by some responsible people in Iowa. Scientific American published a testimonial letter along with the above observation. (R56)

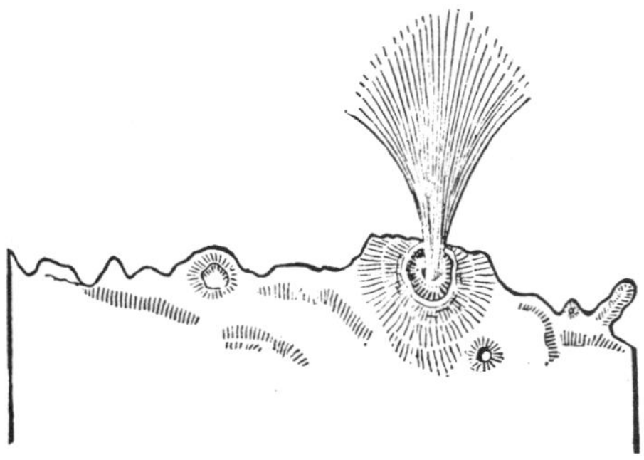

Sketch of a supposed volcanic eruption on the moon. (X21)

X22. November 6, 1880. "On 6th November (1880) we weighed our anchor and sailed out of the Reconcava...On the eve of our departure there stood forth an omen in the sky... Inside the thin crescent of the moon, was one solitary bright star, the only one in the sky." (R30) Another young moon with a probable reflection phenomenon. (WRC)

X23. May 1, 1887. A white patch observed near the lunar Alps. (R19)

X24. November 23, 1887. A gleam of light in the crater of Plato. (R20)

X25. July 15, 1888. Testimony of the Lick astronomer Holden: "I have never been able to understand how Herschel, the keenest of observers, could have been deceived in this (volcano) observation until the night of July 15th of this year, when I was looking at the moon with the great telescope. At the southern extremity of the Alps, in the dark portion of the disk, not far from the terminator, I saw an illumination of the crest of a high peak which was extraordinarily and incredibly bright...No part of this illumination seemed less bright that a first-magnitude star, and, taken altogether, it was the brightest object I have ever seen in the sky. It was apparently ten times as bright as neighboring portions of the moon's surface. Its yellow light was tinged in places with the purple due to the secondary spectrum of the objective; and, viewed as a whole, it presented the appearance of a vast conflagration ---something quite foreign to the brilliant white of the rest of the moon's surface. It would have required no stretch of the imagination to have supposed it to be a tremendous eruption of a range of lunar volcanoes. ...Observations on this and the succeeding nights showed that it was in fact due to a specially brilliant and favorable illumination of a mountain-ridge near the southern termination of the lunar Alps." (R21)

X26. August 12, 1902. "On this date G.S. Jones, using a 6-inch reflector at 250X, remarked at colongitude 18° on the dark side of the terminator and probably near Lambert a brilliant star-like point looking like a third magnitude star and showing one diffraction ring. The point resolved itself into a very brilliant spot as the terminator approached it. Jones observed this mark for at least two hours, and it was consequently probably no impact flare." (R25, R23, R25)

Here follows a hiatus of 46 years!

X27. August 8, 1948. A bright flash. "In carefully keeping the moon in focus with a 50X eyepiece on Sunday, Aug. 8, about 9.40 p.m., I noticed a small bright flash, bluish white then yellow, on the dull part of the moon, somewhat above and to the left of the moon's centre. Duration about 3 sec. This could have been a meteorite in our own atmosphere some 75 or 100 miles away; but it had the appearance of an object striking the moon's surface." (R26)

X28. 1949, during a total eclipse of the moon. A starlike point of light appeared in Aristarchus near the end of the eclipse. It seemed best explained as a reflection. (R28)

X29. November 16, 1953. Transient white spot photographed near terminator. (R35)

X30. April 24, 1955. White flash. "At 19 hrs., 20 mins., U.T., on 1955 April 24, Francis C. Wykes was observing the Moon when he noted a white flash, of short duration, on the unilluminated portion of the disk. The terminator at the time was closely east of Cleomedes, and the position of the flash, according to Wykes' sketch, was in the northern part of the Mare Serenitatis, not far and somewhat east of Posidonius." Colongitude was 300.3°. (R32)

X31. August 26, 1955. "Flares" or flashes. "At 7:51 p.m. CST, while examining the neighborhood of the Apennines, I saw on the dark portion of the moon a bright flare that remained visible for about 35 seconds. It appeared roughly as bright as a 2nd magnitude star does to the naked eye. The terminator region of the moon had been under survey for about an hour, and I am certain that the flare was not present for many seconds before I first saw it." (R31) It seems that this "flare" must have been near the terminator and therefore a possible reflection. (WRC)

X32. September 8, 1955. "Shortly after sunrise on September 8, 1955, I was looking at the moon, high in the sky, through a small 20x telescope. My attention was directed to the Taurus Mountains at the western edge of Mare Serenitatis when, at 7:35 a.m., EDT, I saw two distinct flashes of light, about a quarter second apart, that seemed to come from the edge of these mountains." (R31)

X33. October 10-12, 1955. Seven flashes and streaks of light reported by W. F. Duncan on October 10; three more flashes on the 12th. (R33)

X34. October 28, 1955. William Taylor saw an unusual phenomenon while walking home from work. "He stopped to look at the moon and had been doing this for a few minutes when he suddenly saw a flash and determined it to be in the northeast section. The time was $0^h\ 6^m$, U.T., on October 28, 1955. He says it was on the edge of the dark area known as Mare Vaporum. The light was intense but pinpoint in size and radiated to a large area. It lasted 1/4 second." (R33)

X35. January 17, 1956. A flash was seen by Robert Miles, Woodland, California. "I noticed a flash of white light that caught my eye. At first I thought it could have been a lunar meteor. But it kept flashing on and off.... The light was very bright but changed its color to a very bright blue, like an arc light. It was brighter than the sunlit portions I was looking at." The light flashed for an hour and a half. Location: on the night side of the terminator about 100 miles east of the gap in the mountains on the east boundary of Mare Crisium. (R34) This could hardly be a reflection, considering its duration. (WRC)

X36. November 18, 1956. Several groups of amateur astronomers in the United States scrutinized the moon for transient events during the lunar eclipse. Many were found. A group organized by John P. Bagby recorded 35 bright objects against the moon, five of which were seen simultaneously by three observers. John Mavrogianis and four others saw 25 transient luminous objects, most of which were stationary yellow flashes. A group at Cheyenne reported a fixed light that lasted 4 seconds. (R37)

X37. August 27, 1963. Danny Arthur, observing with a 3-inch telescope, saw a bright light between Mare Crisium and the south end of Palus Somnii. Three others confirmed the sighting. (R38)

X38. July 2-4 and August 2-4, 1965. On these days Project Moonblink recorded starlike flashes of light in Aristarchus. (R44) Project Moonblink employed a special optical device designed to call the observer's attention to light flashes and color phenomena. (WRC)

X39. May 1967. Transient event photographed by Lunar Orbiter IV. "Montes Teneriffe, a group of irregular hills in northern Mare Imbrium, was photographed by Lunar Orbiter IV in May 1967. Conspicuous bright lines approximately follow the contour along the west-facing slopes of two hills near the northwestern end of the group. The line on the southern hill is not interrupted where it crosses shadowed slopes. Both width and sinuous pattern are esstentially identical on two Orbiter high resolution frames exposed 12 hours apart. Shadow differences demonstrate the sun-angle change during the interval. The bright line crossing the shadow is difficult to explain as reflected sunlight; the light may have an internal source. Numerous transient events in the Plato-Montes Teneriffe area have been reported by telescopic observers. Some transient events were described as streaks." (R46)

X40. April 13, 1968. Many points of light during eclipse. "The visual nature of transient lunar phenomena (usually abbreviated to TLP) is not agreed upon by observers. Mrs. W. Cameron of the N. A. S. A. Goddard Space Flight Center noticed a great many star-like points glittering on the moon during the April 13, 1968 eclipse of the moon; they were seen by a group of observers who accompanied her. However, I watched the same eclipse and saw nothing of these star-like points. It is interesting to note that Herschel saw a similar occurrence during an eclipse in 1790." (R47) See X10 for Herschel's observation.

X41. No date given. A bright spot resembling a small star or planet was observed on the shaded portion of the moon at a considerable distance from the terminator. (R22)

X42. General observations. The magnitude of the TLP phenomenon. "Over 1,400 TLP's have been observed and even though a considerable number of these can be rejected as instrumental, atmospheric or physiological artefacts a sufficient residue remain to warrant a detailed physical investigation. Glows, hazes, mists, brief colour changes and temporary obscurations of lunar surface features have been reported. Events seem to be restricted to specific lunar regions, about 300 have been reported from the crater Aristarchus, 75 from Plato and 25 from Alphonsus. They also occur near the boundaries of certain regular maria and near areas rich in rills. The highlands seem to be avoided, but this apparent paucity may be an observational selection effect. Occurrency frequency is not linked with solar activity but does peak when the Moon is at the perigee of its orbit, at times when tidal activity is maximised. Phenomena have been reported from both the sunlit and dark sides of the Moon. The areal extent on the lunar surface is on average 16 km across, this containing brighter spots of 3 and 5 km in diameter. The average duration is about 20 minutes but some have been reported as persisting intermittently for a few hours. No permanent changes have been observed on the lunar surface after a TLP, thus justifying the use of the word 'transient'. Under very favourable seeing conditions TLP's have been seen to twinkle. Colour, if mentioned at all, tends to be described as weak, unsaturated, 'reddish' or 'bluish'." (R54, R43, R49, R53) The above generalizations apply to all TLPs including ALF4 and ALW1.

X43. General observations. Sundry correlations of TLPs.

"The phenomena manifest themselves in five categories, viz., Brightenings, or Darkenings, or as Gaseous, Reddish and Bluish events. Among the hypotheses proposed for their causes are tidal, low-illumination/thermoluminescence, magnetic-tail and solar flare effects. Analyses were conducted to see if different phenomena had different causes. There is some suggestion that they do. As concerns the tidal effects, the strongest peaks are at 0.5 (apogee) for Gaseous and Darkenings phenomena, 0.6 for Reddish events, and 0.7 for Brightenings. Reddish TLP have the strongest correlation with sunrise, while Aristarchus, Plato, Ross D area, and Bluish phenomena have the strongest correlations with solar-flare activity that produced magnetic storms on earth. 'All' observations, the ones labeled 'Best' (probably true anomalies), and Aristarchus, showed minima in the first half and maxima in the last half of the anomalistic (tidal) period. Histograms of several individual sites, including neighboring ones, behave differently, e.g. Aristarchus and Herodotus. When observed data are compared with expected observations (assumed to be evenly distributed) there were various correlations. For the Best data, 12 and 10% of the TLP fall close to perigee and apogee, respectively, and 10% would be expected for each. Seventeen percent occur within one day after sunrise when 3% would be expected; 20% occur while the moon is in the earth's magnetopause where 14% would be expected, and 12% occurred the same day the earth had a magnetic storm where 3% would be expected..... All hypotheses show correlations with some categories and some features. Sunrise correlation is the most frequent correlation. Few correlations involve as many as 50% of the observations. The distribution of all TLP sites is different from and unique compared with deep- and shallow-focus moonquake epicenters. Routine albedo measures reveal unobserved variations which amount to about 10% in nights of observation but 2% of individual albedo observations." (R52) In the above quotation TLP was substituted for LTP to avoid confusion. (WRC)

Another correlation summary. "The periodicities shown by the TLP's and the deep moonquakes and more recently the shallower high-frequency events are tidally related. The statistical periodicity of the TLP's both at local sites such as Aristarchus, Plato and Gassendi and for all sites together was demonstrated in 1965 and is apparently controlled locally by tidal forces just as are the moonquakes. These tidal stress changes are so small that it seems unlikely that they could be the single cause of the events. It appears more probable that these small forces, both moonwide and local tidal-force changes, add their cummulative effect to pre-existing stresses within the moon eventually to breaking point in local areas so that sliding may occur, gas or magma may be released at some points, from point to point within the moon or from the interior to the surface." (R51) Again LTP has been changed to TLP.

X44. General observations. Possible explanations.

Volcanic action. Doubtful because no fresh lava flows have been observed on the moon. Neither have high-temperature infrared sources. (R54)

Luminescence. This effect can be ruled out because the only excitation sources, the solar wind, is too weak. (R54)

Thermoluminescence. Too weak to be seen from earth. (R54) Other investigators still consider it a possible explanation. (R52)

Lightning due to triboelectric charging when dust grains rub together. Some scientists consider this a reasonable explanation. (R54)

Reflection of sunlight at low angles. This remains one of the most popular explanations. (R52)

Meteors in the earth's upper atmosphere

seen end on. (R48)

References

R1. "Remarkable Phenomena in an Eclipse of the Moon," Philosophical Transactions, 82:127, 1792. (X10)

R2. Wilkins, Wm.; "Of an Appearance of Light, Like a Star, Seen in the Dark Part of the Moon, on Friday the 7th of March, 1794," Philosophical Transactions, 84: 450, 1794. (X11)

R3. Maskelyne, Nevil; "An Account of an Appearance of Light, Like a Star, Seen Lately in the Dark Part of the Moon, ..." Philosophical Transactions, 84:451, 1794. (X11)

R4. Fallows, Fearon; "Communication of a Curious Appearance Lately Observed on the Moon," Royal Society, Proceedings, 2:167, 1822. (X13)

R5. "Volcanoes in the Moon," Edinburgh Philosophical Journal, 4:429, 1821. (X12)

R6. "Volcanoes in the Moon," Edinburgh Philosophical Journal, 5:216, 1821. (X12)

R7. "Lunar Volcanoes," American Journal of Science, 1:5:176, 1822. (X12)

R8. Kater, Henry; "Notice Respecting a Volcanic Appearance in the Moon," Royal Society, Proceedings, 2:142, 1821. (X12)

R9. "Curious Appearance Observed on the Moon at the Cape," Edinburgh Philosophical Journal, 7:384, 1822. (X13)

R10. Smythe, Captain; Royal Astronomical Society, Monthly Notices, 3:141, 1836.(X14)

R11. Herschel, Wm.; "Of Three Volcanoes in the Moon," A Popular Display of the Wonders of Nature..., London, 1837, p. 486, (X5)

R12. Rankin, T.; "On a Singular Appearance of the Shaded Part of the Moon on the Evening of March 18, 1847," Report of the British Association, 1847, p. 18. (X16)

R13. Hart, Robert; "On an Appearance Seen in the Moon," Philosophical Magazine, 4:9:238, 1855. (X17)

R14. Hart, Robert; "On an Appearance Seen in the Moon," Royal Astronomical Society, Monthly Notices, 15:89, 1855. (X17)

R15. Hart, Robert; "On a Telescopic Appearance Seen in the Moon," Royal Astronomical Society, Monthly Notices, 15:89, 1855.

R16. Grover, Charles; "Bright Spots on the Moon," Astronomical Register, 3:253, 1865. (X8)

R17. Elger, T.G.E.; "Bright Spot on the Moon," Astronomical Register, 5:114, 1867. (X19)

R18. Sadler, H.; "Aristarchus," English Mechanic, 19:146, 1874. (X20)

R19. Harris, W.H.; "White Patch on the Moon," English Mechanic, 45:219, 1887. (X23)

R20. "Leur Observeé dans le Cirque Lunaire de Platon," L'Astronomie, p. 75, 1888. (X24)

R21. Serviss, Garrett P.; "New Light on a Lunar Mystery," Popular Science Monthly, 34:158, December 1888. (X7, X15, X18, X25)

R22. "The Crescent Moon with a Star within Its Rim," Science, 18:346, 1891. (X41)

R23. Pickering, William H.; "Lunar Phenomena in October," Popular Astronomy, 10:419, 1902. (X26)

R24. Pickering, William H.; "Bright Star-Like Point on the Moon," Popular Astronomy, 10:497, 1902. (X26)

R25. Haas, Walter H.; "Does Anything Ever Happen on the Moon?" Royal Astronomical Society of Canada, Journal, 36:237, 317, 361, 397, 1942. (X5, X7, X8, X18, X26)

R26. Woodward, A. J.; "An Unusual Observation of the Moon," Royal Astronomical Society of Canada, Journal, 42:194, 1948. (X27)

R27. Ogilvy, C. Stanley; "Lights in the Moon," Popular Astronomy, 57:229, 1949. (X3, X5, X11)

R28. Vreeland, Frederick K.; "Lights on the Moon," Popular Astronomy, 57:354, 1949. (X28)

R29. Stratton, F.J.M.; "The Horned Moon with One Bright Star," Observatory, 70: 77, 1950. (X3)

R30. Burnet, Arthur; "The Horned Moon with One Bright Star," Observatory, 70: 159, 1950. (X22)

R31. McCorkle, K.E., and Lambert, W.C.; "Flares on the Moon?" Sky and Telescope, 15:44, 1955. (X31, X32)

R32. "Reported Flash on the Moon," Strolling Astronomer, 9:22, 1955. (X30)

R33. Adams, Robert M.; "Lunar Meteor Research Project," Strolling Astronomer, 9:48, 1955. (X33, X34)

R34. "Another Flashing Lunar Mountain?" Strolling Astronomer, 10:20, 1956. (X35)

R35. Stuart, Leon H.; "A Photo-Visual Observation of an Impact of a Large Meteorite on the Moon," Strolling Astronomer, 10:42, 1956. (X29)

R36. "Herschel's 'Lunar Volcanos'," Sky and Telescope, 15:302, 1956. (X4)

R37. Hars, H.; "The Total Lunar Eclipse of November 18, 1956," Strolling Astronomer, 11:64, 1957. (X36)

R38. "Possible Unusual Lunar Features: A Curious Coincidence," Strolling Astronomer, 17:127, 1963. (X37)

R39. "Lunar Color Phenomena," ACIC Technical Paper No. 12, St. Louis, 1964.

(X4, X5) (ACIC is the U.S. Air Force Aeronautical Chart and Information Center.)

R40. McNarry, L.R.; "Lunar Volcanic Activity," Royal Astronomical Society of Canada, Journal, 58:279, 1964. (X21)

R41. Ashbrook, Joseph; "Coleridge's 'Star within the Moon'," Sky and Telescope, 28:335, 1964. (X3)

R42. Kopal, Zdenek; "On Possible Observations of Luminescence on the Dark Side of the Moon," Icarus, 3:78, 1964. (X5)

R43. Middlehurst, Barbara M., and Moore, Patrick A.; "Lunar Transient Phenomena: Topographical Distribution," Science, 155: 449, 1967. (X42)

R44. Cameron, Winifred Sawtell, and Gilheany, John J.; "Operation Moon Blink and Report of Observations of Lunar Transient Phenomena," Icarus, 7:29, 1967. (X38)

R45. Bispham, K.; "Schroeter and Lunar Transient Phenomena," British Astronomical Association, Journal, 78:381, 1968. (X9)

R46. Dietrich, John W.; "A Possible Lunar Transient Event Photographed by the NASA Lunar Orbiter IV," American Geophysical Union, Transactions, 49:710, 1968. (X39)

R47. Chilton, K.E.; "Transient Lunar Phenomena," Royal Astronomical Society of Canada, Journal, 63:203, 1969. (X6, X40)

R48. Jahn, R.A.; "Transient Lunar Phenomena---A Possible Explanation," British Astronomical Association, Journal, 82: 122, 1971. (X44)

R49. Cameron, Winifred Sawtell; "Manifestations and Possible Sources of Lunar Transient Phenomena (LTP)" The Moon, 14:187, 1975. (X42)

R50. Botley, Cicely M.; "TLPs and Solar Activity, and Other Phenomena," British Astronomical Association, Journal, 86: 342, 1976. (X1, X2)

R51. Middlehurst, Barbara M.; "A Survey of Lunar Transient Phenomena," Physics of the Earth and Planetary Interiors, 14: 185, 1977. (X2, X43)

R52. Cameron, Winifred Sawtell; "Lunar Transient Phenomena (LTP):Manifestations, Site Distribution, Correlations and Possible Causes," Physics of the Earth and Planetary Interiors, 14:194, 1977. (X43, X44)

R53. Westfall, John E.; "A Comprehensive Catalog of Lunar Transient Phenomena," Strolling Astronomer, 27:192, 1979. (X42)

R54. Hughes, David W.; "Transient Lunar Phenomena," Nature, 285:438, 1980. (X42, X44)

R55. Middlehurst, Barbara M., et al; "Chronological Catalog of Reported Lunar Events, " NASA TR R-277, 1968.

R56. Hammer, John, et al; "An Active Volcano on the Moon," Scientific American, 39:385, 1878. (X21)

ALF4 Localized Color Phenomena

Description. Spots and patches of the lunar surface that glow brightly with color. Red is the most common color, but bluish and purplish patches are also on record. Lunar color phenomena last typically 10-20 minutes. They are irregular in shape and measure up to 20 miles in length, where measurements have been made. Most of the events take place near or in craters, such as Aristarchus, where other luminous phenomena also occur (ALF3).

Background. The observation of lunar color phenomena usually requires the use of rather large telescopes. Most reports, therefore, are relatively recent and made by professional astronomers. See also the Background section of ALF3, where the general history of transient lunar phenomena is discussed.

Data Evaluation. Many high-quality observations, with an occasional photo, made mainly by professional astronomers and a few amateurs with good telescopes. Rating: 1.

Anomaly Evaluation. The venting of gases from the lunar crust, possibly correlated with tidal stresses, is not considered anomalous. In the case of lunar color phenomena, it is the emission of visible light that is poorly understood. The energy source(s) and the mechanism for conversion of this energy into light are not known. If lunar color phenomena arise from processes unrelated to gas venting, the situation becomes even more puzzling. Rating: 2.

Possible Explanations. Most astronomers believe that lunar color phenomena begin with the release of gases from the lunar crust and/or the raising of clouds of fine dust. The energy source giving rise to the emission of visible light may be high energy solar radiation (perhaps stored somehow in lunar rocks); triboelectricity; or piezoelectricity. These energy sources may produce a kind of glow discharge. The fluorescence of lunar rocks is a less likely explanation. Rock incandescence due to volcanic action is another outside possibility.

Similar and Related Phenomena. Transient points of light on the moon (ALF3); infrared anomalies (ALF1); earthquake lights (GLD8), especially such colored glows seen on mountains as GLD8-X8; terrestrial mountain-top glows (GLD1); lunar obscurations (ALW1), which may be a nonluminous relative of color phenomena; lunar eclipse color phenomena (ALX4), which are optical in origin.

Examples

X1. 1885. "Observing the moon on the 10th ult., with a 8 1/2-in. refractor, I noticed that the small crater inside Hercules, instead of being filled with black shadow, shone with a dull, deep-red colour. The power I was using was 100.....The hue was quite obvious, and seemed to catch the eye immediately, even in a large field. I turned the telescope to the same spot on the next evening (20th ult.), but could perceive no abnormal appearance." Several powers were tried, but the red color persisted. (R1)

X2. 1931. Goodacre reports a bluish glare on the inner wall of Aristarchus just after sunrise. Telescopic artifacts suspected. (R5)

X3. October 4, 1955. N. A. Kozyrev detects the emission lines of hydrogen from the Aristarchus region. (R5, R10)

X4. November 3, 1958. N. A. Kozyrev observes a reddish glow in Alphonsus followed by the emission of gas. (R8, R10)

X5. January 22, 1959. D. Alter's observation of Aristarchus. "Interior of Aristarchus is a light brilliant blue!! Later in the night it was white again. The observation was made with the Mt. Wilson 60" and a power near 700." (R5)

X6. January 5, 1955. Report from D. A. Logue, Jr. "I saw a strange blue light above the surface of the moon where the night and day meet. I observed this light for more than 30 mins.; it did not move. It appeared like a star, in that the rays of light came from it." (R2)

X7. October 23, 1959. Red glow seen in Alphonsus by N. A. Kozyrev. Anomalous spectrum. (R10)

X8. November 26, 1961. N. A. Kozyrev records another anomalous spectrum; this time from Aristarchus. (R10)

X9. October 29, 1963. The observations of J. A. Greenacre and E. Barr, using a 24-inch refractor. "At 6:50 p.m. I noticed a reddish orange color over the dome-like structure on the southwest side of the Cobra head. Almost simultaneously I saw a small spot of the same color on a hilltop across Schroeter's Valley. Within about two minutes these colors had become quite brilliant and had considerable sparkle (due perhaps to poor seeing).

I immediately called Mr. Barr to share the observation with me. His first impression of the color was a dark orange. After the Wratten filter (deep yellow) was removed, the hue remained, but was brighter and sparkled more. We both agreed at this time the color, seen without the filter, was reddish orange....

No other colored spots were noted until 6:55 p.m., when I observed an elongated streak of pink along the southwest interior rim of Aristarchus. No other hue could be seen on the inside or outside of this crater. Again, Mr. Barr and I observed with and without the filter: the only difference seemed to be a somewhat brighter color in unfiltered light. The colored area along the rim of Aristarchus did not sparkle like the other two spots.

The eyepiece field of view was large enough to have all three areas in sight at the same time. At approximately 7:00 I noticed that the spots at the Cobra Head and on the hill across the valley had changed to a light ruby red, yet their density and sparkle were still sufficient to hide the surface underneath. I had the impression that I was looking into a large polished gem ruby but could not see through it. Mr. Barr at this time thought that the color was a little denser than I had described it, and that it still retained some reddish orange, but less pronouncedly." By 7:15 all spots had disappeared. The spot near the Cobra Head measured about 1.5 by 5 miles. (R3, R4) Note that Aristarchus is also the center of many appearances of bright points of light (ALF3), which are usually attributed to reflections of the sun. In fact, Herschel's famous 1787 observation (ALF3-X5) is similar in some ways to that described above. (WRC)

X10. November 1-2, 1963. Photograph of Kepler region taken by Kopal and Rackham shows red enhancement. (R10)

X11. November 27, 1963. Same observers and telescope as in X9. "In a special search for similar happenings, they observed another colored spot in the same part of the moon on November 27, 1963. This feature, like the others, seemed a light ruby red, according to Mr. Greenacre. It was larger than the previous ones, being about 12 miles long and 1 1/2 miles wide, on the rim of Aristarchus..... The ruby flush was first seen at 5:30 p.m. Mountain standard time, and lasted until 6:45. This 1 1/4-hour duration was longer than for any of the earlier three." Two additional observers at the Lowell Observatory confirmed this event. (R6, R4, R5)

X12. December 28, 1963. A team of Japanese astronomers. "On 28 December, 1963 (UT date) they were carrying out observations in preparation for the lunar eclipse on 30 December. Until 15h 05m UT nothing strange was seen, then at 15.55 one of the observers, Y. Yamada, noted a large distinct pink patch covering the southern part of Aristarchus, which was soon confirmed by the other six observers. It gradually spread toward Herodotus until 16.26, when clouds prevented any further observation. The patch was not markedly brighter or darker than the adjacent area." The location of the red patch is similar to that in X9 and X11. The Japanese were not yet aware of the sightings from Lowell Observatory. (R7)

X13. December 30, 1963. During the lunar eclipse on this date, many observers noted a red glow on the moon's northeast limb. (R10) Actually, lunar eclipses often display unusual color phenomena, most of which seem to associated with light passing through the earth's atmosphere. See ALX4. (WRC)

X14. January 5, 1964. Four English astronomers report a purple-blue patch in Aristarchus. (R5)

X15. August 26, 1964. Project Moon Blink. Red and blue bands in Aristarchus. (R11)

X16. September 20, 1964. Project Moon Blink. Red spots in Aristarchus. (R11)

X17. September 22, 1964. Project Moon Blink. Red area in Kunowsky area. (R11)

X18. October 27, 1964. Project Moon Blink. Red color phenomenon at Alphonsus. (R11)

X19. December 19, 1964. During the lunar eclipse, a region on the edge of Mare Nubium south of Copernicus showed strong anomalous enhancement of radiation. (R10)

X20. August 2, 1965. Project Moon Blink. Pink region in Aristarchus. (R11)

X21. September 11, 1965. Project Moon Blink. Red glows in Aristarchus. (R11)

X22. November 10, 1965. Project Moon Blink. Violet tinge noted at Aristarchus. (R11)

X23. April 1, 1966. Project Moon Blink. Red patch at Alphonsus. (R11)

X24. April 30 through May 2, 1966. "One of us (P.M.) has observed a color phenomenon; reddish glows in the crater Gassendi, 30 April-1 May 1966; it was first seen just before 2200 hours on 30 April by P. Sartory with a blink device using color filters, and was confirmed by T. Moseley. Further color events in Gassendi were seen by several observers between 1930 hours 1 May and 0021 hours 2 May and in September 1966. Details (such as color, areal extent, and duration) were quite similar to those reported in 1963 for events in the Aristarchus region by observers in Flagstaff, Arizona." (R12, R9)

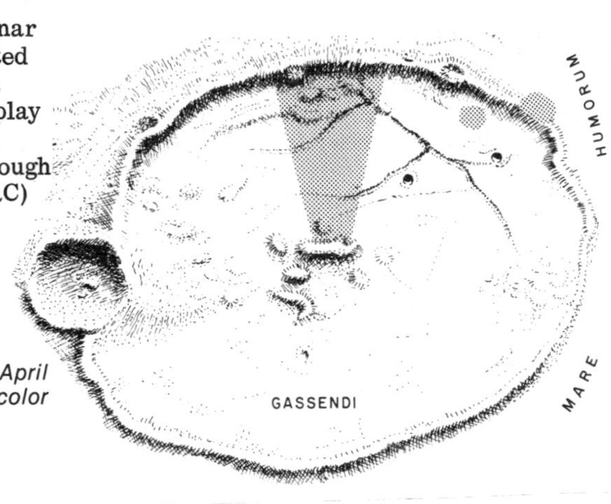

Transient lunar phenomena at Gassendi on April 30 and May 1, 1966. Shaded areas mark the color phenomena. (X24)

X25. September 1966. See details in X24. (R12)

X26. October 31, 1968. K. E. Chilton sees a reddish glow in Eratosthenes. It lasted 5 or 6 minutes. (R13)

X27. April 1, 1969. N. Kozyrev takes spectrograms of Aristarchus when a red spot, 1 to 2 kilometers across, was in evidence. The spot seemed to be the result of the release of molecular nitrogen and cyanic gas. Kozyrev speculated that the gas release might be associated with a terrestrial earthquake the day before; i.e., tidal stresses triggered both the earthquake and lunar gas release. (R14)

X28. General observations. The magnitude of the TLP phenomenon.

See ALF3-X42 for a summary of the results of a cataloging effort. (R19, R21, R23)

Geographical trends in color phenomena. "TLP tend to occur near the boundaries of the regular maria, as has been found by Middlehurst and Moore (1967) and near areas rich in rills, such as the floors of Gassendi and Alphonsus. Much of the most event-prone region is that of Aristarchus, which accounts for about half the reports. It also seems that red TLP and concentrated in the western hemisphere, while colourless events are more common in the eastern. TLP in highlands are relatively rare, but this apparent paucity may be due in part, at least, to observational selection." (R22)

X29. General observations. Sundry correlations of TLPs. See ALF3-X43 for a general summary of all TLP correlations. (R18, R20, R21)

X30. General observations. Possible explanations.

To the list of possible causes of TLPs presented in ALF3-X44, the following should be added:

Color changes due to particular lighting conditions on the lunar surface. This explanation is improbable because the color changes do not recur regularly. (R17)

An "alpenglow" hypothesis, in which sunlight is scattered by dust clouds raised by gas venting. Such a reddening phenomenon occurs in the terrestrial Alps when high altitude dust scatters red light onto the peaks (GEL6). If this hypothesis were true, the color phenomena should be concentrated near the terminator, which is not observed. (R17)

Thermoluminescence of subsurface layers, which acculate energy during many years of irradiation by high-energy particles. These energized layers give off energy (glow) when sporadically exposed and heated. This theory shows some promise. (R17)

Glow discharges in the fine dust raised by lunar outgassing. The motion and separation of the fine dust particles may result in the separation of charge and the buildup of enough potential difference to cause a glow discharge. (R16)

References

R1. Gray, William, jun.; "A Curious Lunar Phenomenon," Knowledge, 7:224, 1885. (X1)

R2. Logue, Daniel A., Jr.; "A Bluish Spot near or on the Moon," Strolling Astronomer, 8:146, 1954. (X6)

R3. Greenacre, James A.; "A Recent Observation of Lunar Color Phenomena," Sky and Telescope, 26:316, 1963. (X9)

R4. "Ruby Color Seen on Moon Again," Science News Letter, 84:386, 1963. (X9, X11)

R5. "Lunar Color Phenomena," ACIC Technical Paper No. 12, St. Louis, 1964. (X2, X3, X5, X11, X14) ACIC is the U.S. Air Force Aeronautical Chart and Information Center.

R6. "Another Lunar Color Phenomenon," Sky and Telescope, 27:3, 1964. (X11)

R7. "Japanese Saw Pink Patch on the Moon," New Scientist, 22:334, 1964. (X12)

R8. "Are There Changes on the Moon?" Sky and Telescope, 28:3, 1964. (X4)

R9. Moore, Patrick; "Color Events on the Moon," Sky and Telescope, 33:27, 1967. (X24)

R10. Middlehurst, Barbara M.; "Lunar Transient Phenomena," Icarus, 6:140, 1967. (X3, X4, X7, X8, X10, X13, X19)

R11. Cameron, Winifred Sawtell; "Operation Moon Blink and Report of Observations of Lunar Transient Phenomena," Icarus, 7:29, 1967. (X15-X18, X20-X23)

R12. Middlehurst, Barbara M., and Moore, Patrick A.; "Lunar Transient Phenomena: Topographical Distribution," Science, 155:449, 1967. X24, X25)

R13. Chilton, K. E.; "Transient Lunar Phenomena," Royal Astronomical Society of Canada, Journal, 63:203, 1969. (X26)

R14. "Earth-Moon System?" Nature, 222:404, 1969. (X27)

R15. Strickland, Zack; "Apollo 12 Seismic Experiment Links Red Lunar Glow to Quakes," Aviation Week, 57, August 10 1970. (X30)

R16. Mills, A. A.; "Transient Lunar Phe-

nomena and Electrostatic Glow Discharges," Nature, 225:929, 1970. (X30)
R17. Jahn, R. A.; "Transient Lunar Phenomena---A Possible Explanation," British Astronomical Association, Journal, 82:122, 1971. (X30)
R18. Cameron, Winifred S.; "Comparative Analyses of Observations of Lunar Transient Phenomena," Icarus, 16:339, 1972.
R19. Cameron, Winifred Sawtell; "Manifestations and Possible Sources of Lunar Transient Phenomena (LTP); The Moon, 14:187, 1975. (X28)
R20. Middlehurst, Barbara M.; "A Survey of Lunar Transient Phenomena," Physics of the Earth and Planetary Interiors, 14:185, 1977. (X29)
R21. Cameron, Winifred Sawtell; "Lunar Transient Phenomena (LTP): Manifestations, Site Distribution, Correlations and Possible Causes," Physics of the Earth and Planetary Interiors, 14:194, 1977. (X28, X29)
R22. Moore, P.; "The Observation of Transient Lunar Phenomena," Philosophical Transactions, A285:481, 1977. (X28)
R23. Westfall, John E.; "A Comprehensive Catalog of Lunar Transient Phenomena," Strolling Astronomer, 27:192, 1979. (X28)

ALF5 Transient, Large-Area Luminescence

Description. The anomalous brightening of large areas of the moon's surface. The enhancement may extend over tens of thousands of square miles. This is evidently not a color phenomenon, although certain portions of the spectrum may be enhanced more than others. Some evidence indicates a correlation with solar flares.

Data Evaluation. Several modern observations using photomultipliers, photographic film, and related instruments confirm the reality of the phenomenon. Rating: 1.

Anomaly Evaluation. Although there is evidence that some events are associated with solar flares, cause-and-effect has not been demonstrated. But at least a reasonable mechanism is at hand. Rating: 3.

Possible Explanations. Solar flares emit energetic particles which arrive at the moon several hours later, causing lunar materials to emit visible radiation.

Similar and Related Phenomena. Small-area bright patches (ALF3) and color phenomena (ALF4).

Examples

X1. November and December, 1883. "On November 4, 1883, I saw on the dark part of the Moon a misty light different from that to be ascribed to the feeble illumination due to light reflected from the Earth. This appearance was seen repeatedly by myself and others in November and December. The weather in January and February was not favorable to observations. But on March 29th last and 30th, the phenomenon was distinctly seen by myself and others. One who had seen it last fall, on looking at it on the 29th, remarked at once that it appeared on a different part of the disc, and, on the 30th, he noticed that the light had narrowed down.

This accumulation of light shifts its place upon the disc; but generally is seen brightest along the dark limb of the Moon. The appearance is that of the light of the Aurora Borealis on the Earth, as it might show to one looking at it from the Moon." (R1)

X2. March 29-30, 1884. See description presented above. (R1)

X3. November 1-2, 1963. "An anomalous increase in surface brightness of the Moon in an area of over 60,000 km^2 around and north of the crater Kepler was observed to occur twice on the night of November 1-2, 1963, on eight photographs secured with the 24-inch refractor of the Observatoire du Pic-du-Midi between 22.35-22.42 U.T. on November 1st, and 00.20-00.35 U.T. on November 2nd, on Kodak 1-F plates exposed through an interference filter of 45 A half-width centered on 6725 A. Control photographs taken through an interference filter of 95 A half-width centered on 5450 A failed to show any such effect. The enhancement in the red was not only observed to recur twice during the same night, but plates taken between 00.20-00.35 U.T. on November 2nd disclosed that the degree of enhancement (resulting nearly in a temporary doubling of surface brightness) increased appreciably within 15 minutes of observation." The authors associated the brightenings with two solar flares that had occurred several hours earlier. They surmised that energetic particulate radiation from the sun caused lunar luminescence. (R2-R5)

X4. December 19, 1964. A 16-inch reflecting telescope, equipped with a photomultiplier confirmed the existence of lunar luminescence during a lunar eclipse. (R6)

X5. March 26, 1970. "Abstract. An anomalous enhancement of brightness of the lunar surface was observed on March 26, 1970, during photometric and polarimetric observations of the Moon made by the 91-cm reflector of the Dodaira Station of the Tokyo Astronomical Observatory. The region near Aristarchus was 0.3 mag. brighter compared with that corresponding to the ordinary brightness variation curve. The polarimetric results also showed evidence quite consistent with the phenomenon of the luminescence (sic). The colour index diminished by about 0.1 mag. The results of the photographic photometry also confirmed the enhancement of brightness over a wide part of the lunar surface that night. This brightning may be related to the large solar flare observed on March 25, 12h UT, 1970, twenty-nine hours before our observations." The author does not consider the evidence conclusive that the lunar luminescence was caused by the solar flare. (R7, R8)

X6. March 13, 1979. "At the recent partial lunar eclipse of 1979 March 13, an anomalous brightening of a region in the umbra was observed. This took the form of a large diamond shape between Mare Serenitatis and the Moon's limb, just after mid-eclipse, and lasted for about ten minutes, indicating a definite unevenness in the radiations illuminating the umbra. A final proof of this theory might be in the correlation of all TLP observations to solar activity; however, this is an extremely difficult task, as observed TLP frequency depends on phases of the Moon, time of year, weather, and other less quantifiable factors. Indeed the strongest correlation so far with TLP frequency appears to be to the Apollo lunar landing missions, probably due to an enormous upsurge in interest in lunar observing at that time." (R8)

References

R1. Haywood, John; "An Auroral Glow on the Moon," Sidereal Messenger, 3:121, 1884. (X1, X2)

R2. Kopal, Zdenek, and Rackham, Thomas W.; "Excitation of Lunar Luminescence by Solar Activity," Icarus, 2:481, 1963. (X3)

R3. Kopal, Zdenek, and Rackham, Thomas W.; "Excitation of Lunar Luminescence by Solar Flares," Nature, 201:239, 1964. (X3)

R4. Kopal, Zdenek, and Rackham, Thomas W.; "Lunar Luminescence and Solar Flares," Sky and Telescope, 27:140, 1964. (X3)

R5. Kopal, Zdenek; "On Possible Observations of Luminescence on the Dark Side of the Moon," Icarus, 3:78, 1964. (X3)

R6. Sanduleak, N., and Stock, Jurgen; "Indication of Luminescence Found in the December 1964 Lunar Eclipse," Astronomical Society of the Pacific, Publications, 77:237, 1965. (X4)

R7. Sekiguchi, Naosuke; "An Anomalous Brightening of the Lunar Surface Observed on March 26, 1970," The Moon, 2:423, 1971. (X5)

R8. Shepherd, James S.; "Lunar Eclipses, Lunar Luminescence and Transient Lunar Phenomena," British Astronomical Association, Journal, 92:66, 1982. (X5, X6)

ALF6 Lightning-Like Phenomena on the Moon

Description. Repeated flashes of light streaking across the moon's dark side.

Data Evaluation. The one modern observation should be given some consideration because it was made by a government scientist and his wife. Rating: 3.

Anomaly Evaluation. Since the moon has virtually no atmosphere, any electrical discharge phenomena would apparently have to occur in lunar dust clouds or vented gases. However, the observed lunar lightning (if that is what it is) covers immense areas; and we know of no lunar process that would raise dust or vent gases over such huge areas. Rating: 2.

Possible Explanations. Large-scale electrostatic discharges caused by triboelectricity or seismoelectric phenomena. Another possibility is that the luminous phenomenon actually takes place in the terrestrial atmosphere and is made more obvious by the lunar backdrop.

Similar and Related Phenomena. Star-like flashes of light on the moon (ALF3); large-area color phenomena, which some have explained as glow discharges (ALF4); flickering light on the dark limb of Venus (AVO9); terrestrial mountain-top glows (GLD1).

Examples

X1. 1715-1746. During a total eclipse of the sun during this period, lightning was seen on the face of the moon. (R2)

X2. June 17, 1931. "During the evening of 17 June 1931, I was working in the yard near our home at Riverside, California, and happened to glance at the moon. It was an unusually fine, clearly outlined new moon, and as I stood looking at it, suddenly some flashes of light streaked across the dark surface but definitely within the limits of the moon's outline. Since this was a phenomenon which I had never seen before, I continued to watch it and saw similar flashes streak across the moon again in a moment or two. Without mentioning what I had seen, I called my wife's attention to the new moon. She admired it. When I asked her to watch it closely to see if she noticed anything strange, she said: 'Oh, yes, I see lightning on the moon,' adding that this appeared to be confined to the moon. We watched it for some 20 or 30 minutes during which the phenomenon must have occurred at least six or seven times." When the observations were reported to Mount Wilson Observatory, they were discounted. (R1)

References

R1. Giddings, N. J.; "Lightning-Like Phenomena on the Moon," Science, 104:146, 1946. (X2)

R2. Botley, Cicely M.; "TLPs and Solar Activity, and Other Phenomena," British Astronomical Association, Journal, 86: 342, 1976. (X1)

ALL THE MOTION OF LUNAR SATELLITES

Key to Phenomena

ALL0 Introduction

Today, the moon has no natural satellites of any consequence. Bits of debris may occasionally occupy temporary lunar orbits, but they are difficult to detect. The important satellites of our satellite are now all artificial. The accurate tracking of these spacecraft has led to the discovery of gravitational anomalies over many of the ringed maria. These anomalies have been explained in terms of subsurface concentrations of mass. Although widely believed to be buried projectiles that originally created the maria, no one is really sure what they are; they could be internally generated, like terrestrial mantle plumes. In any event, the mascons must be explained by any comprehensive theory of lunar evolution.

No direct evidence exists for the past presence of natural lunar satellites. However, paleomagnetism and the arrangement of the maria in bands have led some astronomers to theorize that the moon was once orbited by a group of natural moons that ultimately crashed into its surface. (See ALE21 and ALZ4.)

ALL1 Perturbations of Artificial Lunar Satellites

Description. Large orbital perturbations of artificial lunar satellites that cannot be accounted for assuming a homogeneous outer lunar crust.

Data Evaluation. Precision tracking of lunar satellites by terrestrial networks. Rating: 1.

Anomaly Evaluation. The analysis of the perturbations of lunar satellites permits the construction of a gravimetric map, which suggests the presence of mass concentrations under most lunar maria. The real nature and origin of these concentrations are uncertain. Rating: 2.

Possible Explanations. The mascons could be asteroid-sized bodies buried beneath the maria they created upon impacting the moon. They could also be plumes or plugs of dense material pushed up from below by tectonic forces.

Similar and Related Phenomena. Ancient changes in the lunar magnetic field (ALZ4), which imply that the moon collided with large bodies (possibly lunar satellites) that created the mares and shifted the lunar crust. The arrangement of the maria in bands (ALE9), which suggests the ancient descent and impact of groups of lunar satellites.

Examples

X1. Lunar Orbiter perturbations. "We now report that this new processing of the Lunar Orbiter data has produced unexpected results. A study of local accelerations of the spacecraft resulted in a gravipotential map of the lunar nearside which has revealed very large mass concentrations beneath the center of all five nearside ringed maria (Imbrium, Serenitatis, Crisium, Nectaris, and Humorum). In addition, they were observed in the area between Sinus Aestuum and Sinus Medii (presumably a newly discovered ancient ringed mare), and Mare Orientale. The Urey-Gilbert theory of lunar history has predicted such large-scale high density mass concentrations below these maria, which, for convenience, we shall call masons.....

Even though we have computed both masses and depths for the largest mascons, nevertheless our present quantitative data require further refinements. One may easily compute approximate masses from an assumed depth, such as 50 km. The Mare Imbrium mascon yields numbers on the order of 20 X 10^{-6} lunar masses. A spherical nickel-iron object about 100 km in diameter would be a rough equivalent." (R1)

References

R1. Muller, P. M., and Sjogren, W. L.; "Mascons: Lunar Mass Concentrations," Science, 161:680, 1968. (X1)

ALO ANOMALOUS TELESCOPIC AND VISUAL OBSERVATIONS

Key to Phenomena

ALO0 Introduction

The moon is so close to our telescopes that many surface details are crisp and clear. Further, some features that once seemed enigmatic through the telescope have now been elucidated by spacecraft photographs and the explorations of astronauts. Thus, there are not so many optical or telescopic anomalies for the moon as there are for the other solar system objects. While this section is shorter, the one dealing with geological anomalies is much larger than the corresponding sections for Mercury, Venus, Mars, etc. Better seeing means more accurate classification.

Still, the problems of psychological contrast effects, varying light levels, shifting shadows, and the variation of surface reflectivity with solar angle do not disappear completely. Perhaps the most perplexing of these lunar telescopic anomalies involve the variable spots and streaks of shadow and brightness. Some of these features vary regularly with each lunation; others are confoundingly irregular. Another ephemeral phenomenon is the moon's "annular phase"---that silvery ring rarely seen around the dark limb of the new moon. And, once again, there are "canals"; not the river-like sinuous rilles mentioned in ALE2, but long straight lines akin to those seen (by some observers at least) on Mercury, Venus, Mars and sundry satellites. Apparently the integrating power of the human eye-brain combination is at work here. In fact, the problems of contrast and illusion pervade this entire section.

ALO1 Doubling of Lunar Detail

Description. The long-lived doubling of lunar detail as seen through the telescope. This rare condition may last for over an hour.

Data Evaluation. Only a single report, but it involved two widely separated observers and several different telescopes. There are, however, occasional reports of double and triple lunar crescents seen with the naked eye. Thus, the phenomenon is confirmed on a gross level. Rating: 2.

Anomaly Evaluation. Long-lived image doubling probably involves mirage-action, which results from the interposition of a second layer or region of air possessing a slightly different density than the bulk of the atmosphere. Most terrestrial mirage phenomena are well-understood. The anomalies here are the long-term stability of the phenomenon, given the strong tendency of the atmosphere to mix, and the appearance of the same mirage over a wide area. Rating: 3.

Possible Explanations. Mirage action similar to that creating multiple mirages of terrestrial objects (GEM). Interestingly enough, the use of the telescope often reveals multiple terrestrial mirages unseen by the naked eye.

Similar and Related Phenomena. Multiple lunar crescents (GEM4).

Examples

X1. January 1966. Surrey, England. Report of P.K. Sartory. "Whilst observing the moon one night during January 1966 with a very fine 5-inch Cooke refractor, doubling of the image was seen. The night was clear and seeing very good. It was at first suspected that tube currents were the cause of the effect, but on checking with a 4-inch Maksutov telescope the same effect was noted. The images were quite sharp and steady. Seven photographs were obtained, and all showed the effect quite clearly. It seems, therefore, that the effect is produced outside the instrument since two different telescopes were tested. The doubling of the image persisted for about 1 1/2 hours. A colleague of mine observing about 50 miles away from my station also observed the effect." (R1) Gross multiplication of lunar crescents have been observed. See GEM4. (WRC)

References

R1. Sartory, P.K.; "Doubling of Lunar Detail," Strolling Astronomer, 19:143, 1965. (X1)

ALO2 Variable Spots, Streaks, and Other (Apparently) Optical Phenomena

Description. Dark and light spots and streaks on the moon that appear, disappear, and change shape in a manner that does not seem explicable in terms of the varying angle of illumination. Some of the phenomena vary regularly with each lunation; others seem to follow no laws at all. Most of the light-and-shadow magic occurs around Plato and a few other sites. Cataloged here are passive phenomena only; that is, those involving no intrinsic light sources or the active clouds of vented gases responsible for lunar "mists" and obscurations (ALF2, ALF3, ALW1). See the referenced categories for descriptions of these other types of TLPs (transient lunar phenomena).

Background. The lunar landscape when viewed through the telescope reveals many subtle changes of light and shadow as the sun rises and sets. Ordinary optical laws can explain the bulk of these phenomena when allowance is made the variations of reflectivity with solar angle and type of surface. But a few phenomena do not succumb. This area of lunar research has low priority for most professional scientists, and it has been left mainly to amateur astronomers to record the fascinating nuances of light and shadow on the moon.

Data Evaluation. Over the years, a large corpus of observations of variable spots and streaks and other inexplicable plays of light and shadow have been drawn, photographed, and described in prose. Rating: 1.

Anomaly Evaluation. If only passive mechanisms are involved in these phenomena---no intrinsic light sources and no obscuring clouds of vented gases---we are left with several possibilities, as listed below. None of these mechanisms is in itself anomalous, for they are known also on earth, but we do not know which operate on the moon to cause the various spots and streaks that wax and wane during lunations. Of course, there may be completely unexpected mechanisms at work. Assigning an anomaly rating is difficult. Rating: 3.

Possible Explanations. (1) Changes in the reflectivity and absorption of lunar materials with solar angle; (2) Similar changes dependent upon the surface temperature; (3) Similar changes due to the roughness of the surface; (4) Similar changes due to the deposition of substances vented from the lunar crust, such as volcanic dust and (shades of W.H. Pickering) hoar frost; (5) Sublimation of (4) with surface temperature increases; (5) Chemical changes of the lunar surface due to solar radiation, the solar wind, and other agencies. Here, we should not forget that several older astronomers suggested the presence of lunar vegetation! Finally, some of the phenomena categorized here may really belong in the sections involving intrinsic light sources (ALF2, ALF3) and obscuring matter (ALW1).

Similar and Related Phenomena. Light flashes and bright spots (ALF2); lunar color phenomena (ALF3); mists and obscurations (ALW1).

Examples

X0. General observations. The astronomer William H. Pickering was perhaps the most famous of the investigators of variable lunar spots and streaks. An earlier English amateur, W.R. Birt, actually pioneered the field; but Pickering, because of his radical explanations of the phenomena, grabbed the headlines. The following item reflects Pickering's far-out views.

Referring to Pickering's article in the Century Magazine (R13): "The second, and perhaps more startling, announcement is that there is snow on the moon. He observed that many craterlets are lined with a white substance which becomes very brilliant when illuminated by the sun, and a similar substance is found on the larger lunar craters and a few of the higher mountain peaks. The curious behavior of these patches under different angles of illumination and their change of form have led him to suggest that an irregularly varying distribution of hoar frost may have something to do with the changes observed. The third remarkable deduction refers to the observations of 'variable spots,' which appear to be restricted between latitudes 55° north and 60° south; these spots are always associated with small craterlets or deep narrow clefts, and are often symmetrically arranged around the former. The alterations which these undergo have led him to seek the cause in the change in the nature of the reflecting surface, and the most simple explanation according to him is found in assuming that it is organic life resembling vegetation, but not necessarily identical with it." (R12) This radical interpretations of the data should not detract from the real capabilities of Pickering as an astronomer. (WRC)

Pickering was often called upon to defend his theories rather than his observations: "Since the old question of lunar vegetation has again come up for discussion in the Journal of the B. A. A., I should like to point out once more that there are really two distinct questions involved. One is as to the facts. This is important. The other is as to their explanation, which is of much less consequence. Unfortunately, the interest of the latter, and the discussion which it has involved, has almost completely concealed and smothered the more important question. In denying the explanation, the rejectors have carelessly also denied the facts, without recognizing that they are in a wholly incompetent position to do so. Some of them have even been so careless as to suggest that the darkening areas were shadows, not realizing that shadows are most conspicuous under a low sun on the Moon, and that they disappear completely when the Moon is full. The variable spots do the precise reverse of this. As has often previously been stated, they are inconspicuous, and in some cases invisible under a low sun, but the central ones are strikingly conspicuous at full Moon. In the case of Eratosthenes, to which I have called particular attention, they are perfectly obvious with a 5-inch refractor, and can even be seen with a 3-inch (Popular Astronomy, 27:579, 1919, Plates 44 and 45). Anyone can see them who will take the trouble to turn his telescope upon the Moon. During the three days including the full Moon, however, the novice may at first have a little trouble in the case of this crater, because the shadows have disappeared, and the changes are so marked that the crater itself is rather hard to recognise. The markings with their changing shapes are very easily photographed (Havard Annals, 53, Plates 2, 3, and 4. See also Popular Astronomy, 27:579, 1919, Plates 46 and 47). In the light of these photographs, for anyone to claim that these spots do not exist, or that they do not change their shapes, or that they are not visible at full Moon, is, to say the least, foolhardy." Pickering also mentions a more "delicate" variety of spot that meanders across the face of the Moon erratically, taking a different position each lunation. (R22)

Pickering was not the first astronomer to suggest lunar vegetation. The German astronomer, Schwabe, reported in 1868 that he saw streaks on the lunar mountains that were green at certain seasons. He regarded the streaks as belts of vegetation. (R1) Mare Serenitatis and Mare Humorum often seem so green that vegetation is suggested. (R8) And, of course, Pickering had a longer list of green areas. (R25)

The problem of the variable lunar spots, streaks, and colors was well-summarized by W. H. Haas: "Pickering has observed in a number of lunar regions periodical changes that repeat themselves every lunation. For example, dark areas develop and fade in Eratosthenes, a bright area around Linne varies in size. Messier and Pickering alter their sizes and shapes, and the peaks of Theophilus undergo changes of appearance. Pickering attributed these changes to physical agencies on the moon; i.e., to vegetation, to clouds, to snow, to frost, etc. His explanations demand the presence of a lunar atmosphere, a controversial point. Scientists have long been accustomed to regard the moon as completely dead and naturally received Pickering's observations and theories with much skepticism. That the changes do occur much as Pickering asserted has been established, but the explanation he gave is still not necessarily correct. The majority of the astronomers now ascribe these changes to the variations in the angle at which sunlight is incident upon the lunar rocks. The moon's surface is granulated, a fact proved by the disproportionate increase in brightness of the moon toward the full phase; and it is thought that this granulation causes the area to vary in brightness or in intensity with the ever-changing angle of incident light and thus produces the changes observed. It is between these two theories, incident light and physical change, that the author has attempted to decide. To be sure, there may be other possible explanations; e.g., a chemical change upon the surface of the rocks occurring under a low sun; and it is altogether possible that there is not one explanation for all the changes." (R26)

X1. Plato. Of all the lunar craters, Plato possesses the greatest variety of variable subtle detail. For convenience, we lump these variations into four categories: variable spots, variable streaks, the "sparkling effect", and obscurations. These phenomena occur inside a crater that measures about 60 miles in diameter, with rim walls rising from 3,000 to 5,000 feet above the crater floor. Some peaks on the rim may soar to 7,400 feet. The following descriptions are based on a Sky and Telescope article by Jackson T. Carle, except for the discussion of the "sparkling effect". Additional references are added at the end of each paragraph.

Variable spots. "Since an intensive study of Plato was first organized by the British selenographer W. R. Birt and his coworkers in 1869, some 80 craterlets and spots have been charted. These have never all been seen by one observer or within a limited period of time. Some spots, having been seen easily, apparently disappeared---only to reappear years later and be reported as new discoveries." (R28, R3, R9, R13) See the accompanying figure drawn by Pickering to get an appreciation of how the "ten most prominent spots" in Plato varied in time.

Variable streaks. "As craterlets and spots seem to have appeared and disappeared, so faint streaks and large splotches have shifted on Plato's floor. Some observers connect these two kinds of changes. In 1892, Pickering reported that the bright streaks had altered materially since earlier observations. He sought to explain these markings in terms of his theory of volcanic activity on the moon." (R28, R2, R5, R7, R9, R18, R15, R16)

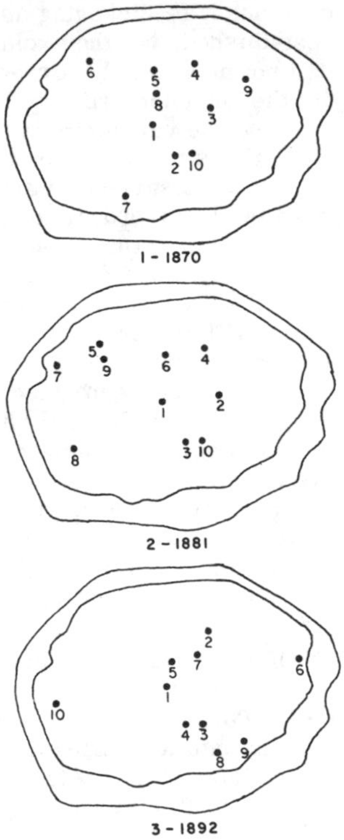

The ten most prominent spots in Plato drawn at different times, by W. H. Pickering, indicating great variability. (X1)

Streaks on Plato's floor. (X1)

Obscurations. The crater of Plato has been called that Great Black Sea, for sometimes it seems to swallow up details one knows should be seeable. Carle graphically describes the third Plato enigma in terms of the thoughts of an astronomer who has been observing the rest of the moon on a fine night when all this orb's detail seems to stand out ---that is, almost all its detail: "Now to Plato. You look, and look again, and see nothing! Yes, the walls are clear and detailed, and that landslip on the east stands out. But the floor appears smooth, flat, and featureless, perhaps with a faint glimmering of something where you know the comparitively central craterlet should be. You see nothing, but this nothing is the something that has puzzled lunar observers for a century, the apparent obscuration of the floor of Plato at times when floor features should be readily visible." (R28, R8, R10, R29) Plato's darkening is a special case of the lunar obscurations described in more detail in ALW1. (WRC)

The sparkling effect. W.R. Birt was evidently the first to report this phenomenon. His first sighting was on November 20, 1871. Henry Pratt made a second observation on January 18, 1872 and reported it thus: "The phenomenon which I witnessed is very difficult to describe. On the dark ground of the floor, or apparently above it, I noticed the play of a continual series of fitful flashes of light, sometimes presenting the form of minute streaks sharply defined, oftener of larger streaks of a more distinct nature, and more frequently still the flashes took the form of spots. I noticed a continual succession of flickering lights, not badly defined as corruscations, if this term does not convey the idea of too much brightness. They were faint, of course, and might have been very easily overlooked even by experienced observers; still, when once perceived they were easily kept in view....Often I saw several distinct specks of light at once; frequently, if the attention was fixed upon it in the endeavour to localise it, would persist in the same place for a second or so, and vanish altogether; and yet its persistence was not one picture, but a rapid succession of images; the same remark applies to the streaks." (R6, R4)

X2. Alphonsus. Variable spots. "Prof. W.H. Pickering has made an extensive study of many variable dark spots upon the floors of various craters. Those in Alphonsus afford one of the best-known examples. These spots commence to darken soon after sunrise, but earlier in low latitudes than it high; some of them attain only a light grey colour, others a dark grey, and other again become intensely black; they attain a maximum, and fade away again towards sunset." (R14, R11, R16)

X3. Mare Tranquilitatus. Variable spots.

"The Mare Tranquilitatis is said to be almost covered by these variable spots, and Prof. Pickering states that the changes may be seen with the smallest telescope, or even with the naked eye; until past the first quarter this area is lighter than the Mare Crisium; it then rapidly becomes the darker of the two until after full moon, when it again becomes lighter." (R11)

X4. Messier. Variable whiteness. "...one of the clearest evidences of hoar frost upon the moon is found in connection with the pair of small craters known as Messier and Messier A. Sometimes one of these craters is the larger and sometimes the other. Sometimes they are triangular and sometimes elliptical in shape. When elliptical their major axes are sometimes parallel and sometimes nearly perpendicular to one another. When the sun first rises on them they are of about the same brilliancy as the mare upon which they are situated, but three days later they both suddenly turn white, and remain so until the end of the lunation. When first seen the white areas are comparatively large, especially that surrounding Messier itself, but it gradually diminishes in size under the sun's rays. By the eighth day little is left outside the crater itself, while at the end of the lunation only the bottoms and interior western walls remain brilliant." (R16)

X5. Eratosthenes. Variable spots. "In the crater Eratosthenes there is a brilliant white area on the summit of the central mountain range. When the sun first rises on it it measures five miles in length by two in breadth. It soon, however, begins to dwindle, and two and a half days later all is gone save two little spots, each about a mile in diameter. They reach their minimum size five days after sunrise, when the smaller is about half a mile in diameter. They then begin to increase, the northern one attaining a length of five miles shortly before sunset. If these markings are due to white quartz, or some similar rock, it is difficult to account for their change in size." (R16, R19-R22)

X6. Littrow B. Variable spots. "About 15 miles N. of this crater there is a conspicuous bright spot, one of many similar objects on the Moon's surface. They vary in size, shape, and brilliancy, but they are all hazy and ill-defined at the edge. This particular spot is about 12 miles in diameter. During the Moon's first quarter, the bright area is much more extensive than it becomes later, and embraces crater B, but only with a slight haze and not so as to hide the crater. Then, in a few days the haze fades away and leaves the spot round and bright. But as sunset approaches, when Littrow is getting near the Terminator, this ill-defined bright spot grows fainter and finally clears away altogether, disclosing several well-defined objects which were before invisible." (R8)

X7. Picard, Variable spots. "There is a white spot to W. of Picard, which Birt considered was in itself of the nature of an obscuration, i.e., this spot itself consisted of material which covered the true surface of the Moon. It often disappeared as sunset approaches, when the true surface presented to the eye a peculiar whiteness, in which were observed towards the end of 1875 two very minute craterlets, emissions from which may have been the cause of the cloudiness and ill definition observed at that time." (R5)

X8. Aristillus. Variable spots. (R20, R21)

X9. Pico. Variable spots. "Some 48 drawings of the mountain, made with a four-inch diameter telescope, show that light and dark areas on the lunar mountain vary from time to time. Mr. Rawstrom states that it is practically impossible to reconcile these changes with the effect of the varying angle of illumination. There are certain markings which actually darken as the lunar mid-day approaches; certain others vary considerably in shape and size during the course of a lunation, the interval between the returns of consecutive new moons. 'Most striking of all, however, are those areas which undergo an irregular change in appearance from one lunation to the next---that is to say, which do not present the same aspect at similar co-longitudes.'" (R23)

X10. Mare Imbrium. Variable spots. "The most conspicuous marking observed is a white area which spreads out from the northeast corner of the mountain (Pico) and extends over a great plain known as the Mare Imbrium, for about 22 miles. Whether this is a haze or a jet of steam of perhaps volcanic origin is unknown." (R23, R24)

X11. Aristarchus-Herodotus region. Variable spots and curious shadows. Of particular interest is a variable white spot that has been observed on rare occasions in the center of Herodotus. The crater Herodotus does <u>not</u> have a central peak, so that the usual explanation employing reflection cannot apply. (R27) In ALF3, bright star-like points of light were generally written off as mountain-peak reflections. (WRC)

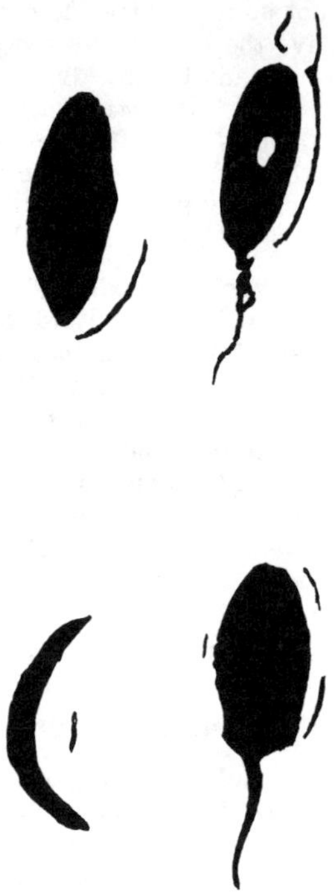

The Aristarchus-Herodotus region drawn by J. C. Bartlett on June 27, 1950, and July 13, 1954. The bright spot in Herodotus is puzzling because the crater has no central peak. (X11)

X12. Shadow phenomenon. May 1, 1909. Good seeing, with high definition of lunar detail. "Under these conditions, while examining the regions near Mersenius then on the terminator, it was suddenly noticed with great astonishment that the broad, irregular, and sharply-indented shadows of the illuminated ridges, etc., in the neighborhood could be distinguished, apparently projected on the dark, unilluminated portion of the lunar surface. Various eyepieces were tried, eye rested, etc., but I could feel no doubt of the objective reality of the phenomenon, some sharp irregularities in the shadow being seen repeatedly. The whole appearance was evasive, and decidedly delicate, but might be within the grasp of a smaller instrument than that employed, care being taken to exclude as much as possible of the illuminated surface from the field. The interior of Mersenius itself was perfectly dark; but the shadow of the eastern wall, only the top of which was illuminated by the rising sun, could be seen projected on the unlighted surface beyond the terminator. The appearance was that of sharply-defined, inky-black shadows projected on a rusty-black background. This effect must almost certainly been caused by a dimly-lighted zone bordering the shadows. At the time, I attributed the phenomenon to reflected light from the bright ridges illuminating the lower ridges; but whatever the cause, the appearance was undoubtedly objective, and the dimly-lighted regions on which the shadows were cast were those which would themselves be shortly illuminated by the rising sun, thus forming a kind of dawn." (R17) This particular lighting phenomenon may be related to the horizon glows detected by spacecraft cameras (ALO11). (WRC)

References

R1. "Lunar Vegetation," Scientific American, 18:280, 1868. (X0)

R2. Birt, W.R.; "Report on the Discussion of Observations of Spots on the Surface of the Lunar Crater Plato," Report of the British Association, 1871, p. 60. (X1)

R3. Birt, W.R.; "Evidence of Recent Changes in the Moon's Surface," English Mechanic, 13:55, 1871. (X1)

R4. "New Phenomenon on Plato," English Mechanic, 14:333, 1871. (X1)

R5. Birt, W.R.; "Report on the Discussion of Observations of Spots on the Surface of the Lunar Crater Plato," Report of the British Association, 1872, p. 245. (X1)

R6. Birt, W.R.; "Lunar Meteorology," English Mechanic, 14:660, 1872. (X1)

R7. Birt, W.R.; "New Objects on the Moon's Surface," English Mechanic, 19:447, 1874. (X1)

R8. "The Green Fields of the Moon," Scientific American, 35:160, 1876. (X0, X1)

R9. Birt, W.R.; "Selenographical," English Mechanic, 33:306, 1881. (X1)

R10. Birt, W.R.; "Selenographical," English Mechanic, 34:110, 1881. (X1)

R11. "Lunar River Beds and Variable Spots," Nature, 51:589, 1895. (X2, X3)

R12. "Changes on the Moon's Surface," Nature, 66:40, 1902. (X0)

R13. Pickering, William H.; "Is the Moon a Dead Planet?" Century Magazine, 64:90, May 1902. (New series vol. 42) (X0, X1, X4)

R14. Saunder, S.A.; "Changes on the Surface of the Moon," English Mechanic, 78:304, 1903. (X2)

R15. "Professor Pickering's Reported Lu-

nar Changes," Franklin Institute Journal, 138:438, 1904. (X1)
R16. Pickering, William H.; "Changes upon the Moon's Surface," Nature, 71:226, 1905. (X1, X2, X4, X5)
R17. Merlin, A. A. C. Eliot; "Curious Lunar Observation---Shadows on Darkness?" English Mechanic, 89:395, 1909. (X12)
R18. Burgess, J. G.; "A Case of Periodic Change on the Moon's Surface," British Astronomical Association, Journal, 25: 183, 1915. (X6, X7)
R19. Pickering, W. H.; "Eratosthenes I, A Study for the Amateur," Popular Astronomy, 27:579, 1919. (X5)
R20. "Confirmation of Pickering's Lunar Observations," Scientific American, 125:159, 1921. (X5, X8)
R21. Christie, W. H. M.; "Changes in the Lunar Craters Aristillus and Eratosthenes .....," Royal Astronomical Society, Monthly Notices, 81:451, 1921. (X5, X8)
R22. Pickering, W. H.; "The Lunar Vegetation," British Astronomical Association, Journal, 37:65, 1926. (X0, X5)
R23. "Change in Lunar Markings," Science, 86:sup 8, July 9, 1937. (X9, X10)
R24. "Changes in Moon Mountain Suggest Presence of Haze," Science News Letter, 32:57, 1937. (X9, X10)
R25. Pickering, W. H.; "Life on the Moon," Popular Astronomy, 45:317, 1937. (X0)
R26. Haas, Walter H.; "The Problem of Lunar Changes," Royal Astronomical Society of Canada, Journal, 32:347, 1938. (X0)
R27. Bartlett, James C., Jr.; "Herodotus: A Light That Failed," Strolling Astronomer, 8:91, 1954. (X11)
R28. Carle, Jackson T.; "Three Riddles of Plato," Sky and Telescope, 14:221, 1955. (X1)
R29. Peterson, Keith; "Plato and Its Mysteries," Strolling Astronomer, 17:99, 1963. (X1)
R30. Robinson, L. J.; "Banded Craters," British Astronomical Association, Journal, 73:33, 1963. (X0)

ALO3 Banded Craters

Description. A class of lunar craters with floors marked by dark bands, which are often geometrically arrayed. The bands vary in darkness and shape as the angle of the sun varies. Banded craters are usually relatively bright. About 200 are known.

Data Evaluation. The dark bands are easily seen through the telescope and on photographs. So far, no data relevant to the banded craters has been found originating in the space program. Rating: 1.

Anomaly Evaluation. The cause of the darkness of the bands is probably not very mysterious. Either superficial surface deposits or regions of fractured rock and disturbed soil would suffice. The major puzzle seems to be the geometric nature of many of the band-systems. Even here, though, the geometry could be the consequence the tendency of rocks to fracture or joint in crude patterns. Rating: 3.

Possible Explanations. The bands may be deposits of volcanic ash, dark dust vented through crustal cracks, or swaths of material laid down during impact events (i.e., like the lunar swirl markings). In other words, the patterns could be either an expression of subsurface fracturing or dynamic processes in crater formation.

Similar and Related Phenomena. Lunar rays (ALE3); swirl markings (ALE5); dark-haloed craters (ALE12); variable spots and streaks (ALO2); lunar lineaments (ALO4).

Examples

X1. General categorizations. "In the most general terms, a banded crater may be defined as a normal crater with one notable peculiarity---the floor and/or walls are scarred by darkish patches of material. These dusky markings may take various forms: from large irregular patches to very regular and linear ray-like areas. Nevertheless, this entire continuum may be organized into the following categories.

Class I. Aristarchus Type. This form of banded crater is probably the most intriquing, for the major bands arrange themselves radial to the centre in a spoke system. This point, while being within the crater, need not necessarily be the geometrical centre of same. The bands, themselves, do not meet at this origin but rather initiate from a dark ring which is quite concentric to the rim of the crater, see Fig. 1A. These craters are very bright; being the brightest crater-like objects on the lunar surface."

The detailed descriptions of the rest of the classes are omitted. The reader can examine the figure to see how the bands are arranged in each class. The author claims to have identified 188 banded craters. He also remarks that the bands display cyclic variations during each lunation. It is this property and the unknown character of the bands which place this phenomenon in the ALO rather than the ALE (geology) category. (WRC)

"As to what constitutes the bands or the dark areas, nothing definite can be said. A single observation by the author found the bands of Aristarchus to be composed of numerous filaments arranged along the axis of the band. It is believed that these 'filaments' are rilles of very small size---this observation is given not as the origin of the dark nature of the bands, but merely as an explanation of the filaments.....

1. These craters all have definite external walls which rise high above the surrounding plain.

2. The walls of banded craters brighten under high solar illumination.....

3. 3. The floors of these craters lie below the surrounding plain.

4. All bright craters, unless intruded upon by another crater, are quite circular.

5. The larger bright craters have central mountain masses.

6. The very bright craters are origins of ray systems." (R2, R1)

References

R1. Abineri, K. W., and Lenham, A. P.; "Lunar Banded Craters," British Astronomical Association, Journal, 65:106, 1955. (X1)

R2. Robinson, L. J.; "Banded Craters," British Astronomical Association, Journal, 73:33, 1963. (X1)

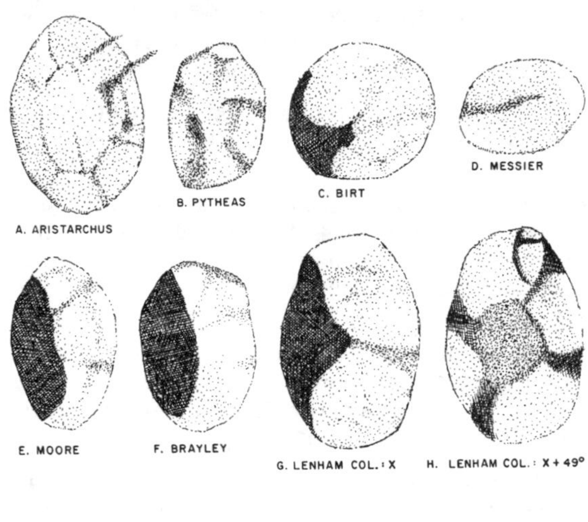

Sketches of several banded craters of different types. (X1)

ALO4 Lunar "Canals" and Lineaments

Description. Linear features seen on the moon with the naked eye, through the telescope, or on close-up spacecraft photographs. At all levels of detail the linear features seem to be integrations of diverse types of lunar features; viz., ridges, valleys, crater chains, etc.

Background. Mercury, Venus, and Mars also display linear features to the terrestrial observer. The Martian canals, of course, fueled much controversy around the turn of the century. Today, the Martian canals, the Venusian spoke system, and linear patterns on Mercury are all ridiculed as illusory. But the phenomenon is so widespread in planetary astronomy that it seems to deserve some explanation beyond being dismissed as merely illusory. The lunar naked-eye canals turn out to be remarkably like the telescopic Martian canals. There is no Percival Lowell around to interpret them as irrigation canals; rather, the claim is made that the eye simply integrates features of diverse origins into nicely geometric lines. Why?

Data Evaluation. The lineaments visible through the telescope and on spacecraft photographs are well-verified. The claims of naked-eye lineaments are rather surprising, but apparently several observers, including W.H. Pickering, have seen them. No studies have been found that try to relate the lineaments seen at various levels of detail. Rating: 1.

Anomaly Evaluation. The "canal" phenomenon actually involves three kinds of potential anomalies: (1) The curious tendency of the human eye-brain combination to create well-defined linear features from diverse markings which seem to dissolve away under detailed scrutiny; (2) The unexplained and unexplored relationships between lineaments seen at various levels of detail; (3) The physical causes of lineaments that are not illusory. There are enough imponderables here to warrant a rather high anomaly rating. Rating: 2.

Possible Explanations. Subsurface fracturing of the lunar crust, jointing, and the surface expression of layered structures; artifacts of the human eye-brain combination akin to geometrical illusions.

Similar and Related Phenomena. Lunar layered structures (ALE7); Mercury's linear markings (AHO4); the Venusian spoke system (AVO6); the Martian "canals" (AMO1).

Examples

X1. Martian canal framework. "Yet knowledge of the existence of an excellent substitute for the Martian phenomena seems to be restricted to a very few; for, though it is by no means generally known, straight dark lines, long and narrow, an exact counterpart of the Martian canals, may be seen with the naked eye on the disc of the Moon every month. The existence of these lines appears to have been announced first about twenty years ago, in 1896, by the Martian observer, Signor Cerulli. But they were not re-observed until their independent re-discovery by Professor W.H. Pickering in 1912. In September, 1917, the writer commenced an elaborate series of observations of these markings; but whilst the two previous observers both employed slight magnifications ---such as are given by opera glasses---he has found the lines quite easy to see with the naked eye. The first evening on which careful observations were made, in September last, three canals were noted, and the following evening eight were drawn. At the present time about fifty of these naked-eye Lunar canals have been mapped. The average Lunar canal appears straight or uniformly curved, of unvarying, though sometimes considerable breadth, and of equal darkness throughout. Various anomalous forms are noticeable---either a variation of darkness as the eye runs from one end to the other, or obvious irregularities in breadth and direction. These are but rare. In intensity one canal varies from another, for they seem to be of all shades of gray, but not usually black. At their intersections there is a notable tendency to the production of round dark spots ('oases'), and at the embouchures of the canals, triangular 'carets' make their appearance on occasion. Several of them are distinctly double---to the writer's eye there is never the slightest doubt upon their gemination. Other cases are more doubtful.

The Lunar canals do not, of course, exist as such on the Moon; they are clearly produced by the integration of delicate shadings, too feeble to present themselves

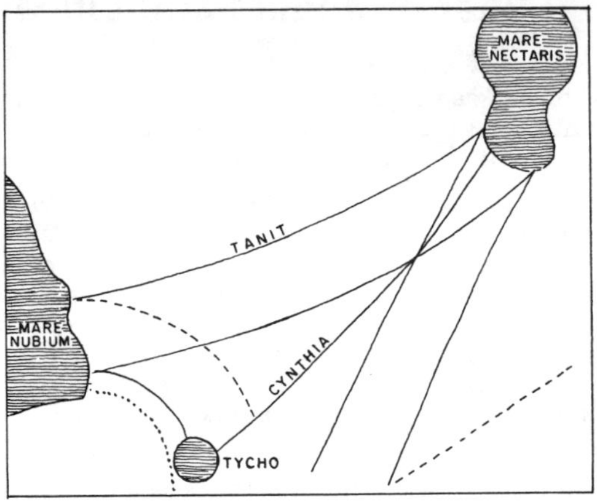

Some reported naked-eye lunar canals. (X1)

separately to the naked eye; but in no case has a purely illusory line deceived the writer into inserting it into his sketches. Every canal drawn has proved on comparison with good photographs of the Moon to lie on the line of various faint but easily identified shadings. But as already indicated, no one of them can be supposed to have any objective existence on the Moon just as it is seen with the naked eye. With higher magnifications---eight diameters, according to Prof. Pickering, four for the writer's eye---the canals vanish and are replaced by complicated detail. (R1-R3)

X2. Telescopic lineaments. "Summary. The lunar lineaments are described and their origin is considered. The components of the lunar grid-system---a system of intersecting lineaments comprised of elongated ridges, joints, crater-chains, and rilles---are described, and methods for producing orthographic charts of the principal global families A, B, C, and D of the grid-system are given. Each family, or system, of lineaments has a well-defined strike and is distinguished from one major regional system of joints (system R_1) which is associated with Mare Imbrium. The global systems display a strong symmetry about the Moon's central meridian. Discussion of system R_1 and a similar regional system of joints associated with Mare Humorum leads to the conclusion that these maria are igneous sinks and were not produced by impact. Examples of dip-slip and strike-slip faults are given." (R4) As in X1. these lineaments are integrated from various types of lunar features. (WRC)

X3. Lineaments in spacecraft photos. "A lunar surface fabric composed of near parallel topographic ridges, valleys, and similar cross structures was noted by many of the early selenographers. More recently, observers using larger instruments, with greater resolving power, have noted, during optimum seeing conditions, that the lunar surface is marked by numerous fine linear structures. It is not surprising to find similar parallel and near-conjugate sets of lineaments well displayed on some of the NASA Lunar Orbiter photographs. The lineaments displayed on the Orbiter photographs are closely spaced, fine, parallel striations that are generally oriented in a NW-SE and NE-SW cancellate design. Locally a tertiary E-W or N-S lineament may also be developed, but it is frequently less distinct and quite often displays a braided appearance. The interspaces between lineaments of the same altitude are frequently higher or lower in elevation than the adjacent area. The difference suggests the possibility that differential movement has taken place along these preferred lines, creating a stepped or seriated appearance to the surface. The cause of the lineaments is not known; however, on Earth, lineaments (straight, parallel) are generally the surface expression of planar discontinuities or breaks in the subsurface rocks such as fractures or joints, faults, shear zones, or stratigraphic layering." (R5)

References

R1. MacDonald, Thomas L.; "Canals on the Moon," English Mechanic, 108:7, 1918. (X1)
R2. MacDonald, Thomas L.; "The Lunar Canals---II," English Mechanic, 108:20, 1918. (X1)
R3. MacDonald, T. L.; "The Lunar Canals ---III," English Mechanic, 108:66, 1918. (X1)
R4. Fielder, Gilbert; "Lunar Tectonics," Geological Society of London, Quarterly Journal, 119:65, 1963. (X2)
R5. Fulmer, Charles V., and Roberts, Wayne A.; "Surface Lineaments Displayed on Lunar Orbiter Pictures," Icarus, 7: 394, 1967. (X3)

ALO5 The Lunar Zodiacal Light

Description. Luminous cones of light on either side of the moon extending along the plane of the ecliptic. Sometimes combined with the normal or solar zodiacal light.

Background. The solar zodiacal light (AZO) is usually seen as an inclined cone of faint light in the west soon after sunset or in the east just before sunrise. Since the axis of the cone lies on the ecliptic, the cone is noticeably slanted in the Temperate Zones. The solar zodiacal light is believed to be light reflected from particulate matter, gas, and/or electrons in orbit around the sun, but usually far from earth and moon.

Data Evaluation. Several careful observations by reputable astronomers are available. It is possible, however, that some observations were confused with lunar pillars (always vertical) or anomalous auroral activity. Rating: 2.

Anomaly Evaluation. True lunar zodiacal light would seem to demand the temporary presence of reflecting matter within the earth-moon system. Such clouds of matter are certainly not normal denizens of the earth-moon system, although they obviously occur in the solar system. In addition, moonlight would not seem intense enough to cause visible reflections from such diffuse matter. Rating: 2.

Possible Explanations. The reflection of moonlight from temporary clouds of gas, electrons, and/or particles in the neighborhood of earth and moon. In this context, it is relevant that some scientists have claimed a correlation between auroral activity, the appearance of high-level clouds, and halo phenomena (GLA12, GLA19); suggesting an influx of matter at very high altitudes.

Similar and Related Phenomena. Lunar halo phenomena; geographically displaced auroras (GLA7); solar zodiacal light (AZO).

Examples

X1. March 6, 1854. Following a discussion of the U.S.-Japan Expedition (1853-1855) during which Jones made a detailed study of the zodiacal light between latitudes 42°N and 53°S. "He (Jones) is very particular in describing what he calls the moon zodiacal light which he witnessed in the tropics. He also witnessed what he termed a joint sun and moon zodiacal light. In his report of one of these observations he says: 'The moon quartered today (March 6, 1854) (lat. 25° 26' N., long. 139° 42' E.). At half past 7 I was astonished to see the zodiacal light fully displayed. It was no doubt a joint sun and moon zodiacal light... My mind was perfectly satisfied that it was clearly a zodiacal light. It differed from the ordinary zodiacal light in not being brightest at its lowest end but was all the way down of a fairly uniform brightness. It was quite distinct. The upper end was lost in the moon's superior light.

The night was very clear.' Naval officers corroborated this and similar observations." (R6)

X2. April 3, 1874. An observation of the astronomer, L. Trouvelot, at Cambridge, Massachusetts. No signs of any auroral activity anywhere. "While going home, I remarked in the east a strange conical light rising obliquely from the top of the roof of a building, behind which the moon, then about 15° or 20° above the horizon, was concealed from view. By going away from the building, the conical light, which closely resembled the tail of a comet, became brighter and brighter as it approached the moon, upon the western limb of which it rested. The base was at least as wide as the diameter of the moon; but it extended beyond, on each side, by a fainter light, which gradually vanished in the sky. The extension of this luminous appendage I estimated to be equal to eight or ten times the moon's diameter. It was not readily visible when the moon was in sight, as the brilliant light of our satellite overpowered its dim brilliancy. The axis of this appendage was found to be coincident, or nearly so,with the ecliptic; and its line prolonged in the west passed a little to the north of Jupiter.

The phenomenon had been observed for about fifteen minutes, when it gradually faded away until it almost totally disappeared five minutes later, although the sky was clear. A quarter of an hour after, the sky was overcast with dense vapors, which continued for nearly an hour.

At 11h 0m the sky had cleared up, and the moon shone brightly. The luminous appendage was still visible, and even appeared more brilliant than before. In order to ascertain whether this appendage was visible only on one side of the moon, or if it was seen on the other side, I went under the piazza of my house, and placed myself in such a position as to have the moon concealed by its upper part, the sky below being visible. As I expected, a similar appendage was observed on the eastern side of the moon, exactly opposite the western one; the axis of both wings, passing through the moon's center, being in the plane of the ecliptic." Later in the night some auroral activity was noticed. (R1, R2)

X3. February 12, 1899. "It was a very clear night, with a rising barometer, on Feb. 12, and the moon was only a day or two old. As it neared the visible horizon, I noticed a faint light extending for some 50° or 60° along the Ecliptic, in a position exactly similar to the Solar Zodiacal Light. It could not have been this because the sun had set some four hours ago, so I must conclude that it was a Lunar Zodiacal Light." (R4)

X4. February 21, 1916. "Dr. W. E. Glanville, of Solomons, Md., reports to the British Astronomical Association that on Feb. 21, 1916, he made a successful observation of the 'moon's zodiacal light,' reported by Jones and other observers. At the time of observation the gegenschein showed on the ecliptic, At 8:15 a distinct luminosity, fainter than the sun's zodiacal light but in the usual shape of the latter when near the horizon, was visible from the moonrise point (over Chesapeake Bay), reaching the southern boundary of Leo up to the gegenschein, which became comingled with the light. The luminosity persisted until moonrise at 8:40. Mr. Gavin Burns, director of the auroral and zodiacal light section of the B. A. A., remarks on this observation: 'As the light of the full moon is only about one-millionth of the intensity of sunlight, it is difficult to believe that the phenomenon described can be due to the light of the moon.'" (R5) To which Glanville responded that the light he observed was unmistakable, and the moon the only source sufficient to produce the effect. (R6)

X5. No date given. Messier, in the Eighteenth Century, described the lunar zodiacal light much as Trouvelot (X2) did. His measurements showed the light extended about 2 1/2° on each side of the moon. (R3)

X6. No date given. M. Hugo described something like the zodiacal light rise about 4° above the lunar disk. (R3)

References

R1. Trouvelot, L.; "The Moon's Zodiacal Light," American Journal of Science, 3:15:88, 1878. (X2)

R2. Trouvelot, L.; "The Moon's Zodiacal Light," American Academy of Arts and Sciences, Proceedings, p. 183, 1877. (X2)

R3. Holden, E. S.; "Moon's Zodiacal Light," American Journal of Science, 3:15:231, 1878. (X5, X6)

R4. Bates, H. E.; "Lunar Zodiacal Light," English Mechanic, 69:85, 1899. (X3)

R5. "Lunar Zodiacal Light," Scientific American, 124:463, 1921. (X4)

R6. Glanville, W. E.; "The Lunar Zodiacal Light," Scientific American, 125:65, 1921. (X1, X4)

ALO6 Distortions of the Lunar Disk

Description. Depressions, large notches, projections, and other departures from circularity of the lunar disk. The deformations are usually seen through the telescope, but are sometimes extensive enough to be visible to the naked eye. They are much larger than the hills, valleys, and other natural features of the lunar rim.

Data Evaluation. A small handful of observations, some rather questionable, all more than a century old. Rating: 3.

Anomaly Evaluation. Very likely, most observations in this category are simply the consequence of optical distortions of the moon's limb, perhaps with some psychological effects (irradiation) thrown in. If other observations resembling X5 could be found, the following anomaly rating would have to be raised. Rating: 3.

Possible Explanations. Small distortions of the lunar disk can probably be attributed to the effect of abnormal refraction in the terrestrial atmosphere, particularly if the moon is near the horizon. Irradiation, a psychological effect operating near bright light sources, could also provide some apparent distortion. Large projections could be the result of halo phenomena or perhaps the lunar zodiacal light (ALO5), assuming it really exists. Observations like X5 have no ready explanation.

Similar and Related Phenomena. The lunar zodiacal light (AOL5); distortions of celestial objects as they approach the horizon (GEL4).

Examples

X1. September 20, 1863. Remarkable depressions noted in the preceding limb of the moon. (R2)

X2. November 3-4, 1864. "Instead of the limb being circular in the neighborhood of the equator it was sufficiently flattened for a spider line to lie along it; the south end clearly coincided with a line from Langrenus to Guttenberg, the north end being nearly identical with the south border of the Mare Crisium." Time of observation: 6 P.M., GMT. Similar flattening observed the following night. (R3)

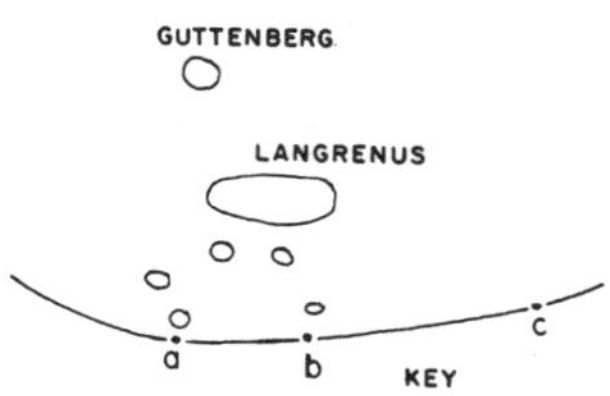

Flattened lunar limb, November 3-4, 1864. (X2)

X3. July 13, 1875. "H. S. M.'s guard-ship Coronation (Champon Bay), July 13 (civil time), in lat 10° 27' 40" N. and long. 99° 15' E., at midnight, the moon bore S. W. by W. magnetic, and its altitude was about 20°, when a prominent projection was seen with the naked eye on the moon's upper limb. The best glasses on board were soon brought to bear upon it.... The protuberance, in colour, was similar to that of the moon." A small protuberance was noted in a different position the following night. (R1)
wer half of the moon. These points gradually moved toward each other along the moon's edge, and seemed to be cutting off or obliterating nearly a quarter of its surface, until they finally met, when the moon's face instantly assumed its normal appearance. When the notches were nearing each other the part of the moon seen between them was in the form of a dove's tail.'" (R5)

X4. November 11-12, 1875. Observations of E. T. Perrott. "8.30 to 9.30 p.m.--Foggy, with occasional glimpses of clear. Flattening of W. limb suspected at ab, pointed to by a semi-ring of small craters W. of Langrenus, the flattening forming the chord of the arc. The flattening not very pronounced, and may be partly optical from irradiation of the bright point, a." Similar flattening reported the following night. (R4) W. R. Birt also saw some flattening on November 12, 1875. (R3)

X5. July 3, 1882. "A singular appearance

of the moon was observed by several residents of Lebanon, Conn., on the evening of July 3. The moon, almost full, was about three-quarters of an hour high. An observer says: 'Two pyramidal luminous protuberances appeared on the moon's upper limb. They were not large, but gave the moon a look strikingly like that of a horned owl or the head of an English bull terrier. These points were a little darker than the rest of the moon's face. They slowly faded away a few moments after their appearance, the one on the right and southeasterly quarter disappearing first. About three minutes after their disappearance two black triangular notches were seen on the edge of the lo-

References

R1. Loftus, A.J.; "Lunar Phenomenon," Nature, 12:495, 1875. (X3)
R2. Key, H.C.; "The Depressions in the Moon's Limb," English Mechanic, 22: 220, 1875. (X1)
R3. Birt, W.R.; "The Depressions in the Moon's Limb," English Mechanic, 22: 272, 1875. (X2, X4)
R4. Birt, W.R.; "The Depressions in the Moon's Limb," English Mechanic, 22: 321, 1875. (X4)
R5. "A Curious Appearance of the Moon," Scientific American, 46:49, 1882. (X5)

ALO7 Bright Diverging Ray above the Moon

Description. Bright vertical rays of light diverging from the top of the moon.

Data Evaluation. Only one report at hand, and that provides little detail. Rating: 3.

Anomaly Evaluation. So little information is available about this phenomenon that it is impossible to assess the scientific significance. It is most probably some sort of halo effect like the solar pillar. Rating: 3.

Possible Explanation. A type of halo phenomenon.

Similar and Related Phenomena. The lunar zodiacal light (ALO5); dark triangles beneath the moon (ALO8); diverging brushes of light from the horns of Venus (AVO8).

Examples

X1. May 2, 1933. "At 0230 G.M.T., an orange-coloured vertical ray was observed projecting from the moon's upper limb at an altitude of about 6 degrees. This ray was very pronounced until it was covered by cloud at 0245 G.M.T. and was not seen again." The weather was clear; auroral displays had been observed in the north 4 hours earlier. (R1)

References

R1. Bone, D.W.; "Lunar Ray," Marine Observer, 11:49, 1934. (X1)

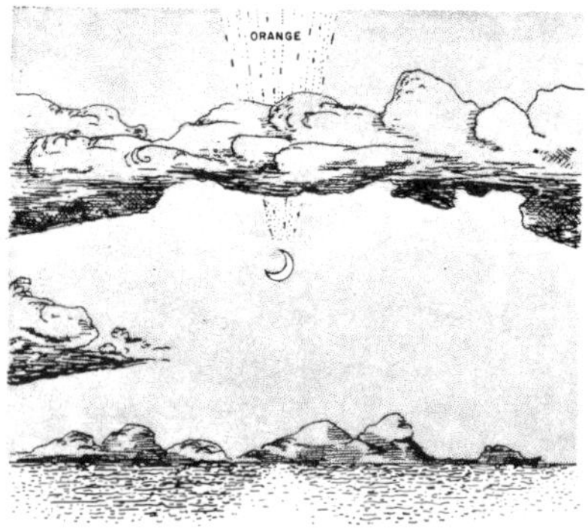

Bright diverging ray above the moon. (X1)

ALO8 Dark Triangular Patches under the Moon

Description. Dark triangular areas beneath the moon seen by mariners. The triangles diverge downward toward the water surface, where they meet the base of the bright triangle of light reflected by the water surface. The observer is at the apex of the bright triangle. Upon looking away from the display for a few minutes and then returning to it, it is found that the dark triangle has disappeared, although it soon returns slowly.

Data Evaluation. A rarely reported phenomenon, although the few reports at hand readily accept the reality of the observations. Rating: 2.

Anomaly Evaluation. This phenomenon is generally written off as due to contrast, but in actuality no one can explain contrast effects in physiological terms. The mystery, then, is not astronomical or geophysical but rather biological Rating: 2.

Possible Explanations. A psychological contrast effect!

Similar and Related Phenomena. Many common optical effects, such as the near-black appearance of white snowflakes seen against a gray sky, or the light edges of black house roofs seen against a gray sky. The "annular" phases of Venus and the moon (AVO2 and ALO9) may be due in part to contrast effects.

Examples

X1. March 15, 1959. "Whilst approaching the Providence N. E. channel a dark patch was observed directly under the moon. The night was mainly clear, with occasional scattered rain showers and the moon three-quarters full. For the main part the horizon was clear cut, emphasised by what appeared to be a light haze, although visibility was good. Directly under the moon the shading was reversed, the sea being lit up by the moon's reflection and the sky very dark. The shaded patch under the moon was directly over Abaco Island and seemed to rise to the moon, tapering away until lost in the night sky.... Note 1. The appearance of a triangle of darkness between the moon and the horizon, contrasting with the triangle of light under the moon which extends from the ship to the horizon is rare but not unknown. No satisfactory explanation can be given but the effect is thought to be physiological, due to contrast, rather than true. As long ago as 1856 an observer noted that the dark triangle disappeared when he screened off the moon and the lower triangle of light. And when, after turning around, he suddenly looked again, the illusion only re-appeared after a few seconds." (R1)

X2. December 21, 1960. "At 1800 GMT (2100 LMT) a dark triangular area was seen directly under the moon---then it its first quarter and at an elevation of about 6°---which extended from the horizon up to an angle of approximately 10°. The sky was cloudless and the visibility excellent, the horizon line being well defined. The observers found, on looking away from the moon and the dark patch for a few seconds, that when they looked back again the shadowy area was not seen for an appreciable interval." (R2)

References

R1. Robert, J. O.; "Contrast of Light Intensity," Marine Observer, 30:193, 1960. (X1)

R2. Chivers, G. F.; "Contrast of Light Intensity," Marine Observer, 31:188, 1961. (X2)

ALO9 Ring of Light around the New Moon

Description. A very bright rim of light around the dark limb of a new moon. The rim seems much brighter than the dark limb, which is illuminated by earthlight. The bright fringe usually disappears when viewed through a telescope---but sometimes it does not.

Background. Venus displays a similar "annular phase", but it possesses an extensive atmosphere, which may refract sunlight for considerable distances. Refraction seems out of the question on the moon.

Data Evaluation. Only a few scattered reports have been uncovered in the literature, but the phenomenon probably occurs rather frequently. To illustrate, the compiler once saw this phenomenon displayed brilliantly near a large city. He did not report it, and no one else did either. Rating: 1.

Anomaly Evaluation. A contrast phenomenon seems the best explanation, but it does seem peculiar that the effect is not mentioned more often, for the appropriate conditions occur frequently. As with other supposed contrast-effect explanations, the anomaly rating really reflects our biological ignorance of the contrast effect itself. Rating: 2.

Possible Explanations. Psychological contrast effect.

Similar and Related Phenomena. The annular phase of Venus (AVO2); the hanging of stars on the moon's limb (ALX0); the rings of light seen around Mercury during transit (AHX2) and around the moon during solar eclipses (ALX11).

Examples

X1. 1882. "I happened to look at the moon about 5 1/2 p.m., and was astonished to see a very fine arc of light completing the circle. I at once applied a power of 80 to a 3-in. telescope, but the arc of light had vanished. In its place however I now saw, quite distinctly, the whole of the dark portion of the disc having a dull, greenish tint. This was, of course, 6d. 6h. after new moon. Looking without the telescope, I again quite distinctly saw the fine bright arc, but again the telescope dispelled it. I now looked with a much higher power, with an ordinary field glass, and with a good binocular, but in every one the bright arc had disappeared, though with the naked eye it still remained. I repeated the experiment several times with the same result." (R1)

X2. March 20, 1912. A distinct rim of very bright golden light made a complete ring around the moon. (R2)

X3. November 7, 1912. "At 6.37 a.m., the moon emerged from behind a cloud, where upon I at once brought out a small telescope from my coat pocket, and, to my delight, perceived a complete circle of silvery light round the moon's disc. In other words, the appearance of the moon was that of a ring, instead of a crescent." (R3) Note that the circle of light did not disappear when the telescope was used in this instance. (WRC)

X4. March 28, 1914. Golden crescent plus a silver rim around the rest of the moon. (R4)

X5. January 17, 1915. The first of a series of observations of this phenomenon by E. A. Stevenson. Most observations were made when the moon was about 2 days old. The rim of light around the dark part of the moon was of almost electric whiteness. (R5)

X6. February 16, 1915. See X5. (R5)
X7. January 7, 1916. See X5. (R5)
X8. February 6, 1916. See X5. (R5)
X9. December 26, 1916. See X5. (R5)
X10. March 14, 1918. See X5. (R5)

X11. December 8, 1918. "I saw very distinctly, round the faintly visible dark part of the Moon, a luminous ring, which seemed different in the nature of its light from the familiar ashen light." (R6) The moon's "ashen light" is simply reflected earthlight. (WRC)

X12. July 31, 1957. "Mr. Craig Johnson writes that on July 31, 1957 at 2h 24m, U.T., using a 4-inch reflector at 91X, he observed a slight ring of light reaching around the north limb of the moon. The ring was just barely brighter than the earthshine and was about 1,000 miles long. Mr. Johnson expresses confidence that the appearance was not an illusion and that it was not wholly due to contrast. The moon's age was 3.9 days, and the seeing was perfect (10 on a scale of 10, while moisture was literally running down the telescope tube). This lunar limb light is nothing new but has been recorded as long ago as Schroeter's time. It has been variously imputed to a lunar atmosphere, to optical effects, and to mountains on the limb." (R9) Again, the ring can be seen through the telescope. (WRC)

X13. General observations. "When the moon is a crescent, either before or after New, a remarkable appearance is often witnessed; in addition to the bright crescent the dark portion of the lunar globe is illuminated by the earthshine, while the dark limb seems fringed with light. To the naked eye the limb appears to be bordered externally with a narrow but inconspicuous fring of light which disappears on telescopic examination, the limb itself then appearing brighter than the remainder of the earthlit portion of the disk. This phenomenon has been remarked for a long time and has been ascribed to contrast---lunar atmosphere or illusion.....

The young earthlit Moon consists of a bright crescent illuminated by direct sunlight and the rest of the disk seen by reflected earthlight. The limb of the crescent appears still brighter owing to shadow concealment, and irradiation causes the bright crescent to appear to be a portion of a sphere of greater radius than the darkened area. The limb of this dark but earthlit surface appears brighter than the rest owing to such large portions of the inner regions being occupied by the in themselves dark plains and the concealment of the low-lying and usually dark areas near the limb. Irradiation and contrast with the dark sky cause an apparent enlargement similar to, but less pronounced than, the bright crescent, and the naked-eye effect is that of a border of light around the dark limb, sometimes intensified at one of the cusps by a narrow strip of directly sunlit elevations. With the telescope the external bright border seen with the naked eye no longer appears, but the true appearance is observed of a bright limb and the darker patchy interior.

Occasionally in addition to the bright limb two or three light patches are observed close to the limb in the earthshine. These have been seen with the naked eye, and may be due to fluorescence as well as to the brilliancy of the earthlight at the time. The actual amount of light reflected by the Earth varies according to the hemisphere turned toward the Moon: it is greater when the great land masses of Asia, Africa or Europe are so situated than when the Pacific Ocean occupies the terrestrial disk seen from the Moon." (R8, R7) But see X3 and X12.

References

R1. "Arc on Lunar Disc," Knowledge, 2: 437, 1882. (X1)

R2. Henderson, Alex. C.; "Complete Golden Ring Round the Young Moon," English Mechanic, 95:232, 1912. (X2)

R3. Henderson, Alex. C.; "Circle of Silvery Light around the Old Moon," English Mechanic, 96:394, 1912. (X3)

R4. Henderson, Alex. C.; "The Young Moon Seen as a Circle Instead of a Crescent," English Mechanic, 99:234, 1914. (X4)

R5. Stevenson, E. A.; "The Young Moon Seen as a Circle," British Astronomical Association, Journal, 28:223, 1918. (X5-X10)

R6. Pickston, J.; "Ring of Light round New Moon," English Mechanic, 108:253, 1918. (X11)

R7. "The Ring of Light All Round the New Moon," English Mechanic, 108:240, 1918. (X13)

R8. Wilkins, H. Percy; "Lunar Crescent and Bright Rim," British Astronomical Association, Journal, 56:91, 1946. (X13)

R9. Johnson, Craig L.; "Lunar Limb Brightening," Strolling Astronomer, 11:118, 1957. (X12)

ALO10 Shortened Lunar Crescents

Description. The failure of a lunar crescent to extend to extend the full 180° required by the geometry of the sun, moon, and terrestrial observer. The cusps are complete, not blunted.

Data Evaluation. A single, century-old observation. Rating: 3.

Anomaly Evaluation. There are not enough details to make a judgment here. One cannot appeal to contrast phenomena here, since irradiation would (and sometimes does) make the cusps seem to extend more than 180°. No explanations are at hand. Rating: 2.

Possible Explanations. None.

Similar and Related Phenomena. Cusp blunting on Mercury (AHO1) and Venus (AVO1); the slight extension of the lunar cusps in the eye of the observer due to irradiation.

Examples

X1. November 8, 1882. "Observing the Moon at 18h 15m Nov. 8, 1882, its age being 27.9 days, I was surprised to find that the 'line' joining the cusps did not pass as usual through the Moon's centre, but a little to the E. of it; in other words, the exterior circumference of the crescent was less than a semicircle." (R1)

X2. January 5, 1883. Same effect noted. (R1)

References

R1. Hopkins, B. J.; "Abnormal Appearance of the Lunar Crescent," Observatory, 6: 247, 1883. (X1, X2)

ALO11 The Lunar Post-Sunset Horizon Glow

Description. Thin lines of light on the lunar horizon that persist for about 2 hours after the lunar sunset. The light is prominent above rocks, crater rims, and other elevated features. Presumably the same phenomenon occurs at lunar sunrise.

Data Evaluation. Numerous photographs from several Surveyor spacecraft emplaced on the moon's surface. Rating: 1.

Anomaly Evaluation. A rather convincing theory identifies the horizon glow with sunlight reflected from electrostatically levitated dust particles. This being so, the phenomenon is more of a curiosity than an anomaly, despite its fascinating features. Rating: 3.

Possible Explanations. As stated above. See X2 for more details.

Similar and Related Phenomena. The terrestrial Alpine Glow (GEL6).

Examples

X1. Frequent occurrence. Photos from a spacecraft on the lunar surface. "Observations of the western horizon shortly after sunset during the Surveyor VIII mission revealed, along the crest of the horizon, a bright line of light similar to that previously reported for the Surveyor V and VI missions. Though not sufficiently well defined to be recognized at the time, the phenomenon also occurred during the Surveyor I mission. Although no sunset observations were made on Surveyor III, it appears that this postsunset phenomenon along the western horizon (and probably the eastern horizon at sunrise) is not an unusual event, but occurs regularly as the natural consequence of some aspect of the lunar environment.

The light was observed for period of time up to about 2 hours after sunset. The center of the solar disk, therefore, is about 1.25 deg below the horizon when the 'afterglow' either stops or the intensity falls below the limits of detection. Pictures of the light from the Surveyor VII mission are shown in Figs. 1 and 2 (photos omitted, but see sketch made from them) when the Sun was centered approximately 0.4 and 1.0 deg, respectively, below the horizon. In Fig. 1, the light intensity permitted normal shutter operation (exposure time, 0.15 sec); the bright line appears to extend only about 2 deg along, and 1/8 deg above, the horizon. The light intensity decreased rapidly; about 1 1/2 hour later, a nominal 1.2-second exposure showed a faint line of illumination extending at least 4 deg along the horizon. A 40-second exposure taken about 2 hours, 40 minutes after sunset, showed no edge of light along the horizon. This last picture (illumination provided by light backscattered from the ridges east of the spacecraft, and by earthlight) provides a valuable comparison of the rocks and horizon geometry with the shape of the bright regions... A particularly striking facet of the phenomenon is the 'mapping,' or shadows in the edge of the light, apparently caused by the rocks extending along and above the lunar horizon line." (R3, R2)

"A patchy glow has been discovered on

Illumination along the western horizon about 15 minutes after sunset, as seen from a Surveyor spacecraft on the moon. (X1)

the moon's horizon at dusk that is believed to be caused by individual dust clouds being continually formed above the hills, crater rims, and rocks which remain sunlit while the shadow of night advances over the lower surroundings.....The leading edge of the 20-mile-wide pattern of cloud patches, which extends from pole to pole at the evening terminator, travels steadily westward at about nine miles an hour, the same rate as the terminator moves. The clouds are present over at least 0.1 percent of the terminator." (R4)

X2. Speculations about the cause of the phenomenon. G. C. I. Rawlings has suggested that some lunar transient phenomena may be related to the terrestrial Alpine Glow (GEL6), in which sunlight is scattered down on the Alps long after sunset by particles high in the earth's atmosphere. (R7) This theory has been combined with electrostatic levitation of lunar dust to explain the Surveyor postsunset glows by J. J. Rennilson and D. R. Criswell. (R4-R6) According to this theory, electrostatic fields are set up by solar X-rays ionizing atoms on the surfaces of sunlit rocks, creating perhaps 500 to 1,500 volts potential difference between these rocks and those in the shadow on the other side of the terminator. Dust clouds about a foot high might be created. Sunlight reflected from these clouds towards the Surveyor cameras constitute the postsunset glow. (R4-R6)

References

R1. Rawlings, G. C. I.; "Lunar Transient Phenomena: Quasi-Alpengleuhn or Twilight-Colour Theory," British Astronomical Association, Journal, 77:309, 1967. (X2)

R2. Allen, L. H.; "The Lunar Sunset Phenomenon," Surveyor Program Results, NASA SP-184, 1969, p. 413. (X1)

R3. Gault, D. E., et al; "Lunar Theory and Processes: Post-Sunset Horizon 'Afterglow'," Icarus, 12:230, 1970. (X1)

R4. "Lunar Terminator Phenomenon," Sky and Telescope, 46:146, 1973. (X1, X2)

R5. Rennilson, J. J., and Criswell, D. R.; "Surveyor Observations of Lunar Horizon-Glow," The Moon, 10:121, 1974. (X2)

R6. Hughes, David W.; "Lunar Duststorms," Nature, 254:481, 1975. (X2)

ALW LUNAR "WEATHER"

Key to Phenomena

ALW0 Introduction

Lunar "weather" is obviously a rather limited subject. Weather as we know it is a medley of atmospheric phenomena; and the moon's atmosphere is in most respects nonexistent. Nevertheless, we do have many observations of lunar clouds and mists. These are probably not composed of water vapor, given the great scarcity of water on the moon; but they may be clouds of gases and dust vented from the moon's soil and upper crust. Besides clouds and mists, astronomers also sometimes find their telescopic command of lunar detail slipping away from them as some lunar medium seems to obscure their targets. Clouds of gas and/or dust, invisible as such, could well mask lunar detail temporarily. Outgassing of the moon is not a common event, but there still seems to be some life in the moon, despite the many old pronouncements of its death!

ALW1 Clouds, Mists, and Obscurations

Description. (1) Apparitions on the moon that resemble in form and movement terrestrial clouds; (2) The veiling or obscuration of topographic detail; (3) The recording of emission spectra typical of gases. There is a definite and unavoidable overlap between these phenomena and the variable patches and streaks described in ALO2.

Data Evaluation. Many reports have been found in all three areas; and there are many more still-to-be-assimilated in the catalogs of transient lunar phenomena. Rating: 1.

Anomaly Evaluation. Clouds, mists, and obscurations are readily explained in terms of sporadic venting of internal gases and/or dust from the lunar crust. The forces that trigger such venting and any energy sources involved are matters of speculation. Rating: 3.

Possible Explanations. Gas and/or dust released or vented from the lunar soil and crust; perhaps triggered by moonquakes and meteor impacts.

Similar and Related Phenomena. Variable patches and streaks (ALO2); lunar color phenomena (ALF3).

Examples

X1. February 13, 1826. Regarding Messier. "Gruythuisen (sic), a very able and conscientious observer, ascertained in 1825 that the western crater was half the size of the eastern, and elongated from the east to the west. On the 18th of February, 1826, a strange fact made itself manifest in the luminous train; the dark band which traversed its center was intermingled with luminous points, 'and I believe,' writes he, 'that I observed that they did not remain always in the same position.' At times a veil or mist appeared to extend over these objects, while that under other circumstance, where they ought to have been less visible on account of the effect of the solar light, they were more so." (R6) Unfortunately, the prose is not very clear here. (WRC)

X2. 1866. In this year, the crater Linne became concealed by a sort of fog or haze that hung over it. When the haze cleared, the prominent crater had been replaced by a smaller one and a white spot. (R16) See ALE22-X2 for a review of the Linne controversy. (WRC)

X3. April 22, 1866. Ptolemaeus. "At the moment, I have an impression that this roughened appearance of the floor is by no means frequent, as I have generally seen the surface (even under very oblique light) perfectly smooth and unruffled, and it is only another of those instances which tend to strengthen a lurking idea always present in my mind of an obscuring medium on the lower levels of the moon." This is the record of a rare clearing rather than a temporary obscuration. (R28)

X4. 1871. A number of dense, dark, cloudy-looking spots of considerable size in the region of Copernicus, Menelaus, and the Apennines. They gave the impression of smoke hovering over the surface, with the lunar surface being visible through the fainter portions. (R1)

X5. November 20, 1871. "On the evening of November 20, 1871, Mr. (Henry) Pratt observed a very remarkable phenomenon on the Mare Frigoris, which he described as one of the most singular and striking of all the local obscurations he had witnessed. The following is an extract from his observing book:---'5.30. On a general survey of Plato and wide neighbourhood, the very peculiar aspect of the Mare Frigoris attracts attention. The appearance can be compared to nothing but a kind of haze, entirely local, hanging around the N.W. foot of the slopes of Plato. It is the more conspicuous, as nothing of the kind is visible either on the part of the Mare Imbrium or on the Mare Serenitatis. The objects on the Mare Frigoris are indistinct, as if veiled.' At 6.30 the appearance was very much modified. At 7.30 very little trace of the veiling was to be seen." (R4, R2, R3)

X6. December 22, 1871. Henry Pratt saw a marked haziness over the northwest floor of Plato. (R4, R3)

X7. April 10, 1873. W.R. Birt fails to detect the lunar feature Picard B (or Peirce A), due to some sort of obscuration, for it is usually visible. (R5)

X8. November 1, 1873. See X7. (R5)
X9, November 28-30, 1873. See X7. (R5)
X10. July 8, 1876. See X7. (R5)

X11. June 16, 1877. W.R. Birt reports that definition on Mare Serenitatis was very bad. (R7) Birt was an enthusiastic student of the moon, as entries during this period prove. (WRC)

X12. November 1, 1878. Quoting Hermann Klein, a German astronomer, on the twin Messier craters. "On the 1st of Nov. 1878, I found to my great surprise that the western ring mountain appeared like a half moon, but the western edge was missing. There was an appearance there was something diffuse, the interior of the crater was like a half shadow. The eastern ring-mountain was sharply defined, complete, and the interior half filled with deep black shadow. On the following day also the western ring mountain was half-moon shaped, full of diffuse shadow. It was entirely impossible to see the western half edge. I am convinced that a kind of fog lay in the interior of the crater at that time, and was spread out over the western half of the ring wall. Any other explanation of the invisibility does not seem to me admissible."

(R28)

X13. April 24, 1882. At a meeting of the liverpool Astronomical Society. "Mr. W.J. Ridd said he had often noticed appearances which could only be accounted for by the existence of a lunar haze. On April 24th, 1882, he had made a note of a strange appearance of shadows, and near Godin and Agrippa of shadows blurred and oscillating. Shadows in Aristoteles steady. The shadows in and around this region, and especially of the highlands west of Agrippa, were misty as though obscured by a fog, which gradually lifted to be again obscured, the intervals from obscuration to obscuration being about 10 minutes, the shadows never became quite clear during the whole time he was watching it. It seemed as though vapour or smoke was being ejected at intervals, and as the vapour cleared away the details became more distinct." (R28)

X14. May 19, 1882. J.G. Jackson, using a 6-inch reflector, saw a peculiar cloud over the western edge of Mare Crisium. It had a misty feathery appearance, and was not less than 100 miles long by 40 to 50 wide. (R11)

X15. July 17, 1882. J.G. Jackson again reports a feathery mist over the western edge of Mare Crisium. (R12)

X16. March 12, 1883. W.H. Davies observed a peculiar blurred appearance to the north of Mare Foaecunditatis (sic), which covered nearly 100,000 square miles of the lunar surface. (R28)

X17. January 27, 1912. An intensely black body, 250 miles long by 50 wide, appeared on the left cusp of the moon. Its shape resembled "a crow poised." (R15)

X18. October 10, 1916. Report of M. Maggini, of the Observatory of Florence, Italy. Craterlets inside Plato, though usually visible under the prevailing illumination conditions, were not detectable. A reddish vapor, resembling nitrous vapor, seemed to be obscuring the craterlets. When the reddish "shadow" disappeared, the missing craterlets were once more visible. (R17)

X19. April 14, 1932. A.V. Goddard's testimony. "I was observing the moon on the night of April 14, at about 10:30 Pacific Time and beheld a most interesting phenomenon. I was using a 16-inch glass and the atmosphere was very steady. A friend and I noticed the unusual absence of all white spots and markings in Plato. At 10:57 P.S.T. in long. 10°E and lat. 51°N a white spot made its appearance and in less than a minute it had spread in a northeasterly direction until it almost reached the rim of the crater. This observation was verified by my friend as I rather doubted my own eyes. White markings are almost always visible in Plato but this is the first time I have ever seen one appear suddenly. It appeared and moved like a cloud of steam but, considering the size of Plato and its rapid motion, this idea seems untenable." (R19)

X20. February 10, 1949. "A lunar phenomenon was reported by F.H. Thornton on February 10, 1949. Using an 18-inch reflecting telescope at his home in Norwich, England, he detected a diffused patch of thin smoke or vapor emanating from a point on the east side of Schroeter's Valley near the Cobra Head and spreading out onto the adjacent plain. Definition was reported as very good in the entire valley except for that area where the patch occurred; there the detail was indistinct and hazy." (R26)

X21. 1955. The Soviet astronomer, N.A. Kozyrev, detects the emission lines of hydrogen gas in Aristarchus. (R26)

X22. November 2-3, 1958. While observing Alphonsus, N.A. Kozyrev notices a reddish obscuration of the crater. Simultaneous spectrograms showed the emission lines of gases. (R21)

X23. November 18, 1958. An observation by two amateur astronomers, one with 20 years of experience, using a 6-inch reflector. "While sweeping the moon's surface, we were startled by what we saw within the crater Alphonsus. A large diffuse cloud completely obscured the crater's central mountain peak and its small craterlet. The cloud was about 20 miles in diameter and irregular in shape ....The diameter of Alphonsus is about 70 miles. The base of the central mountain is approximately six miles in diameter, and the craterlet located on one side of the peak is about three miles wide and five miles long. Two main features attracted our attention: (1) the cloud was large in comparison to the peak that it obscured, and (2) it has a strange diffuse brightness. When looking at lunar features near the terminator, one sees normally mainly contrasting black and white areas. In this case, however, we observed not only the extremely bright white walls of the crater that were being illuminated by the sun and the extremely dark shadow on the crater floor, but also a grayish, diffuse cloud whose brightness was somewhere between the two extremes. The side of the cloud toward the sun was not noticeably brighter than the other. We watched the cloud for 20 to 30 minutes, alternately looking at the moon and Mars. During this period, we saw no change in its size or shape." (R23)

Although the observers involved in the preceding detection of an event in Alphonsus were apparently well-experienced, it is well to add the following cautionary note. "The lunar crater Alphonsus each month shows an illusion familiar to experienced observers. Just after the sun rises, for an hour or so the convex floor of this crater resembles a luminous cloud. It is a trap for the unwary beginner, who is apt to believe he has seen volcanic action there." (R27) But see also the following observation.

X24. November 19, 1958. This event took place in the crater Alpetragius, which is immediately adjacent to Alphonsus, the subject of X23, which occurred a day earlier. "At 22:00 Universal time, I suddenly noted that a portion of the shadow covering the floor of the nearby crater Alpetragius had faded. A few minutes before, the shadow had covered about two-thirds of the interior of the crater, with the central peak a bright spot on a black background. Now, however, about half the shadow had faded, and was replaced by a much lighter shade. I did not see any haze or glow in the crater. At 22:05 UT the shadow gradually darkened, reaching its former state in about 20 seconds." (R22)

X25. October 23, 1959. N.A. Kozyrev detects gaseous emission lines at Aristarchus. Carbon also detected spectrographically. (R25)

X26. November 26 and 28, and December 3, 1961. See X25. Apparently the area on which the emission lines appeared was not larger than several square kilometers. (R25)

X27. October 29, 1963. See X25. (R26)

X28. September 20, 1964. Obscuration in Ross D. (R29)

X29. October 30, 1965. Fuzzy area near Atlas. (R29)

X30. No date given. Mare Crisium. "Now, at various times when studying the floor of M. Crisium, I have noticed waves of light and shade, so that it was difficult at times to see objects with which I was perfectly familiar. Also clouds have passed over the object I have been viewing. These clouds have been seen by other observers, and are mentioned in Neison. That they belong to the moon there can be no doubt, and I gradually come to the conclusion that vapour of some kind still existed on the moon. I had my surmises verified on one particularly fine night---the best I ever had---when I plainly saw a well-defined cloud pass over the object I was copying." (R13)

X31. General observations. Plato. The question of obscurations in this great crater have been discussed in general by W.R. Birt (R8-R10) and others. (R24) Carle (R20) in particular has given a vivid account of the strange blackness that sometimes fills this crater. See ALO2-X1.

X32. General observations. Schickard. P. Moore states that this region is one of several that look blurred during periods of every lunation. (R30)

X33. Unusual explanations of obscurations. D. E. Packer ventured that some of the blurring and obscuration of lunar features may be due to a ring of gas around the earth. (R14) W.H. Pickering believed that snow, hoar frost, and vegetation contributed to lunar changes, including the fuzzing of detail. (R18 and many entries under ALO2)

Lunar clouds, mists, and obscurations comprise just one type of Lunar Transient Phenomena (LTPs). There are also bright, star-like points of light (ALF3) and lunar color phenomena (ALF4). The reader may refer to these sections for additional discussions of possible explanations and references to some massive surveys of TLPs in general, where many more examples can be found of all types of TLPs. (WRC)

References

R1. Tydeman, Edmund M.T.; "'Sun Spots on the Moon," English Mechanic, 14:97, 1871. (X4)

R2. Birt, W.R.; "Lunar Meteorology," English Mechanic, 14:275, 1871. (X5)

R3. Birt, W.R.; "Report on the Discussion of Observations of Spots on the Surface of the Lunar Crater Plato," Report of the British Association, 1872, p. 245. (X5, X6)

R4. Birt, W.R.; "Lunar Meteorology, "English Mechanic, 14:460, 1872. (X5, X6)

R5. "Changes on the Moon's Surface," English Mechanic, 29:25, 1879. (X7-X10)

R6. Flammarion, Camille; "Is the Moon Inhabited?" Scientific American Supplement, 7:2711 and 2696, 1879. (X1)

R7. Dennett, F.C.; "Lunar Mists and Clouds," English Mechanic, 31:470, 1880. (X11)

R8. Birt, W.R.; "Selenographical," English Mechanic, 34:38, 1881. (X31)

R9. Birt, W.R.; "Selenographical," English Mechanic, 34:110, 1881. (X31)

R10. Birt, W.R.; "Lunar Physics," English Mechanic, 34:182, 1881. (X31)
R11. Bridle, M. Herbert; "Cloud West of Mare Crisium," English Mechanic, 35:326, 1882. (X14)
R12. Jackson, J.G.; "Feathery Mist on the Moon," English Mechanic, 35:497, 1882. (X15)
R13. Hardy, Jas. D.; "The Moon---Mare Crisium," British Astronomical Association, Journal, 7:139, 1897. (X30)
R14. Packer, D.E.; "The Earth's 'Crape Ring'," English Mechanic, 81:520, 1905. (X33)
R15. Harris, Frank B.; "Peculiar Phenomenon on the Moon," Popular Astronomy, 20:398, 1912. (X17)
R16. Cook, John A.; "Is the Moon a Dead World," Scientific American, 115:549, 1916. (X2)
R17. "Curious Phenomena in a Lunar Crater," Scientific American, 121:181, 1919. (X18)
R18. Wicks, Mark; "The Moon," English Mechanic, 114:211, 1921. (X33)
R19. Goddard, A. V.; "Unusual Lunar Phenomenon," Popular Astronomy, 40:316, 1932. (X19)
R20. Carle, Jackson T.; "Three Riddles of Plato," Sky and Telescope, 14:221, 1955. (X31)
R21. Kozyrev, Nikolai A.; "Observation of Volcanic Process on the Moon," Sky and Telescope, 18:184, 1959. (X22)
R22. Stein, Raymond J.; "An Unusual Event in Alpetragius," Sky and Telescope, 18:211, 1959. (X24)
R23. Poppendiek, H.F., and Bond, W.H.; "Recent Observations of Possible Volcanic Activity within the Lunar Crater Alphonsus," Astronomical Society of the Pacific, Publications, 71:233, 1959. (X23)
R24. Peterson, Keith; "Plato and Its Mysteries," Strolling Astronomer, 17:99, 1963. (X31)
R25. Kozyrev, Nikolai; "Volcanic Phenomena on the Moon," Nature, 198:979, 1963. (X22, X24, X25)
R26. "Lunar Color Phenomena," ACIC Technical Paper No. 12, 1964. (X20, X21, X27) (ACIC is the U.S. Air Force Aeronautical Chart and Information Center)
R27. "Are There Changes on the Moon?" Sky and Telescope, 28:3, 1964. (X23)
R28. Baum, Richard M.; "Transient Lunar Phenomena: Some Obscure Nineteenth Century Accounts," Strolling Astronomer, 20:155, 1966. (X3, X12, X13, X16)
R29. Cameron, Winifred Sawtell, and Gilheany, John J.; "Operation Moon Blink and Report of Observations of Lunar Transient Phenomena," Icarus, 7:29, 1967. (X28, X29)
R30. Moore, P.; "The Observation of Transient Lunar Phenomena," Royal Society, Philosophical Transactions, A285:481, 1977. (X32)

ALW2 Anomalous Ion Clouds Detected on the Lunar Surface

Description. Clouds of ionized gases that move over the lunar surface. These clouds may be correlated with disturbances of the surface. The gas source(s), ionization mechanism(s), and modes of propagation are all unknown.

Data Evaluation. Three separate clouds were detected by two different instruments left behind by the Apollo-12 astronauts. Rating: 2.

Anomaly Evaluation. Although some suggestions have been made, the origin and motions of these clouds are mysteries. Although the three clouds observed were correlated with the crash of a rocket stage, they did not come from the direction of the impact site. Rating: 2.

Possible Explanations. The impact of the Saturn 4-B stage might have released gases trapped in the lunar soil and crust. Solar wind and ultraviolet radiation could have ionized any gases thus released.

Similar and Related Phenomena. Shock waves from impacts of objects on earth are poor analogies, because the moon does not possess any atmosphere of consequence.

Examples

X1. Circa 1970. "Two lunar surface experiments left on the moon by Apollo 12 astronauts last year have supplied puzzling data on ion clouds resulting from the impact of the Saturn 4-B stage into the moon. About 17 seconds after the impact, the solar wind spectrometer and the lunar ionosphere detector began registering the presence of ionized gas. The gas apparently arrived in three distinct clouds. The particles had energies of 35 to 50 electron-volts. The bulk of them arrived from the north and northeast of the monitoring site, even though the impact was 135 kilometers to the west. The event poses several questions, such as how what was probably a neutral gas became ionized; how the gas traveled from the impact site to the instruments, and how the gas was accelerated to the high energies observed." John Freeman, one of the researchers, surmised that the rocket impact may have released solar wind gas trapped in the lunar soil. Either the solar wind or solar ultraviolet radiation might have ionized the gas. The clouds of gas might have been directed in their motions to some extent by the solar wind and/or interplanetary electric fields. (R1)

References

R1. "Puzzling Ion Clouds," Science News, 98:414, 1970. (X1)

ALX LUNAR ECLIPSE AND OCCULTATION PHENOMENA

Key to Phenomena

ALX0 Introduction

During an eclipse of the moon, the earth's shadow passes across the moon, reducing its brilliancy considerably and, in most cases, turning it coppery-red. The normal coppery color and a handful of interesting anomalies derive from the fact that the earth's atmosphere refracts and scatters sunlight onto the shadowed lunar surface. The variability of the earth-girdling atmospheric lens seems to be the cause of the very dark and very light lunar eclipses, as well as peculiar spots, bands, and arcs of light and shade on the eclipsed moon. Generally, these eclipse-engendered anomalies fall on the curiosity side of the anomaly scale and do not seriously threaten any scientific theories.

The occultation phenomena, though superficially almost trivial, all point in one direction. The projection of stars on the moon's limb, the reception of radio signals from occulted spacecraft, and other similar anomalies involve the apparent bending of electromagnetic waves around (or through) the moon's limb. The occasional presence of a refracting and/or scattering medium on the moon must be considered a possibility. Most of the occultation phenomena are rare and do not always occur when conditions seem most favorable. Such variability does not seem to support the contention that these curious limb effects are all psychological; that is, the products of contrast and illusion. Interestingly enough, some of the lunar occultation phenomena are closely analogous to anomalies occuring during Jupiter's occultation of its satellites and during transits of Mercury and Venus.

ALX1 Very Dark Lunar Eclipses

Description. Total lunar eclipses that are so dark that the moon is completely invisible or just barely visible with a telescope. Usually enough sunlight is refracted by the earth's atmosphere onto the moon to render it easily visible. Dark eclipses usually follow violent volcanic eruptions.

Data Evaluation. Abundant observations of the phenomenon over several centuries. Rating: 1.

Anomaly Evaluation. This phenomenon seems well-explained and is retained in the Catalog for its general interest and relevance to other phenomena in this section. Rating: 4.

Possible Explanations. The sunlight that would normally be refracted onto the moon's face by the earth's atmosphere is absorbed by volcanic ash and other substances in the upper atmosphere.

Similar and Related Phenomena. Eclipse shadow distortions (ALX2); bands and patches on the eclipsed moon (ALX4); unusually bright lunar eclipses (ALX5); dark days on earth (GWD1).

Examples

X1. 1110. An exceptionally dark lunar eclipse, according to the Anglo-Saxon Chronicle. (R12)

X2. 1588. During an eclipse the moon almost disappears completely. Kepler attributed it to mists and smoke in the earth's atmosphere. (R12, R14)

X3. December 9, 1601. According to Kepler, the eclipsed part of the moon was invisible. (R8, R10)

X4. June 15, 1620. Kepler stated that the moon wholly disappeared, although stars of the fifth magnitude were visible in the vicinity. (R1, R3, R4, R8, R10)

X5. December 9, 1620. The moon was invisible to Cysat at Ingolstadt. (R8, R10)

X6. April 14, 1642. Disappearance of moon reported by many observers. Hevelius could not even see it with his telescope. (R1, R8) Dated as April 25 by R4.

X7. May 18, 1761. Wargentin, at Stockholm, stated that the eclipse was so dark that "the moon's body disappeared so completely that not the slightest trace of any portion of the lunar disc could be discerned either with the naked eye or with the telescope, although the sky was clear." (R3, R4, R8, R10, R5)

X8. June 10, 1816. Several observers stated that the moon was invisible. Tomboro had erupted 14 months before; Mayon, two years earlier. (R2, R4, R5, R6, R8)

X9. October 4, 1884. Widely observed dark eclipse. "10h 2m, middle of totality. To the naked eye nothing could be seen but a faint nebulous spot. That the obscurity of the Moon arose from lack of illumination, not from fog or cloud, was seen by the fact that stars of small magnitudes above and below the lunar disk shone as distinctly as on an ordinary dark night." R2-R4, R8)

X10. 1885. Dark lunar eclipse seen in Tasmania. (R5)

X11. October 16, 1902. A very dark eclipse with a broad smear running east-west across the middle of the moon. Mt. Pelee, St. Vincent, and Santa Maria had erupted earlier. (R6-R8)

X12. April 11, 1903. "At the time of the greatest phase, 12.13, and during the retreat of the umbra the eclipsed part of the Moon's disc was entirely of a blackish grey, not a trace of detail being visible." (R5, R6, R8)

X13. March 22, 1913. Very dark and colorless eclipse. (R8)

X14. September 15, 1913. Again very dark and colorless. (R8)

X15. December 30, 1963. Dark eclipse. (R9, R11)

X16. June 25, 1964. Very dark eclipse that coincided with a minimum in solar activity. (R11)

X17. December 18, 1964. "...no colors were seen apart from a very dim rusty cast, and only the part of the Moon's limb closest to the edge of the main shadow remained conspicuous near mid-totality. The rest of the disk could just be seen, but was very obscure. Fourth- and fifth-magnitude stars within a few degrees of the eclipsed Moon were clearly visible to the naked eye."

(R9)

X18. December 30, 1982. A very dark eclipse, due perhaps to the eruption of El Chichon. (R13)

X19. General observations. "Abstract. The moon is visible during total lunar eclipses due to sunlight refracted into the earth's shadow by the atmosphere. Stratospheric aerosols can profoundly affect the brightness of the eclipsed moon. Observed brightnesses of 21 lunar eclipses during 1960-198 1982 are compared with theoretical calculations based on refraction by an aerosol-free atmosphere to yield globally averaged aerosol optical depths. Results indicate the global aerosol loading from the 1982 eruption of El Chichon is similar in magnitude to that from the 1963 Agung eruption." (R12, R14) A comparison of eclipse darkness with known volcanic eruptions also suggests a strong cause-and-effect relationship. (R10)

References

R1. "Eclipses of the Moon," Scientific American Supplement, 4:1485, 1877. (X4, X6, X7)

R2. Johnson, S. J.; "Abnormal Obscurity of the Moon in the Late Eclipse," Royal Astronomical Society, Monthly Notices, 45: 43, 1884. (X8, X9)

R3. Hopkins, B. J.; "The Krakatoa Eruption, and the Lunar Eclipse of Oct. 4, 1884," English Mechanic, 46:580, 1888. (X4, X7, X9)

R4. "Dark Lunar Eclipses," English Mechanic, 77:297, 1903. (X4, X6-X9)

R5. Johnson, S. J.; "A Possible Cause of the Moon's Obscurity on April 11," Royal Astronomical Society, Monthly Notices, 63: 400, 1903. (X7, X8, X10, X12)

R6. Brook, Charles L.; "Dark Eclipses," British Astronomical Asscoiation, Journal, 13:318, 1903. (X8, X11, X12)

R7. English Mechanic, 77:30, 1903. (X11)

R8, Brooks, Edward M.; "Why Was Last December's Lunar Eclipse So Dark?" Sky and Telescope, 27:346, 1964. (X3-X9, X11-X14)

R9. "Did Bali Eruption Darken the Eclipse?" New Scientist, 24:880, 1964. (X15, X17)

R10. Hedervari, Peter; "Great Volcanic Eruptions and the Luminosity of the Moon during Total Eclipses," Strolling Astronomer, 28:158, 1980. (X3-X5, X7)

R11. Meeus, Jean; "Solar Activity and the Eclipsed Moon," British Astronomical Association, Journal, 92:250, 1982. (X15, X16)

R12. Keen, Richard A.; "Volcanic Aerosols and Lunar Eclipses," Science, 222:1011, 1983. (X1, X2, X19)

R13. "December's Dark Lunar Eclipse," Astronomy, 11:75, April 1983. (X18)

R14. "Lunar Eclipses and Volcanoes," Sky and Telescope, 67:512, 1984. (X2, X19)

ALX2 Distortions of the Earth's Shadow

Description. Noncircular, peaked, or jagged edges of the earth's shadow cast upon the moon during a lunar eclipse. The distorted shadow sometimes disappears during the eclipse.

Data Evaluation. Only a few old reports have been uncovered, but they are carefully described and figured in most cases. Rating: 2.

Anomaly Evaluation. This anomaly seems to be a consequence several rather prosaic phenomena, as described below. With so many potential solutions at hand, a low level of anomalousness is implied. Rating: 3.

Possible Explanations. Older astronomers attributed the phenomeon to the irregular edge of the earth; that is, mountain ranges. However, even the highest mountains rise only about 0.1% of the earth's radius, implying that the earth is really too smooth to cast such irregular shadows. Abnormal refraction by the earth's atmosphere is one possible source of shadow distortion. Clouds in the earth's atmosphere may contribute to the distortion on a transitory basis. The phenomenon of irradiation, in which very bright areas seem larger than they really are, could cause apparent deformation of the earth's shadow on the moon, where many bright and dark areas mingle.

Similar and Related Phenomena. Bands and patches on the eclipsed moon (ALX4).

Examples

X1. October 4, 1884. An account from the Belgian astronomer de Boe. "Briefly then, at the stage of the eclipse represented in the annexed slightly exaggerated diagram the arc of the earth's shadow was quite decidedly peaked at a instead of being rigidly circular. Now, at the time of the phase indicated, the moon was on the horizon of the Cordilleras, while when she was half eclipsed she was similarly placed as regards part of the Pacific Ocean. At this time, however, all traces of the peaked appearance had absolutely vanished, and the periphery of the shadow was sensibly circular. Can the curious deformation figured have been the shadow of the Cordilleras, or had it its origin in the form of the lunar surface?" (R1)

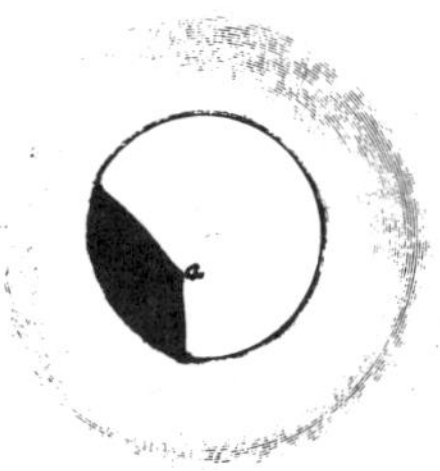

October 4, 1884. Earth's shadow on moon is deformed, peaking at a. (X1)

X2. August 3, 1887. "At 9.30 p.m., local time, at Hamburg, a small cumulus cloud was observed a little distance below the moon, and the darkened part of the lunar surface was taken to be part of the cloud, from its upper edge being flattened. Ten minutes later the cloud had passed away, but the flattened appearance on the moon remained, and it was evident that the earth's shadow was distorted, as seen in the annexed sketch. Several persons noted the peculiarity, which was visible until about 10.30 p.m. in a very clear sky." (R2)

A report from La Tour de Peilz, France. "Here the appearance was certainly unusual; at least I never saw anything like it. The shadow cast on the moon (with a perfectly cloudless sky) was irregular and jagged. I at first thought it was a cloud, but, on looking repeatedly at intervals, I continued to observe the same appearance; allowance being made for the progress of the eclipse." (R3) But at Loch Tay, Scotland, there was no distortion of the earth's shadow. (R3)

X3. October 6, 1903. Greece. "The eclipse was indeed a bright one, as the whole dark border of the moon was very clearly visible..... The border of the earth's shadow protruded from the illuminated disc into the space around the moon, as the annexed figure shows." (R5) (Figure omitted)

X4. General observations. The author doubts that terrestrial mountains could be the cause of the reported irregularities of the earth's shadow on the moon. He suggests clouds instead. (R4)

References

R1. Noble, William; "The Lunar Eclipse of

Strongly flattened shadow of the earth during the eclipse of August 3, 1887. (X2)

October 4," Knowledge, 6:325, 1884. (X1)
R2. "The Lunar Eclipse of August 3," Nature, 36:367, 1887. (X2)
R3. Malet, H. P.; "The Lunar Eclipse of August 3," Nature, 36:413, 1887. (X2)
R4. Hopkins, B. J.; "The Earth's Shadow on the Moon," English Mechanic, 54:350, 1891. (X4)
R5. "The Shadow of the Earth on the Moon as Seen during the Eclipse of Oct. 6, 1903," English Mechanic, 78:448, 1903. (X3)

ALX3 Eclipse Fingers of Light

Description. Radial fingers of light extending over the moon's shadowed surface during a lunar eclipse. In the single example known, the fingers diverge from a point on the edge of the earth's shadow.

Data Evaluation. A single, modern, sketchy observation. Rating: 3.

Anomaly Evaluation. So little information is available that rating the phenomenon is guesswork. Since it may be due to a simple play of light and shade on the variegated lunar surface, a low anomaly rating seems justified. Rating: 3.

Possible Explanations. If the display is as geometrically regular as the sketch suggests, no ready explanation seems to be available. Irregular bands and patches are occasionally observed on the eclipsed moon; and this phenomenon may just be a coincidentally geometrical case of abnormal refraction of sunlight by the earth's atmosphere.

Similar and Related Phenomena. Bands and patches on the eclipsed moon (ALX4).

Examples

X1. July 6, 1963. "Between 2035 and 2300 GMT a partial eclipse of the moon was observed. At maximum eclipse when three-quarters of the moon was in shadow, its surface still remained visible, and fingers of light were seen illuminating the upper section which was in shadow, the appearance being shown in the accompanying sketch." A note following this observation claims to explain this phenomenon but doesn't even present a mechanism for finger formation.

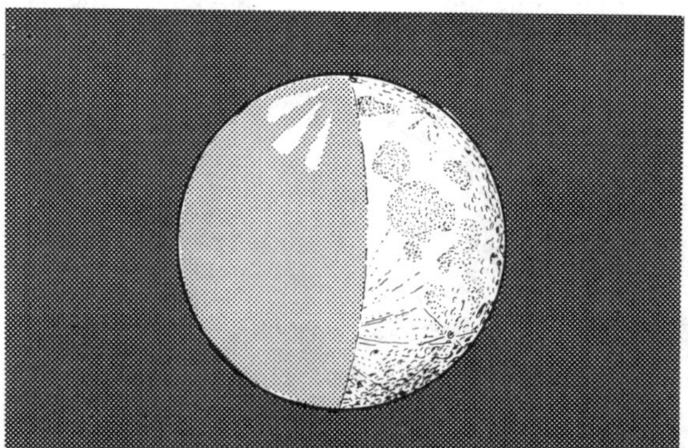

Radiating fingers of light recorded during the partial lunar eclipse of July 6, 1963. (X1)

References

R1. Davies, T.H.; "Partial Lunar Eclipse," Marine Observer, 34:126, 1964. (X1)

ALX4 Bands and Patches on the Eclipsed Moon

Description. Broad bands and patches of gray and various colors appearing on the eclipsed moon. Often remarkably bright and highly colored, these features are usually transient and appear mainly during totality.

Data Evaluation. The astronomical literature contains only a few descriptions of such eclipse phenomena. The December 30, 1963, lunar eclipse, however, produced a brilliant blue band and a red spot that was seen by many. The general phenomenon is well-verified. Rating: 1.

Anomaly Evaluation. The abnormal refraction of sunlight by the earth's atmosphere is in all probability the cause of this phenomenon. No optical principles are challenged, although some minor features remain puzzling. Rating: 3.

Possible Explanations. The eclipsed part of the moon is illumined by sunlight refracted by the earth's atmosphere. The earth's atmosphere, though, is not uniform, being clouded in some places, clear in others, and containing various amounts of impurities. It is in effect an inhomogeneous and dirty lens. Variations in the intensity and coloring of the eclipsed moon are not surprising, although the tendency to produce bands is curious.

Similar and Related Phenomena. Dark eclipses (ALX1); distortions of the earth's shadow on the moon (ALX2); fingers of light on the eclipsed moon (ALX3).

Examples

X1. January 28, 1888. Dr. Dyer called "attention to the fact that about the middle of the eclipse the moon appeared to the naked eye as a fairly bright disc with a dark isosceles triangular shadow on it, having the base to the north. That triangular patch on the moon's disc seems to have been noticed by others. Mr. Backhouse sends us a paper with a diagram showing the triangular patch about in the centre of the disc, with a long description of the physical appearances he noticed." (R2)

X2. October 16, 1902. A very dark eclipse. S.J. Johnson stated that "after the moon was totally eclipsed a broad dusky belt appeared crossing the centre with a dark ruddy patch above and another below this." (R3, R4)

X3. January 29-30, 1953. "2305 to 0140 G.M.T. The commencement of the eclipse was not observed owing to almost stationary Cu covering the moon. During totality a small white patch of light of low brilliancy moved round the north pole of the moon until that phase came to an end at 0030. From that time the white patch increased in area until the end of the eclipse at 0140. During the total phase the face of the moon appeared to be coloured in bands of blue, green, yellow and orange as in the sketch, and stars were visible with the naked eye within 2° or 3° of

Colored bands on the eclipsed moon, January 29, 1953. (X3)

the moon." (R5)

X4. December 30, 1963. "More than 150 reports have been received by Sky and Telescope describing various aspects of the event. It is apparent that the moon's faintness at totality was not the only outstanding feature of the eclipse. At and after second contact, a brilliant blue band was seen at the western edge of the umbrȧl shadow. So vivid was this fringe that some watchers thought the predicted time of second contact was five or more minutes off! Also, during totality a portion of the moon's disk toward the center of the umbra glowed with a reddish tint." (R6)

X5. No date given. Testimony of Lewis Swift. "Starting from near the Moon's center a luminous band extended as far to the east as Aristarchus, passing very near if not exactly over that crater, from near which it bent sharply to the east, somewhat like in form to a huge figure 7, the up and down portion of the character following the curved outline of the lunar limb though distant from it by estimation one-fourth of the Moon's radius. In its curvature to the north it passed, as at the March eclipse, centrally over Plato and ended a short distance beyond it to the west. The band was of a pale drab color and of nearly uniform width throughout its entire length, and on both occasions I compared its width with that of Tycho and found the two equal, so that it must have been fifty miles wide." (R1)

References

R1. Swift, Lewis; "The Colored Lunar Band," Popular Astronomy, 3:269, 1896. (X5)
R2. English Mechanic, 47:52, 1888. (X1)
R3. Brook, Charles L.; "Dark Eclipses," British Astronomical Association, Journal, 13:318, 1903. (X2)
R4. English Mechanic, 77:30, 1903. (X2)
R5. Brown, G.; "Total Eclipse of the Moon," Marine Observer, 24:16, 1954. (X3)
R6. "A Remarkable Eclipse of the Moon," Sky and Telescope, 27:142, 1964. (X4)

ALX5 Very Bright Lunar Eclipses

Description. Lunar eclipses in which the moon's shadowed surface seems to be much brighter than it is during the "average" lunar eclipse. Such eclipses are highly colored; and the brightness may vary during the eclipse. Some very bright eclipses take place during periods of high auroral activity; others do not.

Data Evaluation. While we have good testimony describing a few extremely bright eclipses, these data are all qualitative and often highly subjective. Viewing conditions as well as the observer's opinion of what constitutes an "average" brightness level obviously affect the quality of the data. Rating: 3.

Anomaly Evaluation. It is impossible to determine if an anomaly actually exists here. Very bright eclipses, like the very dark eclipses, could be the natural extreme of a spectrum of eclipse brightness levels. Rating: 3.

Possible Explanations. Luminescence of lunar surface materials stimulated by high energy solar radiation (probably quantitatively inadequate). Enhanced refraction and scattering of sunlight onto the moon by the earth's atmosphere. Unusual transparency of the terrestrial atmosphere and/or the addition of aerosol scattering material plus subjective errors of judgment may combine to account for this phenomenon.

Similar and Related Phenomena. Very dark lunar eclipses (ALX1); transient lunar luminescence (ALF5).

Examples

X1. March 19, 1848. The condensed testimony of several European observers. "I wish to call your attention to the fact, which

I have clearly ascertained, that during the whole of the late lunar eclipse of March 19, the shaded surface presented a luminosity quite unusual, probably about three times the intensity of the mean illumination of an eclipsed lunar disc. The light was of a deep-red colour. During the totality of the eclipse the light and dark places on the face of the Moon could be almost as well made out as in and ordinary dull moonlight night; and the deep-red colour, where the sky was clearest, was very remarkable from the contrasted whiteness of the stars. The Consul at Ghent, who did not know that there was an eclipse, wrote to me for an explanation of the deep red colour of the Moon at 9 o'clock.

Another observer in Devonshire, of the same eclipse says: The appearances were as usual till 20 minutes to 9; at that period, and for the space of the next hour, instead of an eclipse, or the umbra of the earth being the cause of the total obscurity of the Moon, the whole surface of that body became very quickly and most beautifully illuminated, and assumed the appearance of the glowing heat of fire from the furnace, rather tinged with deep red.

Both observers relate that there was at the time a brilliant aurora borealis, and suggest that the luminous appearance of the Moon might arise from the reflection of the light shed by this northern effulgence.

In Ireland, where the same peculiarity was observed, it was noticed that about a quarter of an hour before the end, the appearance was as usual, the eclipsed part was nearly black, and the rest perfectly bright." (R5, R1) Two important points in this series of observations: (1) The unusual brightness did not last for the entire eclipse in some areas; and (2) The presence of an aurora. (WRC)

X2. December 23, 1703. Following a discussion of very dark eclipses. "As an instance of the contrary nature, where the moon has been so strongly illuminated during her presence in the shadow as to admit of the various markings upon her surface being seen with distinctness, and even to lead persons to doubt her being eclipsed, mention may be made of the eclipse on the morning of December 23, 1703, which was observed by various astronomers in the south of France. At Avignon, during the whole duration of the passage through the earth's shadow, 'the moon appeared extraordinary illuminated, and of a very bright red, so that it might have been supposed that she was transparent, and that the sun was behind her globe, and that his rays passed through in the same manner that they are slightly diaphanous.' It is singular, however, that while this was the aspect of the phenomenon at Avignon, different features should have been noted at Montpelier, particularly the total disappearance of the moon, rather quickly towards 6h. 30m. A. M., though the night was as transparent as could have been wished; it was mentioned that the twilight was already very sensible, but that the invisibility of the moon could not be wholly attributed to this cause, since many stars were shining in the same quarter of the sky." (R2)

X3. March 19, 1848. An intensely bright lunar eclipse seen in Europe. An aurora was also seen at the same time. (R2)

X4. September 26, 1950. This lunar eclipse was a thousand times brighter than that of January 18, 1954. (R6)

X5. October 18, 1967. A very bright eclipse. This eclipse occurred three years after the solar minimum and was considered to have been far too bright to conform to Danjon's rule. (R8) Danjon's rule is presented in X6.

X6. Possible correlation of lunar eclipse brightness with solar activity. "In 1920 Danjon, a French astronomer, noticed a relationship between the phase of solar activity and the brightness of a lunar eclipse (Comptes Rendus, 171:1127, 1920). This relationship is now known as Danjon's law and can be stated as follows: in the two years immediately after a solar activity minimum the shadow of the Earth during a lunar eclipse is very dark and has little colour. At the solar activity moves away from minimum the eclipsed Moon becomes brighter and redder until, during the seventh and eighth year after solar minimum the eclipsed Moon is at its brightest and is red, copper-coloured or orange. The brightness curve then falls away very sharply to its minimum value. The maximum phase of the solar cycle passes unnoticed whereas the minimum phase is indicated by a sudden and considerable dimunition in brightness and colour, this change forming a discontinuity." (R6, R3)

If Danjon's law is correct, a source of extra illumination would have to be found that is maximum near the solar maximum. The luminescence of lunar materials when bombarded by solar corpuscular radiation has been shown to be inadequate. Additional scattering from the earth's upper atmosphere due to the additional influx of aersols, as affected by solar corpuscular radiation, has also been shown to be much too small. (R6)

Danjon's original analysis of historical

data on eclipse brightness has been considered flawed by modern scientists; for example, he couldn't fit some of the dark eclipses following volcano eruptions into his scheme. Other studies have, in fact, found little if any connection between eclipse brightness and solar activity. (R4, R7)

References

R1. Forster, Mr.; "Remarkable Appearances during the Total Eclipse of the Moon on March 19, 1848," Royal Astronomical Society, Monthly Notices, 8:132, 1848. (X1)

R2. "Eclipses of the Moon," Scientific American Supplement, 4:1485, 1877. (X2, X3)

R3. "Brightness of the Eclipsed Moon and Solar Activity," Royal Astronomical Society of Canada, Journal, 15:41, 1921. (X6)

R4. Fisher, Willard J.; "The Brightness of Lunar Eclipses 1860-1922," Smithsonian Miscellaneous Collections, vol. 76, no. 9, 1924. (X6)

R5. Inglis, J. Gall; "Lunar Eclipses and Auroras," British Astronomical Association, Journal, 35:276, 1925. (X1)

R6. Hughes, David W.; "Lunar Eclipses and Danjon's Law," Nature, 253:503, 1975. (X4)

R7. Sekiguchi, Naosuke; "Photometry of the Lunar Surface during Lunar Eclipses," The Moon and the Planets, 23:99, 1980. (X6)

R8. Meeus, Jean; "Solar Activity and the Eclipsed Moon," British Astronomical Association, Journal, 92:250, 1982. (X5)

ALX6 Thin Arcs of Light on Rim of Eclipsed Moon

Description. Narrow, crescent-like slivers of light on the edge of the eclipsed moon. These arcs should not be confused with the diffuse patches and bands of ALX4.

Data Evaluation. Only a very few examples of this phenomenon have been found; and each of these differs somewhat in detail. Rating: 3.

Anomaly Evaluation. As with the anomalous bands and patches seen on the eclipsed moon (ALX4), the abnormal bending of sunlight, or perhaps the scattering of sunlight, by the earth's upper atmosphere seems a likely, and not particularly mysterious, mechanism that may account for this phenomenon. Rating: 3.

Possible Explanations. Unusually strong refraction or scattering of sunlight by the earth's atmosphere, especially when the moon is not near the center of the shadow cone. In other words, the shadow cone may be deformed somewhat by special conditions in the earth's atmosphere.

Similar and Related Phenomena. Bands and patches on the eclipsed moon (ALX4); anomalous occultations of planets (ALX7) and stars (ALX8).

Examples

X1. April 11, 1903. Observing the lunar eclipse through a 3-inch refractor, in England, W.W. Magness saw two bright streaks of light, one on either side of the uneclipsed crescent of the moon. (R1)

X2. July 4-5, 1917. Several observers reported that the eclipsed moon was much more luminous around its rim than near the center. Lunar luminescence suspected. (R2)

X3. January 18, 1954. During a total lunar eclipse, a thin sliver of white was evident at one edge of the moon throughout totality. It was surmised that a huge, very clear area of the earth's atmosphere might have refracted enough sunlight to cause the effect, even though theoretical totality existed. (R3)

References

R1. Magness, W.W.; "The Lunar Eclipse," English Mechanic, 77:346, 1903. (X1)

R2. English Mechanic, 107:145, 1918. (X2)

R3. "Lunar Eclipse Total Despite Slim White

Arc," Science News Letter, 65:73, 1954. (X3)

ALX7 Dusky Bands across Planets at Contact Phase of Occultation

Description. The appearance of a dusky band parallel to the moon's circumference across the disk of a planet being occulted by the moon. The band may appear at either immersion or emergence or both. To date, this phenomenon has been noted only with Jupiter.

Data Evaluation. Evidently the phenomenon is quite rare. A few modern observations exist, along with a rather convincing 1892 photograph. Rating: 2.

Anomaly Evaluation. The dusky band crossing the planet is one of several similar phenomena that suggest the presence (probably temporary) of a lunar medium that refracts or scatters sunlight. We do not know exactly what this medium is or where it originates. Rating: 2.

Possible Explanations. Nineteenth century astronomers who saw the dusky band immediately suspected a lunar atmosphere. Today, we recognize that the moon's permanent atmosphere is negligible and are forced to fall back on electrostatically levitated dust clouds and clouds of gas vented from the lunar crust.

Similar and Related Phenomena. The anomalous occultation of stars by the moon (ALX8); the ring of light around the new moon (ALO9); the extended glow of occulted Saturn (ALX12); the radio anomalies associated with occulted spacecraft (ALX10); the ring of light around the moon during a partial solar eclipse (ALX11); postsunset horizon glows (ALO11); the lunar zodiacal light (ALO5); lunar clouds, mists, and obscurations (ALW1).

Examples

X1. January 2, 1857. At least six well-known astronomers reported the following effect. "But the most interesting fact yet remains to be told. The bright border of the moon at this time crosses the soft green face of the planet (Jupiter), not with a clear sharly cut outline like that which had been presented as the disc passed into concealment; it was fringed by à streak or band of graduated shadow, commencing at the moon's edge as a deep black line, and then being stippled off outwardly until it dissolved away in the green light of the planet's face. The shade-band was about the tenth part of the planet's disc broad, and of equal breadth from end to end." (R1)

X2. August 12, 1892. W.H. Pickering, using the Harvard 13-inch telescope, at Arequipa, Peru, photographed the occultation of Jupiter by the moon. The photographs show a distinct dark band at immersion. (R2)

X3. January 13, 1944. In a general discussion of the two occultations of Jupiter seen in 1944 (X3 and X4). "Three kinds of appearances were observed. On January 13th, Hilliard and Campbell observed no dark band at all, while White and Morgan observed a gray band across Jupiter tangent to the moon. At the April 30-May 1 occultation, however, Johnston and Haas observed the same band. The third appearance, observed by me (C. G. Wates) in January and the Hake brothers in April, was that of cusps at the ends of the band, near where Jupiter's limb touched that of the moon. The band was estimated to be three to five seconds in width, and the cusps seemed to me to be about twice the width of the band." (R2, R3)

X4. April 30-May 1, 1944. A report from W.H. Haas. "The occultation of Jupiter by the moon on April 30th-May 1st was observed with the 18-inch refractor of the Flower Observatory, using 150 x. The seeing was poor, as usual in Upper Darby (PA), while the sky was clear, not as usual in Upper Darby. When the planet emerged from behind the bright limb of the moon, a hazy gray band concentric with the lunar limb was seen across the face of Jupiter. The angular width of the band was estimated to be three seconds of arc. This band was about as easy to see as either of the temperate belts on the planet. It certain looked real. I saw nothing similar during immersion at the dark limb. Such a band was seen by at least one other observer at the recent occultation and by at least several observers at the January 13th occultation." (R2, R3)

References

R1. "Fog-Seas of the Moon," Eclectic Magazine, 41:276, 1857. (X1)

R2. Fitzpatrick, Jesse A.; "Jupiter and the Lunar Atmosphere," Sky and Telescope, 3:20, 1944. (X2-X4)

R3. Wates, Cyril G.; "Band and Cusps at Occultation of Jupiter," Royal Astronomical Society of Canada, Journal, 38:258, 1944. (X3, X4)

ALX8 The Hanging or Projection of Stars and Planets on the Moon's Limb

Description. The delayed disappearance of stars and planets being occulted by the moon. The star or planet will either seem to hang (remain stationary) on the edge of the moon for several seconds or it will seem to pass in front of the moon for a short distance, again for several seconds. The star or planet will then either disappear suddenly, gradually fade away, or fade away with a final flare-up before completely disappearing. Either the dark or light limb of the moon may be involved. The reverse process may transpire upon emergence from behind the moon. These phenomena are rare and occur in only a very small fraction of occultations.

Background. This is a delightful optical puzzle which, like the "black drop" phenomenon of solar transits (AHO0), has received very little attention from modern astronomers, who apparently have more challenging things to study. These phenomena have been recorded for over two centuries, and no really satisfactory explanation is at hand. Although the projection of stars on the moon seems trivial on the surface, it could betoken transitory physical phenomena on the moon.

Data Evaluation. A frequently reported phenomenon. Rating: 1.

Anomaly Evaluation. None of the explanations proposed below seems to be completely satisfactory. Superficially, this phenomenon appears to be just a simple problem of optics and perhaps psychological illusion. If it is, why doesn't it occur frequently, for the proper conditions recur every night? On the other hand, if it is due to mirage action involving a transitory release of gases on the moon, we would have to deal with unexpectedly large ventings of gas and/or dust on the moon. Rating: 2.

Possible Explanations. Mirage action (refraction) in transitory lunar gases; simple (?) optical illusion; irradiation; diffraction. The latter suggestions do not really seem to explain all facets of the phenomenon.

Similar and Related Phenomena. Ring of light around the new moon (ALO9); the projection of stars and satellites upon Jupiter's disk (AJX5); the extended glow of occulted Saturn (ALX12); the anomalous reception of radio signals from occulted spacecraft (ALX10); the black drop effect seen during solar transits of Mercury and Venus (AHO0).

Examples

X1. May 4, 1783. "On this evening, when a Dr. Lind and his wife were visiting William Herschel at Datchet, a star was occulted at the dark limb of the moon. Mrs. Lind.... placed herself at a telescope and watched attentively. Scarcely had the star disappeared before Mrs. Lind thought she saw it again, and exclaimed that the star had gone in front of, and not behind the moon. This provoked a short astronomical lecture on the question, but still she could not credit it, because she saw differently. Finally Herschel stepped to the telescope, and in fact he saw a bright point on the dark disc of the moon, which he followed attentively. It became fainter and finally vanished." (R15)

X2. October 23, 1831. Observation of George Innes at Aberdeen, Scotland. "Mr. Innes states that when Aldebaran came within about six seconds of the moon's limb, it passed through the remaining distance with great rapidity, its apparent velocity becoming five or six times as great as before. It appeared to hang on the moon's limb for about five seconds of time, and then suddenly disappeared. In other words, the moon appeared suddenly to swell or extend itself into what was previously the dark space." (R2)

X3. April 4, 1854. The star alpha Geminorum was projected on the moon's bright limb for 4 seconds as it emerged. (R3)

X4. May 19, 1858. Two different observers saw Regulus hang on the moon's limb for 5 seconds upon emerging. (R3)

X5. March 19, 1866. When 31 Arietis was occulted by the moon's dark limb it hung on the limb for several seconds. (R3)

X6. March 28, 1868. "At the occultation of ζ Tauri on March 28th, 1868, I had the good fortune to witness the phenomenon myself, and the event is indelibly fixed in my memory by the surprising beauty of the spectacle as well as by the entirely unexpected nature of the occurrence. The disappearance took place at the dark limb which was distinctly visible, and the atmospheric circumstances were eminently favourable. The star remained on the disk for at least five seconds, and very possibly longer and its distance from the limb was considerable." (R3)

X7. October 14, 1870. The disappearance of ζ Tauri at the moon's bright limb was accompanied by phenomena like those in X6. (R3)

X8. November 10, 1878. Testimony of W. H. M. Christie, as he watched the occultation of 17 Tauri, at the Royal Observatory, Greenwich. "The star is of 4 magnitude, and as it appeared to approach the Moon's limb, I fixed my attention on it, being afraid of losing it in the overpowering light of the Moon. I watched it steadily till it came up to the line of the limb, expecting it to disappear at a slight notch in the limb; but instead of that, the Moon's limb, to my surprise, seemed to recede for some 3 or 4 seconds of time, and the star gradually disappeared in a sort of luminous haze, through which it was seen with more and more difficulty as it advanced. At the instant of disappearance the star was seen apparently perfectly bisected by the limb, if limb it could be called, that is, completely shorn of its rays and half the disk on the one side (towards the Moon), and intact on the other. (R4, R5)

X9. December 10, 1885. A star disappeared at the dark limb of the moon, which was visible. Fifteen seconds before extinction, the star was suddening dimmed for 10 or 12 seconds, and then appeared to brighten a bit before final disappearance. (R8)

X10. April 11, 1891. "I observed the immersion of a star in Taurus, 6.5 or 7 magnitude, 60° from the north point, 10° from the vertex. The star disappeared at 8h 10m 1s, Central standard time; re-appeared and disappeared again about one second after the first disappearance." Seeing was not steady, and the phenomenon was attributed to the action of the earth's atmosphere. (R6)

X11. May 21, 1897. The star emerged at the moon's dark limb and hung there for a few seconds. (R8)

X12. August 18, 1900. Same as X11.

X13. September 3, 1900. Saturn was occulted at the moon's dark limb, which was invisible. A last little patch of light parallel to the moon's limb seemed to linger strangely. (R8)

X14. September 30, 1905. H.P. Hollis was watching the moon occult stars. Some of these stars hung on the moon so long "that the observer got tired of waiting." (R9)

X15. April 22, 1913. M.E.J. Gheury was timing the occultations of stars by the moon with a stopwatch. Carefully watching the spot where an occulted star was expected to reappear, he suddenly saw a star at the correct spot but much fainter than he expected. After a second, the star suddenly brightened. (R10)

X16. November 6, 1922. "The occultation of Aldebaran on 1922 November 6 was observed with a 3 1/4-inch refractor X 200 under the most favourable conditions of seeing and transparency. I expected to see the star disappear at the instant at which its spurious system was precisely bisected by the Moon's limb. Instead, however, the star appeared to pass completely on to the Moon's surface, and only disappeared when its first diffraction ring was completely within and, as far as I could judge, just past internal contact with the bright limb. The disappearance then took place instantaneously without any previous diminution in brightness." (R11)

X17. January 28, 1928. E. Prain, at Dundee, Scotland, was watching the moon occult the star V. Piscium. "...the contact came, and the star---correctly speaking, the light of the star---not only remained for some seconds on the edge of the moon, but it grazed about two degrees past the edge as if the star were between the moon and the earth---it was not a 'transit meteor' this time---then began to diminish and vanished. I will never forget the sight." (R12)

X18. July 17, 1938. "On July 17, while walking down the King's Road, Chelsea, with two others (unaware of the predicted occultation of Mars), we saw the Moon above the house-tops, with a planet in close proximity. We walked to the Embankment, to get a clear view, and watched it for some time with the naked eye. It was evident that an occultation was about to take place. But instead of disappearing at the dark limb of the Moon, the planet appeared to enter the dark field of the Moon's surface and to continue to travel across it, until it vanished on reaching the illuminated edge of the Moon. We discussed this mysterious phenomenon as it occurred. It lasted several minutes. We could only account for it by some optical illusion, which we presumed would be a matter of common knowledge and frequent occurrence, to be easily explained by astronomers." Two other similar observations were made elsewhere. The moon was at half phase. (R13)

Another group, eight miles away, knowing of the occultation and watching it carefully saw no projection of Mars on the moon at all. (R14)

X19. November 18, 1956. "Mr. Paul J. Nemecek of Whittier, Calif. reports on a curious occultation which he witnessed at 6h 33m, U.T. as follows:'Moon occulted fourth stellar magntiude star near Pleiades. Star remained on limb about 5 seconds before disappearing behind limb. Star faded slowly away in magnitude as it hovered at limb.'....Mr. Nemecek also writes of seeing a halo around the star before it disappeared. Another observer in Missouri also saw an anomalous occultation at this time. (R16)

X20. Attempts at explanation. Several half-hearted attempts to explain this phenomenon appear in the literature. None is very successful. (R1) Suggestions include optical illusion (R2); unsteadiness of the earth's atmosphere (R7); and irradiation and diffraction (R11)

References

R1. Powell, Professor; "On Attempts to Explain the Apparent Projection of a Star on the Moon," Report of the British Association, 1846, p. 5. (X20)

R2. Airy, G. B.; "On the Apparent Projection of Stars upon the Moon's Disk in Occultations," Royal Astronomical Society, Monthly Notices, 19:207, 1859. (X2, X20)

R3. Plummer, John J.; "On the Apparent Projection of Stars upon the Moon's Disk in Occultations," Royal Astronomical Society, Monthly Notices, 33:345, 1873. (X3-X7)

R4. Christie, W. H. M.; "Note on a Phenomenon Seen in the Occultation of a Star at the Moon's Bright Limb," Royal Astronomical Society, Monthly Notices, 39: 198, 1879. (X8)

R5. Christie, Mr.; "On a Phenomenon Seen in the Occultation of a Star at the Moon's Bright Limb," Astronomical Register, 17:11, 1879. (X8)

R6. Parkhurst, J. A.; "Peculiar Star Occultation," Sidereal Messenger, 10:252, 1891. (X10)

R7. Davis, Herman S.; "Anomalous Occultations," Observatory, 24:417, 1901. (X20)

R8. Tupman, G. L.; "Anomalous Occultations," Observatory, 25:56, 1902. (X9, X11-X13)

R9. Hollis, H. P.; "Anomalous Occultations," English Mechanic, 81:60, 1905. (X14)

R10. Gheury, M. E. J.; "Evidence of an Atmosphere on the Moon," Observatory, 36:268, 1913. (X15)

R11. Waterfield, R. L.; "The Projection of a Star upon the Moon's Bright Limb," British Astronomical Association, Journal, 33:250, 1923. (X16, X20)

R12. Prain, Ernest; "A Remarkable Occultation: Evidence of a Lunar Atmosphere?" English Mechanics, 3:348, 1928. (X17)

R13. Thackeray, A. D.; "Optical Illusion during the Occultation of Mars," British Astronomical Association, Journal, 48: 126, 1938. (X18)

R14. "Optical Illusion during the Occultation of Mars," British Astronomical Association, Journal, 48:179, 1938. (X18)

R15. Haas, Walter H.; "Does Anything Ever Happen on the Moon?" Royal Astronomical Society of Canada, Journal, 36:374, 1942. (X1)

R16. Hars, H.; "The Total Lunar Eclipse of November 18, 1956," Strolling Astronomer, 11:64, 1957. (X19)

ALX9 Post-Eclipse Changes of Surface Features

Description. Temporary enlargment, brightening, or other changes of lunar features following lunar eclipses. The only fairly-well-verified phenomenon in this category is the apparent enlargement of the white area surrounding the crater Linne. This curious effect lasts for the better part of an hour.

Data Evaluation. The enlargement of Linne has been seen several times by various different observers. However, there have been no observations found within the last 50 years. Rating: 2.

Anomaly Evaluation. The explanation very likely involves a simple change of surface properties with temperature or exposure to solar radiation. Rating: 3.

Possible Explanations. Thermal alteration of the lunar surface properties of reflectivity and absorption by physical rearrangement or surface chemical change. Another possibility is the temporary rearrangement of surface materials by the same electrostatic mechanism blamed for the post-sunset horizon glows (ALO11). Also, temporary luminescence should not be forgotten, although there doesn't seem to be any adequate stimulus here. Whatever the cause, the Linne area is unique, for the phenomenon has not been seen elsewhere.

Similar and Related Phenomena. Historical changes in the Linne area (ALE22-X2); the post-eclipse brightening of Jupiter's moon Io (AJX6).

Examples

X1. December 27, 1898. Linne enlarged by 0".5 for about 30 minutes after the crater reentered sunlight. (R1)

X2. December 16, 1899. Possible enlargement of Linne by 0".14 after lunar eclipse. (R1)

X3. October 16, 1902. "Linne was so strikingly enlarged that Pickering doubted its identity when it first emerged from the umbra. The amount was 2".75 and there was no decrease in size for 45 minutes after the shadow had left. (R1)

X4. April 11, 1903. "Pickering observed Linne to be enlarged by 0".55, but 0".35 of this amount was gone 75 minutes after emersion. Saunder obtained an enlargement of 0".4 or 0".5. Wirtz observed a similar increase and further found the spot growing rapidly larger before the shadow covered it." (R1)

X5. August 14, 1905. Although Linne did not enter the umbra, it seemed to increase to 1".24. (R1)

X6. February 8, 1906. Frost and Stebbins announced that the lunar eclipse had enlarged Linne by about 1". (R1)

X7. July 16, 1935. Haas reported that Linne was enlarged perhaps by 1".5. (R1)

References

R1. Haas, Walter H.; "Does Anything Ever Happen on the Moon?" Royal Astronomical Society of Canada, Journal, 36:398, 1942. (X1-X7)

ALX10 Reception of Radio Signals from Occulted Spacecraft

Description. The detection on earth of spacecraft radio transmissions while the spacecraft is occulted by the moon. Such signals persist for almost a minute after theoretical occultation.

Data Evaluation. The phenomenon has been observed at least twice, with one instance being carefully monitored. However, confirming data are desirable. Rating: 2.

Anomaly Evaluation. Radio waves can be transmitted for short distances around the curvature of an atmosphereless, ionosphereless planet by the phenomena described below. So, these anomalous radio transmissions, which seem quite startling at first, are probably not really so remarkable. Rating: 3.

Possible Explanations. The processes of scattering and diffraction can transmit radio waves along the curved surface of a planet for some distance. Some radio waves may also be refracted through the lunar limb itself. If lunar ion clouds, as mentioned in ALW2, are present, reflection of radio waves (and refraction, too) are possible, although the radio frequencies used in the cases noted are quite high for these mechanisms to be very important.

Similar and Related Phenomena. The projection of stars and planets on the moon's limb (ALX8), which is an almost perfect optical parallel of this radio phenomenon.

Examples

X1. December 15, 1972. Following the unexpected detection of radio signals from the Apollo 15 spacecraft Endeavour, while it was occulted by the moon, scientists paid special attention to the occultation of the spacecraft America. The transmitter used for communication between the command module and the lunar landing party was 259.7 MHz. In this experiment, the 150-foot radio telescope of the USAF Cambridge Research Laboratory, near Hamilton, Massachusetts, was employed.

"Occultation of the signal occurred at 0143:36 GMT, December 15, 1972. The signal did not immediately go to a zero level; it seemed to go to zero at approximately 0144:02 GMT and reappeared at 0144:12 GMT, persisted for 52 s, and finally disappeared at 0145:04 GMT. This behavior indicates a dependence of the post-occultation signal on the exact topography of the lunar surface beneath the spacecraft." The height of the spacecraft was 96 kilometers at the time of occultation.

Several explanations were considered, including the refraction of waves and the propagation of lunar surface waves. The latter possibility was considered the most promising. (R1)

References

R1. Salisbury, W. W., and Fernald, D. L.; "Post-Occultation Reception of Lunar Ship America Radio Transmission," Nature, 242:601, 1973. (X1)

ALX11 Bright Ring around Moon during Partial Solar Eclipse

Description. The appearance of a bright ring around that part of the moon that is not superimposed on the solar disk during a partial eclipse of the sun.

Data Evaluation. Supposedly a well-known phenomenon, but only one rather vague reference has been located. Rating: 3.

Anomaly Evaluation. As in the case of the ring of light seen around the new moon (ALO9), most scientists would probably opt for an explanation embracing contrast effects; that is, they would suggest that the rim of light is not a real physical phenomenon but rather a perceptual artifact. In X1 below, however, convincing photographs are mentioned, which implies physical reality. In sum, if the phenomenon is physically real, we have no obvious physical explanation on an atmosphere-less moon; if it is "only" a contrast effect, we do not understand the physiology of the situation. Rating: 2.

Possible Explanations. A contrast phenomenon;

Similar and Related Phenomena. Ring of light around the new moon (ALO9); ring of light around Mercury during transit (AHX2); the annular phase of Venus (AVO2); the hanging and projection of stars and planets on the moon's limb (ALX8).

Examples

X1. General observations. From the deliberations of the Royal Astronomical Society. "Mr. Raynard said, with regard to the line of light which had been spoken of as existing round the moon's limb, there can be no doubt as to its presence on the photographs of partial phase eclipses. The line was observed over and over again in 1860. Mr. de la Rue and the Astronomer Royal had both noticed it, and referred it to an effect of contrast. After the 1869 eclipse in America Dr. Curtiss noticed the band of increased brightness along the moon's limb in the partial phase photographs taken at Des Moines, and he had made some interesting experiments upon them, which showed that the band of brightness had an actual existence and was not merely due to contrast." (R1)

References

R1. "Royal Astronomical Society," Scientific American Supplement, 3:1069, 1877. (X1)

ALX12 Extended Glows of Occulted Planets

Description. Glows seen at the edge of the lunar disk emanating from occulted planets still well behind the moon's limb.

Data Evaluation. Two modern observations are on record so far. Rating: 2.

Anomaly Evaluation. Both of the two potential explanations mentioned below deviate significantly from the current scientific pictures of the planets involved. Rating: 2.

Possible Explanations. The scattering or reflection of the light of the occulted planets by some unrecognized medium surrounding the moon; or the presence around the planets involved (Jupiter and Saturn) of extended atmospheres or scattering media.

Similar and Related Phenomena. The reception of radio signals from occulted spacecraft (ALX10); the hanging and projection of stars and planets on the moon's limb (ALX8); the ring of light around the new moon (ALO9); the bright rings around the moon during partial solar eclipses (ALX11).

Examples

X1. October 19, 1968. "Harold Brock of Akron, Ohio, observed the occultation of Jupiter by the moon from near La Pryor, Texas, where the event was grazing. As the planet skirted the moon's south limb, it passed behind two lunar mountains, the first peak lower and narrower than the second. Mr. Brock, observing with a 4 1/4-inch telescope at about 150x, noted an unusual phenomenon just as the planet was completely covered by the first peak. 'I became aware of an arc of light over the planet's position and just above the moon's dark limb. It reminded me of the glow of a distant city. The light was brightest near Jupiter's assumed position, becoming fainter away from the limb. As the occultation deepened, the arc diminished. But it remained sufficiently bright so that I could follow the planet's progress behind the moon's limb. Just before Jupiter reappeared from behind the first peak, the glow became more visible, but it vanished the moment Jupiter's disk began to emerge. The effect was repeated behind the second peak.'" Others observed the same effect. (R1)

X2. October 17, 1973. "The observation upon which this report is based was made by the visual observer at a time between 1 s and 2 s before the appearance at the dark lunar limb of the first pin-point of light from Saturn's rings. At this time, the observer's eye was attracted to a faint glow which he probably first saw by averted vision but which delineated the dark limb and which appeared like seeing 'a campfire on the other side of a treeless hill on a dark night.' The extent of the glow was estimated afterwards to have been a little greater than the east-west extent of the rings. The pin-point of light from the rings which appeared in the centre of the faint glow rapidly brightened and became bigger as the rings became visible following the third contact. The glow increased slightly in width and showed easily detectable Moon limb curvature on its sharp edge but at about 4 s beyond third contact the glow was indistinguishable against the increasing ring brilliance. Superficial explanations for this glow can be quickly dismissed. If internal reflection within the telescope of light from the bright lunar surface caused the glow, it was certainly a coincidence that it not only appeared in the exact position where the rings showed themselves but also was aligned with the direction and position of the dark Moon edge. Mist in the Earth's atmosphere could have produced a hazy glow just as the first light of the rings appeared but this would almost certainly have been circular from the initially observed edge of the rings. This must remain a distinct possibility but it is at least as tempting to associate this glow with the planet and a tenuous outer atmosphere surrounding Saturn and its rings which is normally invisible against the planet's brightness." The observation of Brock (X1) was then mentioned as supportive of this theory. (R2)

References

R1. "Unexplained Occultation Phenomenon," Sky and Telescope, 37:122, 1969. (X1)

R2. Reed, G., et al; "Extended Glow Preceding Reappearance of Saturn during a Lunar Occultation," Nature, 247:447, 1974. (X2)

ALZ THE ENIGMA OF LUNAR MAGNETISM

Key to Phenomena

ALZ0 Introduction

One of the greatest surprises in space exploration has been the presence of significant magnetic fields on the moon and around Mercury---both planet-sized objects that seemed very poor candidates for hot, fluid cores operating in a dynamo mode. Venus, large and hot, would not have startled anyone if it possessed a magnetic field, but it apparently does not have a self-generated field. Mars does not either. There seems to be no rhyme or reason to the magnetism of the planets.

When the results of the Apollo Program were analyzed, the magnetometers carried on lunar satellites and placed on the surface by astronauts found the lunar magnetic field to be patchy; that is, nonuniform, more a montage of individual magnetic anomalies. The lunar rocks returned to earth revealed various degrees of remanent magnetism. They suggested the past presence of a strong general lunar magnetic field rather than the present weak, patchy field. Another discovery of potentially great import has been that the directions of magnetization of the lunar magnetic anomalies have three or four preferred alignments. The implication here is that ancient lunar poles wandered---assuming the moon once actually had poles.

One group of lunar scientists believes that the moon indeed did have magnetic poles at one time and that its strong magnetic field was generated internally by dynamo action, just like the earth's is today. Another faction favors impact magnetization by comets and perhaps other solar system objects. In impact magnetization the intrinsic magnetic field of a comet is strongly compressed upon impact and induced magnetism in the materials surrounding the impact site. Obviously, the story of lunar magnetism is far from finished.

ALZ1 Anomalous Features of Lunar Magnetism

Description. The presence on the moon of surface magnetic anomalies and remanent magnetism in lunar rock samples.

Background. Because the moon is small, of relatively low density, and spins very slowly, it was considered certain that no general lunar magnetic field existed. Spacecraft, such as Explorer 35, confirmed that the moon's magnetic dipole moment was indeed very tiny. The subsequent discovery of patchy surface magnetism and remanent magnetism in lunar rocks came as great surprises, for they implied the past presence of a fairly strong general magnetic field.

Data Evaluation. Lunar magnetic anomalies have been measured repeatedly by three different methods (see X2). Scientists have compiled many measurements of remanent magnetism in returned lunar samples, although there is some disagreement as to how much of this remanent magnetism is due to an intrinsic lunar field and how much to externally impressed fields. The consensus is that a rather large intrinsic lunar magnetic field did exist at one time. Rating: 1.

Anomaly Evaluation. The origin of the moon's general magnetic field and the surface concentrations of magnetism (Magcons) constitute two major lunar mysteries, although it is conceivable that a single theory could explain both. The major challenge posed by lunar magnetism is the strong implication that the moon once possesses a molten core which, through dynamo action, created a substantial magnetic field. Rating: 2.

Possible Explanations. General lunar magnetism might have been self-induced through dynamo action, or externally induced by a much closer earth, the solar magnetic field and/or the nebular magnetic field prevailing at solar system formation. The first explanation is considered the most likely. The local magnetic anomalies may be the consequence of: (1) the confinement of the moon's field to an outer shell that "leaks" at spots due to thermal action, meteor impact, or some other disturbance; or (2) induction by the magnetic fields accompanying comets or other colliding objects.

Similar and Related Phenomena. The unexpected magnetic field of Mercury (AHZ1); the general puzzle of planetary magnetic field generation, given the apparent inadequacy of the dynamo theory (EM).

Examples

X1. The general magnetic field. The magnetometer on the Explorer 35 lunar satellite measured the global magnetic dipole moment of the moon as less than 10^{20} gauss cm^3, implying a mean surface magnetic field of less than 2 gammas. (R9, R4, R5) In comparison, the earth's surface magnetic field is about 60,000 at the north magnetic pole. This negligible lunar magnetic field was just what scientists expected, because it was thought certain that the moon never had a core dynamo like the earth. However, the situation was complicated by field measurements at the surface and the magnetic properties of lunar rock samples (X2 and X3 below). (WRC)

X2. Local magnetic anomalies. "Local magnetic anomalies have been measured, however, by three independent techniques: (1) surface magnetometers showed values of 36 gamma at the Apollo 12 landing site; 43 gamma and 104 gamma at two positions near the Apollo 14 site; less than 5 gamma at the Apollo 15 site; 121, 125, 180, 231 and 313 gamma at various positions near the Apollo 16 site; (2) the magnetometers aboard the Apollo 15 and 16 subsatellites recorded anomalies as large as 1 gamma at a flight height of about 100 km apparently associated with large basins on the back side of the Moon; (3) limb compression effects detected by the Explorer 35 satellite reveal many magnetic sources most of which are located in the lunar highlands; more detailed results were obtained by the Apollo 15 subsatellite due to its lower flight height." (R9, R1, R4, R5, R16) See X4 for implications of these localized magnetic fields or magcons, as some have called them. (WRC)

X3. Magnetic properties of lunar samples. "From the magnetic viewpoint, three classes of material have been returned from the Moon, namely, mare basalts, breccias and soils. It has been well established that the mare basalts carry a remanent magnetisation which seems in part to be due to cooling in the presence of a field of a few thousand gamma. In general the basalts have a natural remanent magnetisation (NRM) of intensity about 10^{-5} e.m.u. g^{-1} consisting generally of a stable remanent magnetisation of about 2 x 10^{-6} e.m.u. g^{-1} and a larger soft component which may or may not be of lunar origin." Lunar rocks with such magnetic properties might account for some of the weaker surface fields measured by the magnetometers emplaced by the Apollo astronauts (X2); however, the anomaly at the Apollo 16 site of over 300 gamma would require a remanent magnetization of about 2 x 10^{-4} e.m.u. g^{-1}. Lunar rocks with this level of magnetization are rare. (R9, R3, R5, R6)

A caveat was introduced by C.E. Helsley: "Abstract. Partial thermal remanence experiments on lunar igneous rocks indicate that the magnetization of lunar rocks is not a normal single component thermoremanent magnetization. The magnetization therefore may not have been acquired at the time of initial cooling of the rock and thus should be used cautiously in making estimates of the intensity of the ancient lunar magnetic field." (R7)

X4. Potential explanations. Most scientists feel that the magnetic data reviewed above imply an ancient global magnetic field of several thousand gammas. There are three theories for the origin of this original field: (1) The moon was much closer to earth and was magnetized by the earth's strong field. However, tidal action would cause a moon near enough to the earth for magnetization to recede very quickly. It would not stay close enough to earth for a long enough time to be magnetized to its present level; (2) The moon might have been magnetized by the solar field, but this field seems too small; (3) The moon might have once possessed an internal dynamo and generated its own field. Today, though, the moon is thought to spin too slowly for dynamo action. However, we do not really know the moon's history---it may have spun more rapidly, especially it originated elsewhere and was captured by earth. (R5)

One possibility is that the moon's magnetism is confined to an outer shell. "The observations of the altitude dependence of the lunar magnetic field indicate that the lunar crust is magnetized to a depth of roughly 10-100 km, at least on the farside of the moon. On the other hand, there is no evidence for a global dipole moment as would be expected from a lunar dynamo, or from a uniform magnetization of the lunar crust imposed by an ancient external field. Runcorn has recently suggested that this dilemma can be resolved if the crust was magnetized by an ancient lunar dynamo so that the magnetization is proportional in strength and orientation to the ancient lunar field. Such a magnetized layer has no external field. In the framework of this model the magnetic features are due to changes in the crustal magnetization caused by variations in the mineral properties or due to erasure, say by local heating or impact. Large well-resolved features such as the one near Van de Graaff can then be used to infer the orientation of this ancient lunar dynamo. Using the Van de Graaff data, one obtains an ancient dipole moment in the moon's equator along the moon earth-line. This suggests that the lunar rotation axis may have once had a quite different selenographic position." This sort of model explains the lack of a general field as well as the localized concentrations. Runcorn later developed a theory of lunar polar wandering (ALZ4). (R12)

Magnetic theory predicts that a magnetic shell, with its magnetic components along the lines of force produced by a central dynamo, would have zero external magnetic field. (R13)

At one time, S.K. Runcorn proposed that the lunar dynamo was sustained by the energy of decay of superheavy elements. (R15) No traces of such superheavy elements have been found to date. (WRC)

Some of the magnetic anomalies may have been caused by cometary impacts, as amplified in ALZ3, where magnetic anomalies are correlated with swirl markings. (WRC)

References

R1. "Apollo 12's Magnetometer Indicates a Lunar Core," New Scientist, 46:103, 1970. (X2)

R2. Strangway, D.W., et al; "Magnetic Properties of Lunar Samples," Science, 167:691, 1970. (X3)

R3. Helsey, Charles E.; "Magnetic Properties of Lunar Dust and Rock Samples," Science, 167:693, 1970. (X3)

R4. Dyal, Palmer, et al; "Apollo 12 Magnetometer: Measurement of a Steady Magnetic Field on the Surface of the Moon," Science, 169:762, 1970. (X1, X2)

R5. Driscoll, Everly; "The Magnetic Moon: How Did It Get That Way?" Science News, 101:346, 1972. (X1-X4)

R6. Nagata, T., et al; "Lunar Rock Magnetism," The Moon, 4:160, 1972. (X3)

R7. Helsley, Charles E.; "The Significance of the Magnetism Observed in Lunar

Rocks," The Moon, 5:158, 1972. (X3)
R8. Hammond, Allen L.; "Lunar Research: No Agreement on Evolutionary Models," Science, 175:868, 1972. (X4)
R9. Strangway, D.W., et al; "Lunar Magnetic Anomalies and the Cayley Formation," Nature (Physical Science), 246: 112, 1973. (X1-X3)
R10. "The Moon Is Less Magnetic Than We Thought," New Scientist, 64:715, 1974. (X4)
R11. "Was the Early Moon a Permanent Magnet?" Nature (Physical Science), 243:57, 1973. (X4)
R12. Russell, C.T., et al; "On the Origin of the Ancient Lunar Magnetic Field," American Geophysical Union, Transactions, 56:387, 1975. (X4)
R13. "The Mystery of the Moon's Magnetism Deepens," New Scientist, 65:548, 1975. (X4)
R14. Runcorn, S.K.; "The Origin of Lunar Paleomagnetism," Nature, 275:430, 1978. (X4)
R15. Runcorn, S.K.; "The Ancient Lunar Core Dynamo," Science, 199:771, 1978. (X4)
R16. Runcorn, Keith; "The Moon's Deceptive Tranquillity," New Scientist, 96: 174, 1982. (X2, X4)
R17. Runcorn, S.K.; "Lunar Magnetism, Polar Displacements and Primeval Satellites in the Earth-Moon System," Nature, 304:589, 1983. (X4)

ALZ2 Anomalous Demagnetization Behavior of Lunar Samples

Description. Nonreproducible remanent magnetization values measured when demagnetizing lunar samples at a specific alternating field.

Data Evaluation. Reported to be a general property of lunar rock samples, but only a single report has been found so far. Rating: 2.

Anomaly Evaluation. The cause of the nonreproducibility is unknown, although one theory has been proposed below. Perhaps the most significant factor here is the potential undermining of the validity of the techniques used to determine the magnetic properties of lunar samples. Rating: 2.

Possible Explanations. The presence of multidomain "supergrains" in the minerals.

Similar and Related Phenomena. All magnetic measurements made on the moon and samples returned from it (ALZ).

Examples

X1. Experimental results. Abstract. "We present here an extended investigation of lunar olivine basalt sample 15535.28 which dramatically illustrates unorthodox alternating field (AF) demagnetization behavior as also observed to varying degrees in many lunar rocks. This behavior is characterized by nonreproducible remanent magnetization values upon demagnetization at a given peak AF. In addition, the direction of the remanence in 15535 following AF demagnetization is roughly confined to a particular plane. This behavior is shown to be an intrinsic property of certain magnetic carriers present in the sample.....It is suggested that the above behavior is due to the presence of a few planar, multidomain grains representing a local mineral fabric. These 'super-grains' do not demagnetize---neither by AF demagnetization to 1000 Oe nor by heating to moderate temperatures---and hence, if present in a given lunar sample, they could easily render any paleointensity determination of little value." (R1) Here is another caveat applying to all lunar magnetic measurements. (WRC)

References

R1. Hoffman, K.A., and Banerjee, S.J.; "Magnetic 'Zig-zag' Behavior in Lunar

Rocks," Earth and Planetary Science Letters, 25:331, 1975. (X1).

ALZ3 Correlations of Magnetic Anomalies with Surface Features

Description. The geographical correlation of lunar magnetic anomalies with impact basins and the dark-and-light swirl markings. In addition, some concentrations of swirl markings, which are also anomalously magnetic, are located approximately at the antipodes of large impact basins.

Background. The bright feature Reiner gamma was correlated early with an intense lunar magnetic anomaly. Reiner gamma also turns out to be one of the enigmatic swirl markings. At present the situation is difficult-to-decide; are the swirls basin ejecta, cometary debris, or seismic disturbances from antipodal impacts?

Data Evaluation. The data involved in these correlations come from spacecraft photographs of the moon's surface and indirect measurements of magnetic anomalies by the Apollo 16 subsatellite. All data seem to be of good quality. Rating: 1.

Anomaly Evaluation. Several unknowns are associated with this anomaly. Whether the swirls are cometary debris or basin ejecta, we are unsure as to how they acquired their magnetism. The possible role of the antipodal convergence of seismic waves from impact shocks is unexplored. The most significant implication of the proposed cometary origin of lunar magnetic anomalies is that all lunar magnetic anomalies might be the consequence of cometary or meteoric impacts, thus permitting lunar scientists to reject the hypothesis of core dynamo action. Rating: 2.

Possible Explanations. The swirls could be cometary debris or indigenous lunar materials magnetized by the impact compression of cometary magnetic fields. Impact shock waves could also converge on the lunar antipodes, disturbing the crust and causing leakage of magnetic flux from the postulated magnetic shell.

Similar and Related Phenomena. Mercury's swirl markings (AHE3); all other lunar magnetic anomalies (ALZ); effects of antipodal earthquakes on earth (GQG1).

Examples

X1. Reiner gamma correlation. "Abstract. Previously unmapped Apollo 16 subsatellite magnetometer data collected at low altitudes over the lunar near side are presented. Medium-amplitude magnetic anomalies exist over the Fra Mauro and Cayley Formations (primary and secondary basin ejecta emplaced 3.8 to 4.0 billion years ago) but are nearly absent over the maria and over the craters Copernicus, Kepler, and Reiner and their encircling ejecta mantles. The largest observed anomaly (radial component ~21 gammas at an altitude of 20 kilometers) is exactly correlated with a conspicuous light-colored deposit on western Oceanus Procellarum known as Reiner γ. Assuming that the Reiner γ deposit is the source body and estimating its maximum average thickness as 10 meters, a minimum mean magnetization level of $5.2 \pm 2.4 \times 10^{-2}$ electromagnetic units per gram, or $\sim$ 500 times the stable magnetization component of the most magnetic returned sample, is calculated. An age for its emplacement of $<$ 2.9 billion years is inferred from photogeologic evidence, implying that magnetization of lunar crustal materials must have continued for a period exceeding a billion years." (R4, R1, R3, R8) The strong Reiner gamma correlation led to much speculation about the origin(s) of magnetic anomalies. Were they due to magnetic

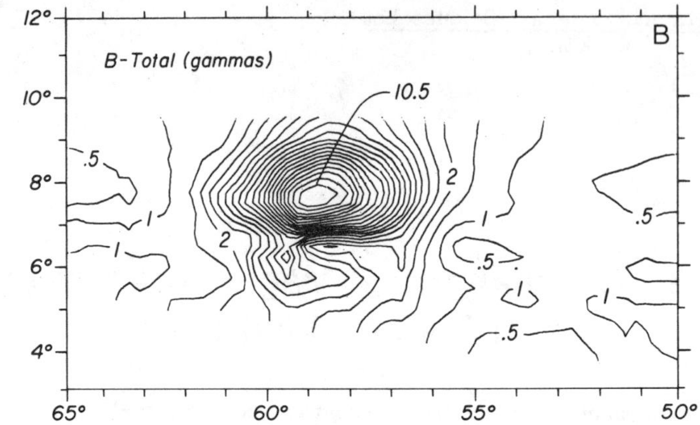

The Reiner gamma magnetic anomaly. A swirl marking is also located in the same area. (X1)

basin ejecta or perhaps induced by impacting comets? Reiner gamma also marks the location of a lunar swirl marking, as discussed in X3 below. (WRC)

X2. Correlation with basin ejecta. Abstract. "By the method of electron reflection, we have identified seven well-defined magnetized regions in the equatorial belt of the lunar far side sampled by the Apollo 16 Particles and Fields subsatellite. Most of these surface magnetic fields lie within one basin radius from the rim of a ringed impact basin, where thick deposits of basin ejecta are observed or inferred. The strongest of the seven magnetic features is linear, at least 250 km long, and radial to the Freundlich-Sharonov basin. The apparent correlation with basin ejecta suggests some form of impact origin for the observed permanently magnetized regions."(R2, R1)

X3. Correlation with lunar swirl markings. "Although cometary impacts have long been assumed for inner Solar System bodies such as the Moon and Mercury, there has been little evidence for distinguishing such an event from a meteoroid impact. We propose that the enigmatic bright and dark swirls which cross portions of the lunar farside may be best explained by an impacting comet complex. The strong magnetisation of at least one swirl (Reiner γ) and the close association of other swirls with lunar regions containing strong magnetic anomalies suggest that these features were magnetised by impact. We argue that the swirls are young deposits, implying a recent cometary impact ($< 10^8$ yr ago) and ruling out an active lunar dynamo as the source of their magnetising field. Rather, this field may have been of cometary origin, perhaps amplified during the impact." (R6, R5, R8)

The paper of Schultz and Srnka (R6) brough a response from L. L. Hood. "Although we agree that many of the swirls are probably relatively strongly magnetized deposits of impact-generated material, we consider that meteoroid impacts occurring over a span of time comparable to the age of the Solar System can explain their existence at least as well as can recent comet impacts." Hood notes that many medium-amplitude magnetic anomalies are well-correlated with basin ejecta. Furthermore, four concentrations of swirls are antipodal to impact basins, suggesting a secondary origin such as a convergence of seismic waves. These four regions are also anomalously magnetic. (R7)

Schultz and Srnka have answered Hood's critique is some detail, pointing out that the correlation of swirls and magnetic anomalies is intriguing but not really too convincing. Reiner gamma, for example, does not have any reasonable antipodal basin candidate. (R7)

References

R1. "Magnetic Deposit on the Moon," Eos, 60:28, 1979. (X1)

R2. Anderson, K. A., and Wilhelms, D. E.; "Correlation of Lunar Far-Side Magnetized Regions with Ringed Impact Basins," Earth and Planetary Science Letters, 46: 107, 1979. (X2)

R3. Hood, L. L.; "Sources of Lunar Crustal Magnetic Anomalies," Eos, 60:298, 1979. (X1)

R4. Hood, L. L., et al; "The Moon: Sources

of the Crustal Magnetic Anomalies," Science, 204:53, 1979. (X1)
R5. "Magnetism Is Patchy on the Moon," New Scientist, 85:925, 1980. (X3)
R6. Schultz, Peter H., and Srnka, Leonard J.; "Cometary Collisions on the Moon and Mercury," Nature, 284:22, 1980. (X3)
R7. Hood. L. L., et al; "Cometary Collisions on the Moon and Mercury," Nature, 287:86, 1980. (X3)
R8. Hood, L. L.; "The Enigma of Lunar Magnetism," Eos, 62:161, 1981. (X1-X3)

ALZ4 Correlated Directions of Magnetization of Lunar Magnetic Anomalies

Description. The tendency of the directions of magnetization of lunar magnetic anomalies to be aligned in a small number of preferred directions.

Data Evaluation. The directions of magnetization of the lunar anomalies have been derived from data from the Apollo 15 and 16 subsatellites. Rating: 1.

Anomaly Evaluation. The small number of preferred directions of magnetization for the lunar magnetic anomalies requires some sort of explanation. The major theory at hand, that of S.K. Runcorn, points to polar wandering caused by the impacts of large, ancient lunar satellites. Prevailing lunar theory has not yet assimilated Runcorn's hypothesis, which of necessity assumes an ancient fluid iron core. Impact magnetization is now preferred in many quarters. Rating: 2.

Possible Explanations. The preferred alignments of lunar magnetic anomalies may be due to the past existence of several different magnetic poles on the moon. Such polar wandering may have been caused by the impacts of decaying lunar satellites of large size, which caused the lunar crust to shift over the core.

Similar and Related Phenomena. Apparent polar wandering on the earth (EM) and Mars (AME13); alignments of impact basins (ALE23).

Examples

X1. Analysis of subsatellite data. Abstract. "Magnetism was discovered in the returned Apollo samples from the Moon and parts of the lunar crust were found to be magnetized from satellites. The origin of this magnetism has been a mystery because it is known that the Moon today, unlike the Earth, has no magnetic field. The strength of the magnetic field responsible for the magnetization 4×10^9 years ago is twice the present Earth's field and decayed exponentially to about 1/5th of this value at 3.2×10^9 years. From palaeomagnetic directions conclusive evidence has been obtained that the Moon had a magnetic field at least between 4.2×10^9 and 3.2×10^9 years ago and it is concluded that this was generated in a small iron core. Determinations by Coleman, Russell and Hood of the directions of magnetization of the crust from the Apollo 15 and 16 subsatellite magnetometers has been an important step forward. I find that the pole positions calculated from this data fall into three or four bipolar groups along different axes inclined to the present axis of rotation. The antipodal groupings suggest reversals of the lunar dynamo. I identify the different axes with different times of magnetization, i.e. Pre-Nectarian 4.2×10^9 years, Lower Nectarian 4.0×10^9 years, Upper Nectarian 3.9×10^9 years and Imbrian 3.85×10^9 years. I explain successive reorientations of the Moon with respect to its axis of rotation by the creation of the large basins which tend to move towards the pole. The striking association of the multi-ring basins with the palaeoequators of corresponding age excludes the possibility that they are high velocity impacts of asteroids in a heliocentric orbit and gives the first observational evidence that there were satellites around the Moon. The directions of the incoming body obtained from studies of asymmetry of the multi-ring basins supports this view." (R3, R1, R2, R4, R5)

References

R1. Runcorn, S. K.; "Lunar Polar Wandering," Eos, 62:818, 1981. (X1)

R2. Runcorn, Keith; "The Moon's Deceptive Tranquillity," New Scientist, 96:174, 1982. (X1)

R3. Runcorn, S. K.; "Lunar Palaeomagnetism, Polar Wandering and the Existence of Primeval Lunar Satellites," Meteoritics, 18: 389, 1983. (X1)

R4. "The Moon's Ancient Magnetism," Sky and Telescope, 65:506, 1983. (X1)

R5. Runcorn, S. K.; "Lunar Magnetism, Polar Displacements and Primeval Satellites in the Earth-Moon System," Nature, 304:589, 1983. (X1)

AM INTRODUCTION TO MARS

Key to Categories

AME MARTIAN GEOLOGICAL ANOMALIES
AMF LUMINOUS PHENOMENA ON MARS
AML THE CURIOUS SATELLITES OF MARS
AMO MARTIAN TELESCOPIC ANOMALIES
AMW ATMOSPHERIC PHENOMENA ON MARS

Of all the other planets that circle the sun, we know Mars the best. Though not as conspicuous in the night sky as Venus, Mars has no opaque atmosphere to hide its surface details from terrestrial telescopes and spacecraft cameras. Thanks to the Mariner and Viking spacecraft, we now have high-quality, close-up photos of the Martian surface. One would think that all the mysteries of Mars would have melted away under such scrutiny. But new mysteries have replaced some of the old ones, and many of the old enigmas persist down to the present day.

Martian mysteries fall into three broad categories: (1) permanently visible surface features; (2) transient Martian phenomena that resemble in some ways the transient lunar phenomena (TLPs) of ALF3 and ALF4; and (3) the rather odd Martian moons. Whether permanent or transient, the features and events seen on Mars emphasize that this is a dynamic planet with still-active atmospheric and geological processes, although the planet's pulse is admittedly rather slow. There have been and may still be in the future: glaciers, rivers, floods, volcanos, and quakes---almost the full geophysical repertoire. Then there are those spacecraft photos showing Martian phenomena with no terrestrial analogs, such as planet-wide dust storms, pedestal craters, and layered polar ice. There may not be, as Percival Lowell once asserted, a race of Martians maintaining a life-sustaining canal system of immense proportions, but Mars today is no less intriguing to the planetary scientist than it was to Lowell.

AME MARTIAN GEOLOGICAL ANOMALIES

Key to Phenomena

AME0 Introduction

The Mariner and Viking orbiters and landers sent back crisp, close-up photographs of the Martian surface that contrasted strongly with the fuzzy images seen through telescopes. Seen closely, the shadowy markings drawn by telescopic astronomers turned into craters, valleys, channels, volcanos, scarps, and a variety of other geological features---some earth-like, some not. In the last few years the whole scientific vision of Mars has changed.

As a whole, the planet, like the moon and Mercury, was found to possess hemispherical asymmetry; that is, its northern topology was quite different from that observed in the south. There is also evidence of polar wandering, with suggestions that the poles may once have been located near the present equator. Finally, there exists an immense bulge---the Tharsis bulge---that seems to affect the geology of almost one-third of the planet. Mars was obviously a very active planet in the past.

Another characteristic of Martian geology is the widespread evidence of water action, from the huge outflow channels and layered deposits of Valles Marineris to the erosion around the the pedestal craters. Mars, in fact, may still harbor considerable water in the form of permafrost and in liquid form in aquifers. The Solis Lacus region seems abnormally wet and may betoken the presence of water at or very near the surface. Many areas of chaotic terrain suggests surface collapse when subterreanean water gushed out, possibly when permafrost suddenly melted. If Mars really did once have lakes and extensive ice cover, it was a radically different planet from the arid object we see today.

For over a century, scientists have wondered about the possibility of life on Mars. Our close-up photography reveals no vegetation or anything we can construe as being alive. The life-detection experiments on the two Viking landers produced only inconclusive results. In the Martian soil there may be an agent that promotes the synthesis of organic compounds. The safest interpretation of these data is that the reactions are abiological. For this reason this subject resides in the geology section, there being no section on Martian biology---yet!

AME1 The Martian Channels

Description. Broad channels on the Martian surface that resemble terrestrial arroyos and Washington's Channelled Scablands. Superficially, these features seem to be eroded by water or some other fluid. One channel type is typically hundreds of kilometers long, ten wide, and originates in chaotic terrain. A second type is typically tens of kilometers long, one wide, and originates in networks flowing out of featureless, desert-like regions. Most planetologists believe the Martian channels were carved over a billion years ago.

Background. Even in Lowell's time, Mars was considered an extremely dry planet. The discovery by spacecraft of the impressive channels came as a great surprise. The presence on Mars of large quantities of water in the past and perhaps even today, locked up in permafrost and aquifers, has important consequences for the search for extraterrestrial life.

Data Evaluation. Large numbers of high-quality photographs of the Martian surface from the Mariner and Viking spacecraft. Rating: 1.

Anomaly Evaluation. Initially, the Martian channels were highly anomalous because they did not fit the model of Mars most scientists subscribed to prior to the acquisition of close-up spacecraft photos. Now, however, the model of Mars has been modified to include large reservoirs of water in the past and possibly the present. The original source(s) of this water and the triggering mechanism(s) that release it from its reservoirs are not known with any certainty. Although water and ice are considered the most likely eroding agents, no one knows for certain. Rating: 3.

Possible Explanations. The most popular theory is that the Martian channels were eroded by water and ice released from underground reservoirs (permafrost and aquifers) by such stimuli as subterranean thermal activity and meteor impacts. The water probably came from the minerals that originally formed Mars; viz., water of hydration; but one cannot exclude the infall of solar system ice in meteoric or cometary form.

Similar and Related Phenomena. The highly geometric and largely illusory Martian canals (AMO1); the lunar sinuous rilles (ALE2); terrestrial arroyos and Channelled Scablands (ET).

Examples

X1. General descriptions. Two of the many general portrayals of the Martian channels are quoted below. Others may be found in R2, R4, R5, R15, R16.

"The channels seem to be of two types. One type is large---hundreds of kilometres long, and typically ten kilometres wide---and originates in depressed areas of jumbled hills called chaotic terrain. Geomorphologists studying these large channels conclude that they resulted from a sudden catastrophic release of water, estimated from the channel structures to be from 10^6 to 10^8 cu. m per second. Release of this water from underground reservoirs caused collapse of the ground, explaining the chaotic terrain. Data from the Viking landers and Earth-based observations suggest that the Martian soil still contains this water, both as absorbed water in mineral grains and as a perma-frost layer of unknown depth. Several mechanisms have been proposed that might cause the rapid release of this ground water, including: (1) local heating by geothermal activity, (2) liquefaction caused by stresses that raise the pore pressure in the soil above the hydrostatic load pressure, (3) breakout from old buried layers of water-carrying rock (aquifers), due to impact or other disturbances. Although the slow, uniform erosion of river beds, as proposed by Hutton and Lyle, prevails on Earth in most regions, catastrophic erosion of channels by major floods is not uncommon in frozen or arid regions. For instance, a catastrophic release of dammed-up water that then flowed across the Channelled Scablands in Washington State produced features similar in scale and form to those found in the large channels of Mars.

This analysis of the large channels of Mars, contrary to our first impressions, may not require any radical revision of the Martian climate in early history. This many cubic kilometres of water is trapped in permafrost and occasionally breaks out; it simply gushes out across the landscape of loose dust and rubble, eroding a channel until it empties into a plain where it simply evaporates. Mars' climate is so cold now that ice would tend to form and perhaps protect the water from too rapid evaporation or spontaneous boiling; broken ice slabs in parts of the channel would, in fact, aid the erosion, according to terrestrial observations. On the other hand, an ancient, warmer Mars, with higher air pressure, would make this scenario more believable. A warmer Mars would keep the water liquid and more pressure---such as 3000 Pa to 10 000 Pa or more---would reduce the evaporation rate, allowing the channels to act more like flash floods on Earth.

The second type of channel on Mars is small, typically tens of kilometres long and up to a kilometre wide, forming dendritic networks that present a different problem.

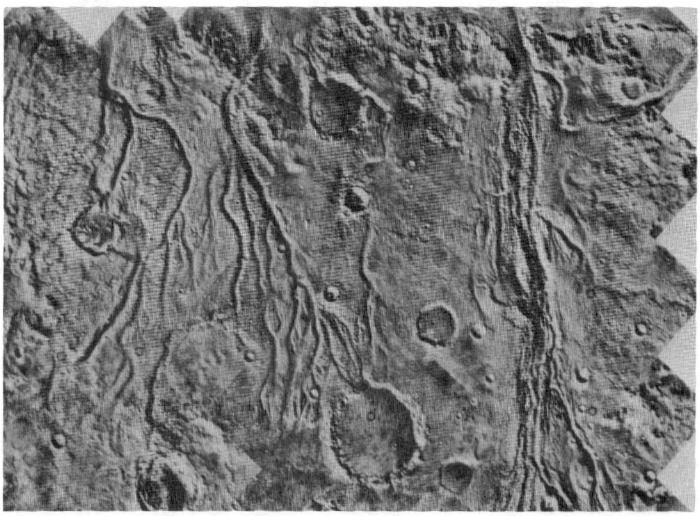

Martian channels just west of Chryse Planitia photographed from Viking Orbiter 1. (X1) (Courtesy NASA)

These appear exclusively in older crater terrain, suggesting that they formed some 10^9 years ago, but not within the past 10^8 to 10^9 years. They do not emanate from collapsed chaotic terrain, but seem to form networks flowing out of featureless deserts. Following the discovery of the smaller channels in Mariner 9's photographs of 1972, some researchers suggested rainfall as a mechanism for creating these features, implying, of course, a much denser and wetter atmosphere as well as a warmer climate. However, recent studies of these features suggest that they may grow primarily by sapping---a process of headward erosion in which the dendritic pattern fans outward from the original source. Again, this could involve a permafrost layer. If a fault or some other disturbance exposes buried ice which then melts and flows away, the sapping process will gradually eat upstream into the terrain creating the channelled features. Although one could imagine this process working in an environment almost like that on Mars today, erosion by flowing water clearly seems to indicate a warmer climate with a higher atmospheric pressure some 10^9 years ago when these channels were forming." (R11)

Here follows an abstract of a more technical paper on the channels. "All Martian channels and valleys visible at a resolution of 125 to 300 meters between 65°N and 65° S were mapped at a scale of 1:5,000,000 and the maps then digitized. Correlations of valley presence with other surface features show that almost all the valleys are in the old cratered terrain, preferentially in areas of low albedo, low violet/red ratios. and high elevation. The networks are open, the individual drainage basins are small relative to Earth, and large distances separate the basins, features which all suggest an immature drainage system. The simplest explanation of the correlations and the restriction of valley networks to old terrain is that the channels themselves are old, and that the climatic conditions necessary for their formation did not prevail for long after the decline in the cratering rate around 3.9 billion years ago. Two types of outflow channel are distinguished: unconfined, in which broad swaths of terrain are scoured, and confined, in which flow is restricted to discrete channels. The outflow channels have a wide range of ages and may form under present climatic conditions. Fretted channels are largely restricted to two latitude belts centered on 40°N and 40°S, where relatively rapid erosion along escarpments results from mass wasting. They probably form by enlargement of preexisting channels by escarpment retreat." (R12)

X2. Possible eroding agents. Most students of Mars prefer scenarios based on water and ice as eroding agents, as in the descriptions in X1. There are many studies supporting this viewpoint, including analyses of permafrost, aquifers, evaporation of ice, the erosive power of ice streams, and climatic changes that might induce the release of water. (R1, R2, R4, R9-R11, R14-R17) One rather unusual speculation has the Martian water tied up in carbon dioxide hydrate, $CO_2 \cdot 6H_2O$. (R3)

In addition to water/ice, the following agents have been proposed: hydrocarbons (R7, R8); lava (R18); and wind-and-sand (R6, R13)

References

R1. Sagan, Carl, et al; "Climatic Change on Mars," Science, 181:1045, 1973. (X1, X2)

R2. Hartmann, William K.; "Geological Observations of Martian Arroyos," Journal of Geophysical Research, 79: 3951, 1974. (X1, X2)

R3. Milton, Daniel J.; "Carbon Dioxide Hydrate and Floods on Mars," Science, 183:654, 1974. (X2)

R4. Sharp, Robert P., and Malin, Michael C.; "Channels on Mars," Geological Society of America, Bulletin, 86:593, 1975. (X1, X2)

R5. Malin, Michael C.; "Age of Martian Channels," Journal of Geophysical Research, 81:4825, 1976. (X1)

R6. "Wet Worlds and Dry," Science News, 113:198, 1978. (X2)

R7. Yung, Y. L., and Pinto, J. P.; "Primitive Atmosphere and Implications for the Formation of Channels on Mars," Nature, 273:730, 1978. (X2)

R8. "Martian Surface in Good Spirits," New Scientist, 79:19, 1978. (X2)

R9. Wallace, David, and Sagan, Carl; "Evaporation of Ice in Planetary Atmospheres: Ice-Covered Rivers on Mars," Icarus, 39: 385, 1979. (X2)

R10. Carr, Michael H.; "Formation of Martian Flood Features by Release of Water from Confined Aquifers," Journal of Geophysical Research, 84:2995, 1979. (X2)

R11. Hartmann, William; "The Watery Past of Mars," New Scientist, 82:1083, 1979. (X1, X2)

R12. Carr, Michael H., and Clow, Gary D.;

Some channels are associated with collapse features like those shown in the above photo. (X2) (Courtesy JPL)

"Martian Channels and Valleys: Their Characteristics, Distribution, and Age," Icarus, 48:91, 1981. (X1)

R13. Cutts, James A., and Blasius, Karl R.; "Origin of Martian Outflow Channels: The Eolian Hypothesis," Journal of Geophysical Research, 86:5075, 1981. (X2)

R14. Lucchitta, Baerbel K., et al; "Did Ice Streams Carve Martian Outflow Channels?" Nature, 290:759, 1981. (X2)

R15. Mars Channel Working Group; "Channels and Valleys on Mars," Geological Society of America, Bulletin, 94:1035, 1983. (X1, X2)

R16. Carr, Michael H.; "The Surface of Mars: A Post-Viking View," Mercury, 12:2, January/February 1983. (X1, X2)

R17. Nedell, Susan S., and Squyres, Steven W.; "Geology of the Layered Deposits in the Valles Marineris, Mars," Eos, 65:979, 1984. (X2)

R18. Baird, Alex K., Clark, Benton C.; "Did Komatiitic Lavas Erode Channels on Mars?" Nature, 311:18, 1984. (X2)

AME2 Systems of Lineaments on Mars

Description. Systems of lineaments that are aligned in the same directions over wide areas of the Martian surface. The lineaments may be ridges, valleys, channels, the walls of polygonal craters, sand dunes---in short anything with a linear appearance.

Data Evaluation. The Mariner and Viking photographs of the Martian surface are employed in lineament analysis. Rating: 1.

Anomaly Evaluation. Regional systems of lines, even though composed of different kinds of geological features, tend to reveal basic crustal jointing patterns and past tectonic activity. Lineaments themselves are not anomalous, being found on many planet-size objects. The anomaly in this instance is our ignorance of the precise origins of the several systems of Martian lineaments. Rating: 3.

Possible Explanations. Cooling and contraction of the Martian crust; the drying of wet sediments; the cooling of lava flows; the cracking of the crust due to the stresses from spin deceleration and/or polar wandering; the cracking due to the impact of large meteors and other objects. See also X5.

Similar and Related Phenomena. The famous Martian canals (AMO1), which have been seen by some astronomers through the telescope and which are poorly correlated with the lineaments seen in close-up photographs; naked-eye lineaments seen on the moon (ALO4).

Examples.

X1. Global system of lineaments. The photographs from the Mariner 4, 6, and 7 flights were analyzed for linear features, such as polygonal crater walls, linear rilles, linear ridges, linear albedo boundaries, and linear scarps. When these features are plotted, they demonstrate the existence of a well-developed, planet-wide system of lineaments. This system of fractures might be the consequence of changes in the planet's rate of rotation, polar wandering, or similar stresses. (R1)

X2. Radial and concentric lineaments. The analysis employed in X1 also led to the discovery of two systems of radial and concentric lineaments; one associated with the Hellas basin, the other with the south polar basin. (R1)

X3. Networks of dark filamentary markings. Abstract. "Some Mariner 9 B-frames show networks of criss-crossing rectilinear albedo markings typically 10 km long by 100 m wide. This paper discusses the location, variability and possible nature of these dark filamentary markings. Although not common on Mars, the markings are concentrated in at least two areas: Depressio Hellespontica and Cerberus/Trivium Charontis. In Depressio Hellespontica their emergence coincided with a general darkening of the region which occurred 2 months after the end of the 1971 global dust storm. The very definite criss-cross pattern of many of these markings suggests that they may be controlled by joints. It is unlikely that the markings are linear dunes." (R3)

X4. Polygonal trough systems. "Polygonally fractured terrain in the northern plains of Mars was first seen on the Mariner 9 B-frames. However, the extent and detailed morphology of this terrain was not known until high resolution Viking photographs of the northern latitudes were obtained. These photos revealed extensive systems of crudely polygonal troughs hundreds of meters wide, with average spacings of 5 to 10 km. Most of the troughs are concentrated in three areas centered at 45°N, 15°W (southeastern Acidalia Planitia), 36°N, 256°W (northwestern Elysium Planitia), and 49°N, 233°W (Utopia Planitia, near the Viking 2 landing site." These polygons are probably due to crustal cracking in response to tension, which may have been generated by cooling and contraction, or the drying of water-saturated sediments, or the cooling of lava. (R5, R4) The sides of the polygons naturally contribute to the global lineament studies, like X1. (WRC)

X5. General observations. "Surficial linear features of Mars seem to fall into four generalized categories. Using a width to length ratio of 1:10 or smaller for the surface features, and making analogies based on earth and lunar features, the four categories are as follows: (1) Uplifted linear features--ridges; (2) Depressed linear features--graben-like valleys or canyons and rilles; (3) Superposed linear features--mainly erosional/depositional; (4) Arcuate linear features--alignment and superposition of crater rims. Origin of the features appears to be quite diverse. Among possibilities are: (a) Aeolian processes due to wind and dust storms; (b) volcanic processes including fissure eruptions; (c) Former fluvial processes; (d) Superposition of meteoritic impact crater rims; (e) Grabens and horsts resulting from rifting of Martian crust; (f) Folding of Martian crust; (g) Differential melting of water ice and CO_2 ice which are possibly rich in fine sand and silt; (h) Moraines related to ice lobes that are advancing toward equatorial regions." (R2)

References

R1. Binder, Alan B., and McCarthy, Donald W., Jr.; "Mars: The Lineament Systems," Science, 176:279, 1972. (X1, X2)

R2. Ablordeppey, Victor K., and Gipson, Mack, Jr.; "Preliminary Classification of Linear Surface Features of Mars," Eos, 54:347, 1973. (X5)

R3. Veverka, J.; "Variable Features on Mars. VII. Dark Filamentary Markings on Mars," Icarus, 27:495, 1976. (X3)

R4. Pechmann, James; "The Origin of Polygonal Trough Systems on Mars," Eos, 58:1182, 1977. (X4)

R5. Pechmann, James C.; "The Origin of Polygonal Troughs on the Northern Plains of Mars," Icarus, 42:185, 1980. (X4)

AME3 Pyramidal Structures

Description. Geometrically sculpted pyramidal land forms on the Martian surface, with triangular and polygonal bases. The bases are typically several kilometers across.

Background. The Martian pyramids have been seized upon by some as evidence of intelligent life forms on Mars.

Data Evaluation. The pyramidal structures can be discerned on close-up spacecraft photos, particularly by the geometrical shadows they cast. Rating: 1.

Anomaly Evaluation. The explanation of such regular geometrical features is not simple, but wind-faceted mountains do exist on earth, so that at least one reasonable geological process shows some promise. However, the regularity of all sides of the Martian pyramids and their confinement to one small area are very puzzling. Rating: 2.

Possible Explanations. Wind/sand erosion.

Similar and Related Phenomena. Wind-faceted mountains in Peru.

Examples

X1. Permanent surface features. Abstract. "Triangular and polygonal pyramid-like structures have been observed on the Martial surface. Located in the east central portion of Elysium Quadrangle (MC-15), these features are visible on the Mariner 9 photographs, B-frames MTVS 4205-3 DAS 07794853 and MTVS 4296-24 DAS 12985882. The structures cast triangular and polygonal shadows. Steep-sided volcanic cones and impact craters occur only a few kilometers away. The mean diameter of the triangular pyramidal structures at the base is approximately 3.0 km, and the mean diameter of the polygonal structures is approximately 6.0 km. The observed Martian structures tend to line up suggesting joint or fault control. However, they do not appear to be controlled by the visible faults. The structures appear to be either wind-faceted volcanic cones and blocks or solidified blocks which have been rotated in semiconsolidated lava." (R1)

References

R1. Gipson, Mack, Jr., and Ablordeppey, Victor K.; "Pyramidal Structures on Mars," Icarus, 22:197, 1974. (X1)

AME4 Ice Layering in the North Polar Regions

Description. Discrete, thin ice layers extending over several thousand square kilometers in the north polar region of Mars. The layers are surprisingly constant in thickness laterally and from layer to layer.

Background. Rhythmic layering has also been observed in the Valles Marineris region, but it is impossible at present to associate this with the polar ice layering.

Data Evaluation. Viking 2 orbiter photographs display the layering clearly. Rating: 1.

Anomaly Evaluation. We can only speculate about the source of the water and the periodic stimulus. The regularity of this feature betokens a series of climatic pulsations of unknown origin. Rating: 2.

Possible Explanations. Climatic variations, perhaps augmented by periodic dust storms. The regularity of the layering may be a consequence of periodic orbital variations. Another possibility is the periodic escaping of water from reservoirs, possibly associated with periodic bombardment by comets, to which some attribute periodic terrestrial extinctions of life.

Similar and Related Phenomena. Rhythmic layering of sediments on Mars (AME19); the episodic, planet-wide dust storms on Mars.

Examples

X1. Permanent surface features. "The layered deposits contribute more than any other feature to the geological distinctiveness and significance of the polar regions of Mars. We suspect that these layered deposits are exposed at the surface in all areas mapped in Fig. 1 (not reproduced) where frost is absent. They are manifested on slopes by a parallel striping of the surface; however, in only a fairly small number of locations can the topographic and geological nature of the exposures be clearly discerned. In one of these areas, selected so that shading and shadowing dominates locally over brightness variations caused by albedo, a terraced slope can be recognized. With the use of arguments presented previously for features viewed in the south polar region, a strong case can be made that erosion of a layered deposit accounts for the origin of the terraced slope. The lateral continuity of terraces and their uniformity in height suggest that the individual layers that gave rise to the terraces extend as continuous thin sheets over areas of several thousand square kilometers. Moreover, these individual sheets are not only quite constant in thickness, but the thicknesses of different sheets are rather similar to one another." Rhythmic climatic changes, perhaps involving dust deposition, may account for the accumulation of these uniformly layered structures.

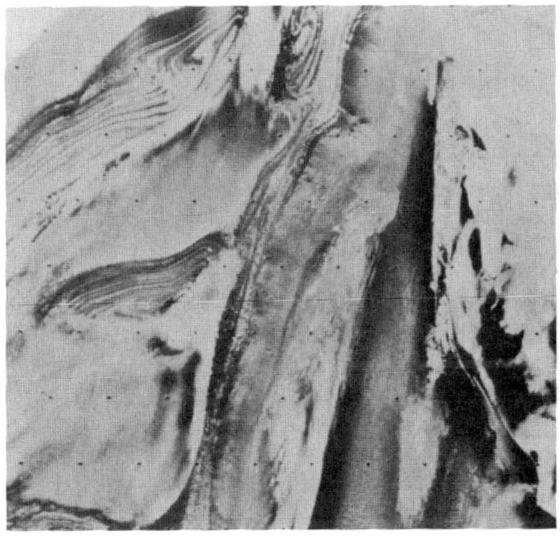

Layered deposits in the north polar region of Mars. The regular banding may indicate cyclic climatic conditions. Photo area is 65 by 100 kilometers. (X1) (Courtesy NASA)

References

R1. Cutts, James A., et al; "North Polar Region of Mars: Imaging Results from Viking 2," Science, 194, 1329, 1976. (X1)

AME5 The "Searchlight" Areas

Description. Long bright patches with straight parallel or slightly diverging sides found in the north polar region. These features may maintain their linear trends for over 100 kilometers irrespective of local topography. The searchlight areas seem translucent because albedo features can be seen through them.

Data Evaluation. Photographs taken by the Viking Orbiters. Rating: 1.

Anomaly Evaluation. Ice sheets are to be expected in polar regions, but the Martian searchlight areas are so large, straight, and unperturbed by local geology that some special explanation seems in order. Unfortunately, the phenomenon is so bizarre that no convincing theories are at hand. Rating: 2.

Possible Explanations. The searchlight areas seem to be thin sheets of translucent ice, but there seem to be no explanations of their origin, size, and geometry.

Similar and Related Phenomena. None.

Examples

X1. Permanent polar features. "One class of phenomena is informally known as 'searchlight patterns'. These bright elongate patches are bounded by abrupt margins that sustain a linear trend for 100 km or more, maintaining a parallel or slightly diverging aspect. Wind erosion or deposition seems an obvious explanation for these features, although the exact mechanism is unknown. Linear dark markings also form angular patterns on the ice; Fig. 5a includes an example of two almost identical patterns separated by about 3 km. We have considered the possibility that lateral motion of the upper portion of a single feature caused this apparent replication, but insufficient information is known about the topography to pursue this idea in any detail." It is also puzzling that the searchlight features transect major topographical features without changing direction. (R1) The searchlight features seem like translucent covers of very thin ice, for albedo features can be seen through them. (R2)

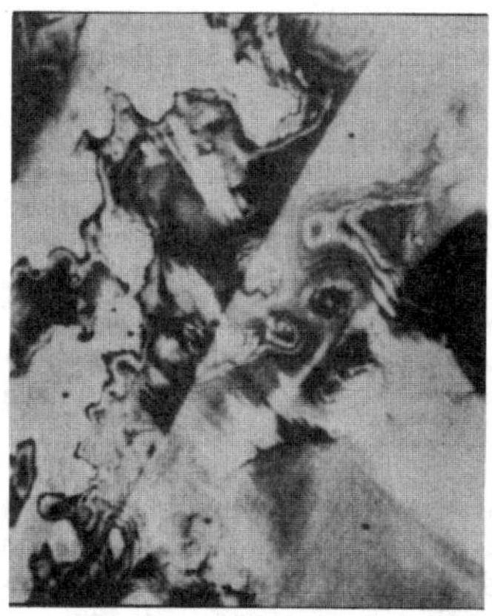

A "searchlight" area in the north polar region. (X1) (Courtesy NASA)

References

R1. Cutts, James A., et al; "North Polar Region of Mars: Imaging Results from Viking 2," Science, 194:1329, 1976. (X1)

R2. "Viking: Polar Dunes and Captured Moons," Science News, 110:276, 1976. (X1)

AME6 Evidence for Surface and Subsurface Ice

Description. Martian geological features that resemble those found in terrestrial polar and glaciated regions. Permafrost action seems widespread, and some features of the Martian channels seem ice-worked.

Data Evaluation. The Mariner and Viking spacecraft photographs are excellent, but it should be pointed out that claims of ice action are only implied by analogy with terrestrial features. Mars is an alien planet, and alien geological forces may be at work. Rating: 1.

Anomaly Evaluation. The geological evidence for widespread ice action on Mars implies a large inventory of water---much more than can be accounted for in today's atmosphere and past escape to space. Current thinking places this water under the Martian surface. Actually, the existence of this water is in accord with the present theory of Mars' origin, so no real anomaly exists. Mars should have water, and it does---underground. In addition, the occasional release of some of this water, its freezing, and subsequent erosion of landforms is quite understandable. Rating: 4.

Possible Explanations. Much of Mars' water inventory, as estimated from the prevailing theory of the planet's formation, could still be underground as ice and permafrost and in aquifers in liquid form. Martian thermal activity, climatic changes, and meteor impacts could cause temporary melting and release of some of the water. Collapse features could result. The freed water could freeze and create temporary ice rivers and glaciers.

Similar and Related Phenomena. Polar-like features near the present Martian equator (AME11); the Martian channels (AME1), which show evidence of ice erosion; apparent ice-formed features on the moon (ALE4).

Examples

X1. Evidence for permafrost. Abstract. "The outgassing history of Mars and the prevailing temperature conditions suggest that ground ice may occur to depths of kilometers over large areas of the planet. The presence of permafrost is also indicated by several topographic features that resemble those found in periglacial regions of the earth. East of Hellas and in the Protonilus and Nilosyrtis regions there are features that resemble those formed on earth by gelifluction, the slow creep of near-surface materials aided by freeze-thaw of ground ice. In the south part of Chryse Planitia there are irregular depressions that resemble thermokarst features, and the pattern of tributaries to the equatorial canyons is suggestive of a sapping process that would result from the melting of ground ice. The morphology of ejecta around fresh Martian impact craters is distinctively different from that around lunar and Mercurian craters. Such differences could be ascribed to the presence of ground ice in the target materials. The convergence of these different observations supports permafrost conditions not only at present but also for much of the planet's history." (R1)

"The observed features that are most likely to reflect ground ice are thermokarst-like pits and debris flows. Landforms with ambivalent origins include polygonally patterned ground, lobate ejecta blankets, craters with central pits, and curvilinear features. The most persuasive morphological evidence for ground ice is thermokarst pits and debris flows; the thermokarst pits are primarily located in the volcanic regions of Tharsis and Elysium. The association of ice-related features with these volcanic areas suggests that these forms are not directly latitude dependent. Activation by orbital variations could produce periodic, multiple episodes of melting that are dependent upon latitude. The presence of ice-related features in both hemispheres and the equatorial region of Mars indicates that ground ice may be---or have been---present over the entire planet, as predicted by the cryosphere model." The authors also remark that about 90% of the planet's original inventory of water cannot be accounted for unless one assumes it is underground. (R2, R4)

X2. Evidence for surface ice sculpturing. Abstract. "Many landforms in Martian outflow channels have characteristics that suggest sculpture by glaciers, ice streams, or ice sheets. Viking Orbiter and terrestrial satellite images were examined at similar resolution to compare features of the Martian outflow channels to features produced by the movement of ice on earth. Many resemblances were found. They include the anastomoses, sinuosities, and U-shaped cross profiles of valleys; hanging valleys; and linear scour marks on valley walls, grooves and ridges on valley floors, and streamlining of bedrock highs. The question of whether ice could have moved in the Martian environment is investigated." The author concludes that episodic, temporary masses of ice or slush could have formed in the Martian channels and caused the sculpturing seen on the Viking photos. (R3)

References

R1. Carr, Michael H., and Schaber, Gerald G.; "Martian Permafrost Features," Journal of Geophysical Research, 82:4039, 1977.

R2. Rossbacher, Lisa A., and Judson, Sheldon; "Ground Ice on Mars: Inventory, Distribution, and Resulting Landforms," Icarus, 45:39, 1981. (X1)

R3. Lucchitta, Baerbel K.; "Ice Sculpture in the Martian Outflow Channels," Journal of Geophysical Research, 87:9951, 1982. (X2)

R4. Squyres, Steven W., and Carr, Michael H.; "The Distribution of Ground Ice Features on Mars," Eos, 65:979, 1984. (X1)

AME7 Anomalous Hillocks and Ridges

Description. An immense number of hillocks, 100-500 meters across, and a system of parallel ridges, hundreds of kilometers long, on the northwestern flank of the Martian volcano Arsia Mons.

Data Evaluation. Mariner and Viking spacecraft photographs. Rating: 1.

Anomaly Evaluation. The hillocks and ridges do not seem to have any terrestrial parallels, and no convincing explanation(s) are at hand. Rating: 2.

Possible Explanations. The hillocks and ridges might be landslide features.

Similar and Related Phenomena. The famous Mima Mounds on earth occur in large numbers in some localities but are very much smaller than the Martian hillocks.

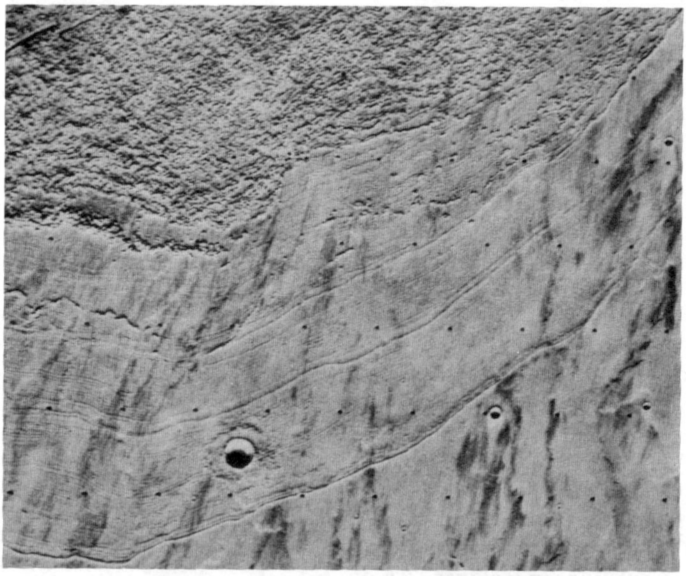

Puzzling hillocks and parallel ridges near the Martian volcano Arsia Mons. (X1) (Courtesy NASA)

Examples

X1. Permanent surface features. "No one knows what chain of events led to the strange appearance of this northwestward extension of the flanks of the huge volcano Arsia Mons in the Tharsis uplands. Countless hillocks, mostly 100 to 500 meters across, cover the flank's edge, which is surrounded by parallel ridges that run for hundreds of kilometers, undisturbed by craters, flow features or even variations in surface brightness. One hypothesis is that the hillocks were formed by a huge landslide, perhaps assisted by gravity since the surrounding plains slope downward 0.5° to the northwest. (Ash-flow deposits are deemed unlikely, due to the lack of signs that any of the material was blasted into place from vents in the volcano.) The ridges may be folds or 'reverse faults' caused by the drag of the landslide over the underlying terrain, which would transmit an outward pressure perhaps capable of passing beneath surface features." (R1)

References

R1. "Strange Hillocks and Ridges on Mars," Science News, 113:43, 1978. (X1)

AME8 Martian Surface Asymmetry

Description. The north-south hemispherical asymmetry of Martian surface features. The southern hemisphere is old, heavily cratered, and at a higher elevation than the northern hemisphere, which is young and lightly cratered. The average difference of elevation between the two hemispheres is about 3 kilometers.

Data Evaluation. Mariner and Viking spacecraft photographs. Rating: 1.

Anomaly Evaluation. The fact that the surfaces of all large objects in the inner solar system, except for cloud-shrouded Venus, are hemispherically asymmetrical is unquestionably one of the great enigmas of the solar system. On the moon and Mars, the asymmetry is not restricted to crater populations; rather the average differences in elevation amount to more than a kilometer. It is not known whether these dichotomies are the consequence of a short-lived episode of bombardment, the result of internal planetary forces, or due to some catastrophic event akin to earth-moon fission. Rating: 2.

Possible Explanations. The impacts of very large meteors; large-scale thinning of the northern hemisphere crust by internally generated forces, perhaps related to those associated with terrestrial continental drift; planetary fission, such as proposed by some for the earth-moon system.

Similar and Related Phenomena. The hemispherical asymmetries of the moon (ALE1) and Mercury (AHE1); the earth's Pacific Basin, which some propose was the scar left behind after earth-moon fission.

Examples

X1. Permanent surface features. Abstract. "Mars is divided into two fundamentally different geological provinces of approximately hemispherical extent. The more southerly province is heavily cratered, contains relatively old geological units, and superficially resembles the lunar and Mercurian highlands. The northern province is relatively lightly cratered and contains younger geological units, including extensive plains, volcanic edifices, and volcanic calderas. Except for the Tharsis and Elysium regions and other large volcanoes, most of the younger, northern province consists of lowlands, which lie an average of 3 km below the highlands. Lowlands occupy about one-third of Mars. They are separated from the highlands by a distinct scarp or by a sloping transitional zone as much as 700 km wide in which highland materials have been disrupted and partly replaced by lowland deposits. The transition is characterized by a variety of landforms unknown on other planets. The highlands and lowlands are in isostatic equilibrium across the transitional zone. No generally accepted explanation for the cause of the highland-lowland dichotomy has been proposed although thinning of the lithosphere in the northern hemisphere by internal processes has been suggested. Large impacts, which were a feature of early Solar System history, have also been suggested as an important cause of such early planetary-surface heteorgeneities. Subcircular lowlands such as Argyre, Hellas, and Isidis Planitia and the plains south of Tharsis occupy previously recognized impact basins. We propose here that the largest expanse of lowlands also results from formation of a large impact basin early in Mars' history that has profoundly shaped the planet's surface." (R3) Similar descriptions of the planet's dichotomy are to be found in R1 and R2.

References

R1. "The Mystery of the Hemispheres," Science News, 105:241, 1974. (X1)

R2. Metz, William D.; "Update on Mars: Clues about the Early Solar System," Science, 183:187, 1974. (X1)

R3. Wilhelms, Don E., and Squyres, Steven W.; "The Martian Hemispheric Dichotomy May Be Due to a Giant Impact," Nature, 309:138, 1984. (X1)

AME9 The White Rock

Description. A 14 x 18 kilometer white feature inside a crater near the Martian equator. Its composition is unknown, although it is always referred to as a "rock". It is the only such feature found on Mars so far.

Data Evaluation. Viking spacecraft photographs. Rating: 1.

Anomaly Evaluation. Since the true composition of the "rock" is not available, it is difficult to assess its anomalousness. If it were salt or ice, it would represent a serious anomaly; but we can probably assume that it is only an "ordinary" white rock of some kind. Rating: 3.

Possible Explanations. Since the white rock is situated inside a crater, it may be crustal rock exposed by the excavation; however, it is strange that no similar situations have been found. Even though the site of the white rock is near the equator, an exposure of subsurface ice should not be ruled out. Several widespread terrestrial rocks are white: marble, limestone, chalk, salt, etc.; but the conditions necessary (i.e., life) for their production on Mars probably never existed.

Similar and Related Phenomena. None.

Examples

X1. Permanent feature. "This strange, white, horseshoe crab-shaped feature found inside a Martian crater is to be studied by scientists working with the NASA Viking mission. The feature, nicknamed 'white rock', is the only one of its kind on Mars. Its composition is unknown, but it is unlikely to be either ice or snow because it lies near the equator. The 'white rock' is about 18 km long and 14 km wide, and is at the bottom of a crater located at 8° South, 335° West." (R1-R3)

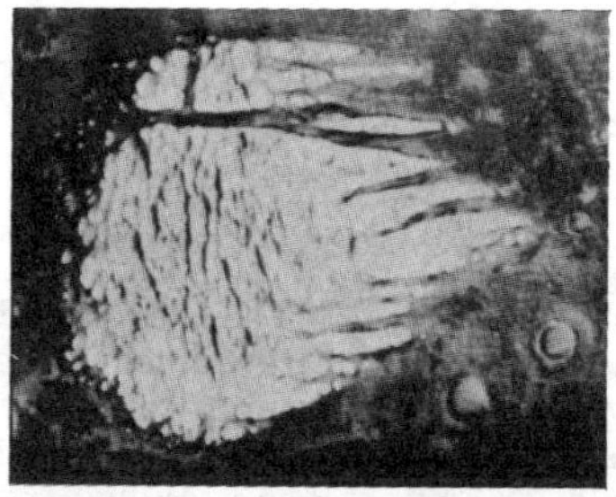

The white rock, as photographed by Viking Orbiters. (X1) (Courtesy JPL)

References

R1. "NASA Is to Investigate Martian 'White Rock'," Nature, 276:434, 1978. (X1)
R2. "Coming: The Daily Planet," Science News, 114:323, 1978. (X1)
R3. "A Martian Mystery," Astronomy, 7:64, January 1979. (X1)

AME10 Anomalous Frost on the Martian Surface

Description. The winter appearance of a thin coating of water-frost on the Martian surface in the absence of preceding, planet-wide dust storms. The first appearance of frost on Mars followed a huge dust storm; and it was theorized then that such storms were necessary to provide sufficient nuclei in the atmosphere.

Data Evaluation. Photographs by the Viking 2 Lander show obvious frost coatings during the winter seasons. Rating: 1.

Anomaly Evaluation. The Viking observation of frost formation without a preceding dust storm implies either that there is normally sufficient dust in the atmosphere for frost formation, or that other suitable nuclei are present, or that frost can form directly on the surface. None of these possibilities is particularly anomalous. Only minor shifts in viewpoints would be required. Rating: 3.

Possible Explanations. The prevailing theory is that water and carbon-dioxide ice crystallize on dust particles in the atmosphere. Thus burdened, the particles fall to the surface, where the carbon dioxide sublimes, leaving behind dust and water ice. The fall of true snow might also be possible. Since frost formation occurs both with and without preceding dust storms, the storms are evidently not necessary in the production of atmospheric nuclei. Frost formation directly on the surface---as seen on earth---is another possibility.

Similar and Related Phenomena. Terrestrial snowfall without clouds, which is not uncommon in very high latitudes (GWP1). Terrestrial clouds usually contain many more ice crystals than dust nuclei, indicating that on earth, too, dust may not be the primary nucleating agent (GWC15).

Examples

X1. Apparent annual occurrence. Photos from the Viking 2 Lander. "Lander 2's cameras have revealed a new layer of water frost on the Martian surface at the Utopia Plains landing site. It is Martian winter again, and a thin layer of frost can easily be seen in the photos. The new frost layer poses a scientific puzzle to members of the Viking team: in September 1977, Viking Lander 2 found frost on the surface during the Martian northern winter. (That was 1 Martian year or almost 2 earth years ago.) Scientists associated that frost collection with a major dust storm that had obscured the planet's surface before and during that period. But recent observations have shown no dust storms on Mars this year---in fact, the atmosphere is clearer than scientists have seen it since Viking arrived in 1976. So no one is certain just what triggers the appearance of frost." (R1) The dust particles had been considered essential to the formation of frost, but the rate of frost formation was the same with and without the dust storm. (R2)

References

R1. "Viking, Three Years Later," Eos, 60:635, 1979. (X1)
R2. Eberhart, Jonathan; "Mars: The Wet Look," Science News, 116:108, 1979. (X1)

AME11 Polar Features near the Equator

Description. The occurrence near the present Martian equator of geological features typical of the Martian polar regions, such as curved valleys, pedestal craters, and layered terrain.

Data Evaluation. Mariner and Viking spacecraft photography. The data are high quality; it is their interpretation that is controversial. Rating: 1.

Anomaly Evaluation. The confirmation of large-scale shifting or drifting of the Martian crust relative to the axis of rotation would indicate, as with terrestrial continental drift, the existence of unknown driving forces, internal and/or external. Rating: 2.

Possible Explanations. Mars' present equatorial region contains areas which were once near the Martian regions, but they were catastrophically relocated by the impact(s) of large meteors. Slow, terrestrial-type continental drift could have accomplished the same effect over a much longer time span. Alternatively, the equatorial "polar" features could be the consequence of much more extensive ice cover in the past.

Similar and Related Phenomena. Terrestrial polar wandering and continental drift (ET); the remarkable dichotomy in Martian topography (AME8), which is divided into north/south hemispheres rather than east/west hemispheres like on Mercury (AHE1) and the moon (ALE1), and is suggestive of a 90° crustal shift.

Examples

X0. Evidence of small-scale polar wandering. Unique quasi-circular structures in the Martian polar regions have been interpreted as proof of polar wandering on the scale of about 15° during the past 10^8 years. The large volcanic structures near the planet's equator and its gravitational roughness are also considered as evidence of past polar wandering. (R1)

X1. Large-scale polar wandering evidence. "Three spots on the surface of Mars, all of them within 15° of the equator, show signs of once having been at the poles, according to Peter H. Schultz and A. B. Lutz-Garihan of the Lunar and Planetary Institute in Houston. The implication of these possible 'paleopoles,' says Schultz, is that redistribution of the planet's mass by internal activity shifted the crust. Viking spacecraft photos from orbit, he says, show the regions to have curved valleys like those in the present polar caps, 'pedestal craters,' whose shapes suggest that they formed in now-vanished ice, and signs of laminated terrain reminiscent of the present caps' familiar layering, which could indicate cyclic climate changes. The present poles represent a stable position established after the formation of the huge Tharsis rise, the scientists report, while the proposed paleo-poles existed after, during and before a no-longer-evident mass redistribution associated with the volcanic Elysium region, a fourth of the way around the planet. If crustal volcanics are due to a mantle plume that is fixed relative to the Martian spin axis, the authors suggest, the proposed polar wandering can also account for the northwest trend and age sequence of the line of huge volcanoes along the Tharsis ridge." (R2)

References

R1. Murray, Bruce C., and Malin, Michael C.; "Polar Wandering on Mars?" Science, 179:997, 1973. (X0)

R2. "The Poles of Old Mars," Science News, 119:216, 1981. (X1)

AME12 Ice Cap Melting Correlated with Solar Activity

Description. The correlation of the rate of melting of the Martian polar caps with the number of sunspots. This effect is superimposed on the normal seasonal melting.

Data Evaluation. A single, second-hand account of one astronomer's analysis. So far as is known, this phenomenon has never been confirmed by anyone. Rating: 3.

Anomaly Evaluation. Since the thermal effect on Mars of increased solar activity should be negligible, the demonstration that a substantial melting effect really does occur would be contrary to theoretical expectations. Rating: 2.

Possible Explanations. Increased radiative heating would be an acceptable mechanism, but it doesn't seem powerful enough. Possibly, the Martian climatic cycle is controlled in other ways by solar activity, as may also be the case on earth (GWS).

Similar and Related Phenomena. Terrestrial weather correlated with solar activity (GWS).

Examples

X1. Cyclic phenomenon. "Recent observations made by M. Antoniadi, the well-known authority on Mars, show that the melting of the polar caps of Mars proceeds at different rates in different years, and when the statistics of the shrinkage of the caps is compared with those of sunspots it is found that there is a very slow melting when there are few spots, and a rapid melting when the sun is at maximum activity. This seems to be a striking confirmation of the view of Abbot and others that the output of radiant energy from the sun is greatest at spot maximum and least at spot minimum. The effect is masked in the case of the earth by secondary atmospheric phenomena, but on Mars, with its rarer atmosphere, the response to solar fluctuations seems to be more direct. M. Antoniadi has carried his comparisons as far back as 1856, and finds that the association holds good 'as a rule,'" (R1)

References

R1. "Martian Ice Caps and Sunspots," Scientific American, 115:187, 1916. (X1)

AME13 Excess of Grazing-Incidence Craters

Description. The presence on Mars of approximately ten times as many grazing-incidence craters as is predicted for bombardment of heliocentric projectiles. The alignments of these craters also shift with geological age.

Data Evaluation. Mariner and Viking spacecraft photography. Rating: 1.

Anomaly Evaluation. The presence of the seemingly excessive number of grazing-incidence craters can be easily explained by assuming the past presence of a ring of Martian satellites. Neither is it surprising that Mars may have retained a complement of satellites after the planet formed. However, the shifting of the crater axes implies crustal shifting, which is a radical notion for Mars. Rating: 2.

Possible Explanations. The grazing-incidence craters were excavated by impacting Martian satellites over a long period of time, during which the Martian crust shifted underneath the ring of satellites.

Similar and Related Phenomena. The alignments of lunar maria (ALE9-X3); Martian polar features near its equator (AME11); the dichotomy of Martian surface features (AME8).

Examples

X1. Permanent features. About 5% of the Martian craters are elliptical with butterfly-wing patterns of ejecta perpendicular to the long axes of the craters. This type of elongate crater is made by projectiles hitting the planet at grazing incidence. However the 176 grazing-incidence craters identified on Mars represent ten times the number that would be expected if the projectiles were all in heliocentric orbits. Furthermore, "when the 176 Martian grazers were grouped according to their age (determined on the basis of the degree to which they had been eroded), it became apparent that the major axes of craters of roughly the same age often fall on the same great circle and that the great circles traced by the axes of older craters are increasingly inclined toward the Martian equator. In other words, the axes of newer craters are roughly east-west with respect to the Martian geographic poles and the axes of older craters are increasingly north-south with respect to the poles." The surmise of the scientists who made this study (P. H. Schultz and A. B. Lutz-Garihan) is that the projectiles that made these elliptical craters were actually in orbit around Mars in a plane perpendicular to the planet's axis of rotation. The axis of cratering changed with time because the Martian crust shifted due to internal changes. (R1)

References

R1. "Star Grazers," Scientific American, 249:88, October 1983. (X1)

AME14 Unidentified Active Ingredient in Martian Soil and the Possibility of Life

Description. Chemical experiments that indicate the presence of unknown, highly oxidizing substances and an agent that promotes the synthesis of organic chemicals.

Background. The results of the Viking life-detection experiments on Mars were widely proclaimed by the media to be negative. Actually, the results were controversial and inconclusive, with some scientists contending that the results of the Viking experiments are best explained by the presence of lifeforms.

Data Evaluation. Three life-detection experiments on each of the two Viking landers. Several successful runs were made with each experiment on samples of the Martian soil. The data, however, are not clear-cut; in fact, they are very puzzling. Rating: 3.

Anomaly Evaluation. Given the absence of organic material in the Martian soil, such as would be required to provide a foundation for the evolution of life-as-we-know-it, the verified presence of Martian life would indeed be startling. The anomalousness is underscored by the scarcity of water and oxygen and the harshness of the climate. If Martian life were the only possible interpretation of the Viking experiments, we would have a first-class anomaly. But given the inconclusiveness of the experiments, one can safely fall back on a mysterious, non-biological, "active" ingredient. Rating: 2.

Possible Explanations. An unknown, "active", but non-living ingredient of the Martian soil that mimics the chemistry of lifeforms.

Similar and Related Phenomena. Terrestrial biological phenomena.

Examples

X1. Results of the Viking biological experiments. Abstract. "The essential findings of the three biological experiments aboard the two Viking Mars landers are reviewed and compared. All three of the experiments yielded significant data in repeated tests of Martian surface samples. Some of the results are consistent with a biological interpretation, although there are serious reservations in accepting this conclusion. Most of the findings, however, are inconsistent with a biological basis. The combined data suggest the presence of several classes of oxidants on Mars and these would account for most of the observations. An explanation for the apparent small synthesis of organic matter in the pyrolytic release experiment remains obscure." (R2) Several scientists, however, are emphatic that the Viking experiments do indicate the presence of life on Mars. (R1, R3)

This positive outlook was described by R. Lewis: "Each of the three experiments was designed to detect evidence of microbial presence in the soil of Mars by a different method. Two based on gas exchange and on carbon assimilation by microorganisms yielded interesting results, but these were interpreted as the products of some unearthly, inorganic, chemical reaction in the soil when it was moistened with water or with a nutrient solution. The labelled release experiment sought to detect evidence of metabolism in the release of radioactive gas from a 0.5 cu. cm soil sample after it was fed a nutrient 'soup' labelled with radioactive carbon-14. Four cycles of the experiment were run at one of the two Mars landing sites and five at the other. Radioactive carbon dioxide appeared in each cycle when the soil was not heated to sterilisation temperature and failed to appear when the sample was heated before testing. This is the experiment that holds the most promise of a sign of life on Mars. Although these results would be considered incontrovertible evidence of biological activity in a terrestrial soil, they were not deemed conclusive on Mars because of a suspicion that some unknown agent in the soil was mimicking a biological response. The suspicion was based in part on the surprising chemical reactions displayed by the gas exchange and carbon assimilation experiments. They were attributed to an inorganic agent that reacted in some ways like a biological one. In larger part, however, doubt was based on the failure of a gas chromatograph mass spectrometer to detect organic molecules in the soil in a molecular analysis experiment. The instrument probed to ranges of parts per million to parts per billion. Consequently, it was difficult for most experimenters to believe that life could exist in a soil that seemingly was barren of organic material from which life could have arisen." Basically, there is on Mars an agent capable of rapidly recomposing organic chemicals. Some scientists prefer to believe that this unknown, active agent is really a form of life. (R1)

Results of an experimental study of possible nonbiological explanations of the Viking data. Abstract. "The Viking Labeled Release (LR) data obtained on Mars satisfy the criteria established for a biological response. The importance of the issue, especially when viewed against the harsh environment on Mars, requires careful consideration of possible nonbiological reactions that may have produced false positive results. A 3 1/2-year laboratory effort to investigate possible chemical, physical, and physicochemical agents or mechanisms has been concluded. Among nonbiological possibilities, hydrogen peroxide, putatively on Mars, emerged as the principal candidate. When placed on analog Mars soils prepared to match the Viking inorganic analysis of the Mars surface material, hydrogen peroxide did not duplicate the LR Mars data. When other materials were used as substrate, hydrogen peroxide could be made to evoke the type of responses obtained by the LR Mars experiment. However, essential criteria concerning the formation, accumulation, and preservation of hydrogen peroxide to qualify it as the active agent on Mars have not been met and new data show it to be essentially absent from the Mars atmosphere. The presence of a biological agent on Mars must still be considered. This interpretation of the LR results is strengthened by a recent report that the Viking organic analysis instrument (GCMS) failed to detect organics in an Antarctic soil in which the LR instrument had demonstrated the presence of microorganisms." (R5)

References

R1. Lewis, Richard; "Yes, There Is Life on Mars," New Scientist, 80:106, 1978. (X1)

R2. Klein, Harold P.; "The Viking Biological Experiments on Mars," Icarus, 34: 666, 1978. (X1)

R3. Levin, Gilbert V., et al; "Viking Mars Labelled Release Results," Nature, 277: 326, 1979. (X1)

R4. Levin, Gilbert V., and Straat, Patricia Ann; "Recent Results from the Viking Labeled Release Experiment on Mars," Journal of Geophysical Research, 82: 4663, 1977. (X1)

R5. Levin, Gilbert V., and Straat, Patricia A.; "A Search for a Nonbiological Explanation of the Viking Labeled Release Life Detection Experiment," Icarus, 45:494,

1981. (X1)
R6. Strand, Linda J.; "The Search for Life on Mars: Shots in the Dark," Astronomy, 11:66, December 1983. (X1)

AME15 Lack of Water-Ice at the Southern Polar Cap

Description. The apparent scarcity of water ice in the southern polar regions, despite the presence of sufficiently low temperatures and atmospheric water vapor.

Data Evaluation. Mariner and Viking spacecraft spectroscopy. Rating: 2.

Anomaly Evaluation. Since the atmosphere of Mars does contain some water vapor, and this vapor does freeze out on the northern polar cap during that hemisphere's winter, the same process should also occur at the southern polar cap, which is the colder of the two polar regions. The anomaly seems to be in the planetary distribution of water vapor. Rating: 3.

Possible Explanations. The water vapor from the melting northern polar ice may not reach the southern polar region in large quantities due to wind circulation patterns.

Similar and Related Phenomena. The mixing of northern and southern air masses on earth is sluggish.

Examples

X1. Permanent situation. "Another Martian mystery is the polar caps. The residual northern cap was discovered early in the Viking mission to consist almost entirely of water ice, but scientists waited in vain to see vapor given off from the north to freeze out in the south as the seasons reversed. Last week, colloquium participants reported that the residual southern cap seems to be dominated by frozen carbon dioxide. Scientists are developing hypotheses to explain the difference, but they are complex." The southern cap is normally colder than the northern one due to the planet's inclination. Carbon dioxide can freeze there, but so can water, and it seems strange that it doesn't. (R1)

References

R1. "Mars: How Little We Really Know," Science News, 115:53, 1979. (X1)

AME16 Anomalously Wet Areas

Description. Areas of the Martian surface where water appears to be close to the surface. The Solis Lacus region is the major such spot identified so far.

Data Evaluation. The visual observations of clouds, fogs, and frost from earth and orbiting spacecraft. The detection of concentrations of water vapor by spacecraft instrumentation. The radar reflectivity of the surface. Rating: 2.

Anomaly Evaluation. Geologically speaking, the presence of abnormally wet areas on a planet with subsurface water is not surprising. The presence of oases in terrestrial deserts mark places where favorable topography and water table come together. It seems reasonable that similar situations occur on Mars. Rating: 4.

Possible Explanations. We know little of the Martian water table and aquifers, but Martian oases could well be created the same way oases are in terrestrial deserts. Solis Lacus, in this view, is simply a region where the Martian water table is close to the surface. Subsurface water may become more obvious when storms strip away sand and dust.

Similar and Related Phenomena. Terrestrial desert oases.

Examples

X1. Apparent permanent feature. While most of Mars is extremely arid, "remote sensing data indicate that the Solis Lacus region may contain a reservoir of weakly-bound H_2O that extends to within a few cm. of the surface. This was the site of the origin of the great 1973 dust storm, and reflectance spectra revealed the Solis Lacus clouds to be more hydrated than the surrounding surface material. Also spectral vidicon images revealed a huge deposit of ground fog and frosts that was apparently pulsed from the central Solis Lacus region at the onset of the storm: several tens of mg. of weakly bound H_2O was apparently driven from the upper few cm. of topsoil by perihelion solar heating. This points to an anomalous reservoir of H_2O in this region that extends to within a few cm. of the surface and replenishes the large evaporative losses of the highly desiccating environment. The Viking IRM maps show thermal inertia and pre-dawn residual temperature anomalies in Solis Lacus that are consistent with such a reservoir. A striking anomaly was also revealed by the Viking Water Vapor Mapper. Laboratory studies indicate that the H_2O should occur as a salt brine at $\gtrsim 1$ cm. depth that undergoes diurnal freeze-thaw throughout most of the year. Oxidants that destroy organic matter at the Viking sites should not form in the moist soils of Solis Lacus, and with daytime temperatures as high as $20^{\circ}C$ it would appear that Solis Lacus may be an oasis on an otherwise hostile planet" (R2, R1) See AME14 for Viking soil analyses. (WRC)

Earth-based radar provides more support for water at Solis Lacus. "It appears certain now that water exists on the surface of Mars both as ice and as liquid. A recent report by S.H. Zisk and P.J. Mouginis-Mark indicates that liquid water exists in Mars' southern hemisphere at the area of Solis Lacus (Lake of the Sun). Interestingly enough, the results were obtained from radar data from earth-based stations rather than from Viking spacecraft. The reason for this is that the data from the Viking Orbiters showed a rough, evidently solid, frozen surface. Images from the Viking spacecraft were obtained during the Martian winter, but the earth-based radar measurements were obtained in the Martian summer. The conclusion is that the ice melts and ponds with a saturation depth of 1 m or so. Needless to say, the radar images of the area appeared extremely smooth, consistent with a liquid surface." (R3, R4)

References

R1. Eberhart, Jonathan; "Mars: The Wet Look," Science News, 116:108, 1979. (X1)

R2. Huguenin, R. L., et al; "Mars: An Oasis in Solis Lacus," Eos, 60:306, 1979. (X1)

R3. "Water on Mars," Eos, 61:1233, 1980. (X1)

R4. Zisk, S. H., and Mouginis-Mark, P. J.; "Anomalous Region on Mars: Implications for Near-Surface Liquid Water," Nature, 288:735, 1980. (X1)

AME17 Spectroscopic Evidence of Vegetation

Description. Absorption bands in the reflection spectrum of Mars that closely resemble those of some types of terrestrial vegetation.

Data Evaluation. A single study of Martian spectra recorded using a terrestrial telescope.

Rating: 3.

Anomaly Evaluation. The similarity of the Martian absorption bands to those of terrestrial vegetation is probably only fortuitous. The Mariner and Viking missions produced no evidence of anything like Martian vegetation. However, given the altitudes of the orbiters and the small fraction of the surface examined by the two landers, vegetation of some sort cannot be completely disproved. Nevertheless, it is more likely that we have here a case of mistaken identity. Rating: 3.

Possible Explanations. The absorption bands recorded may actually be due to molecules in the Martian and/or terrestrial atmospheres or minerals on the Martian surface.

Similar and Related Phenomena. The Martian springtime wave of darkness (AMO3); the Viking life-detection experiments (AME14).

Examples

X1. Spectroscopic evidence. "Abstract. A new test for the presence of vegetation on Mars depends on the fact that all organic molecules have absorption bands in the vicinity of 3.4 μ. These bands have been studied in the reflection spectrum of terrestrial plants, and it is found that for most plants a doublet band appears which has a separation of about 0.1 μ and is centered about 3.46 μ. Spectra of Mars taken during the 1956 opposition indicate the probable presence of this band. This evidence and the well-known seasonal changes of the dark areas make it extremely probable that vegetation in some form is present." (R1)

References

R1. Sinton, William M.; "Spectroscopic Evidence for Vegetation on Mars," Astrophysical Journal, 126:231, 1957. (X1)

AME18 Apparent Lack of Extensive Surface Erosion

Description. The generally well-preserved appearance of Martian craters, channels, and other geological features despite the obvious presence of strong winds and frequent dust storms.

Data Evaluation. Mariner and Viking spacecraft photographs. Rating: 1.

Anomaly Evaluation. The well-preserved appearance of Martian surface features is incompatible with the known wind-and-dust environment and the estimated ages of the geological features. Rating: 2.

Possible Explanations. The windblown dust and sand may not be as erosive as wind-tunnel experiments suggest due to clumping of the particles. Also, the Martian dust storms may seem spectacular from afar, they may not transport as much dust as their appearance suggests. Much more radical explanations would have the Martian surface much younger than 3.5 billion years, or the present Martian erosive environment much more recent.

Similar and Related Phenomena. The Martian pyramidal structures (AME3), which are supposed to be wind-eroded mountains; the lack of dust and meteoric material on the moon (ALL14); an episode of accelerated crater erosion (AME20).

Examples

X1. Apparent lack of erosion. "...although the planet is still geologically active (as indicated by young volcanic features and faults), the pace of that activity has slowed with time. The densely cratered terrain dates from before the decline in impact rates over 3.5 billion years ago; the Tharsis bulge, and most of the fractures around it, most of the volcanic plains, most channels, and many of the volcanoes date from the first half of the planet's history. This conclusion is somewhat surprising, since many of these ancient features are well preserved; the annual global dust storms, despite their apparent violence, must have little erosive power. They probably simply move previously eroded loose material around; their effect on cohesive rocks must be negligible, because otherwise the old landforms would be destroyed or at least substantially modified. Our estimates are that the erosion rate is no more than a billionth of a meter per year." (R1) This conclusion is based on the assumption that most craters were excavated about 3.5 billion years ago and the apparent pristine character of these supposedly old features. (WRC)

X2. Laboratory simulation of Martian conditions. R. Greeley and his colleagues have used a wind tunnel at NASA's Ames Research Center to simulate erosive conditions on Mars. Atmospheric pressure on Mars is only a few thousandths of that at sea level on earth, so Martian winds must possess very high velocities to transport dust particles. These high wind speeds should produce a highly erosive environment. "Based on these factors, and on wind patterns derived from Viking lander data, the researchers came up with some surprising results. They calculate that Mars should be eroded at rates of up to 2 centimeters per century. But if this were the case, they note, the craters visible at the Viking sites (which are hundreds of millions of years old) should have been worn away long ago." Greeley supposes that Martian sand may exist in cohesive clumps which break apart on impact, thus reducing their erosive power. (R2)

References

R1. Carr, Michael H.; "The Surface of Mars: A Post-Viking View," Mercury, 12:2, January/February 1983. (X1)

R2. "The Windblown Planet Mars," Sky and Telescope, 68:507, 1984. (X2)

AME19 Layered Deposits

Description. The presence of extensive layered deposits, apparently of sedimentary origin. The deposits in the Valles Marineris, the only examples reported in the literature surveyed, are up to 5 kilometers high, with layers 170-220 meters thick, extending up to 50 kilometers.

Data Evaluation. Mariner and Viking spacecraft photography. Rating: 1.

Anomaly Evaluation. Assuming that the layered deposits were laid down in standing water, the implication is that water was once abundant on Mars and that it played a large role in shaping Martian topography. Such a fluvial period in Martian history is no longer considered a radical idea, although the origin of the event and the disposition of the water are not well understood. Rating: 3.

Possible Explanations. Sedimentary layers were deposited by a series of inundations during a planet-wide fluvial period. Sediment deposition by wind is another possibility.

Similar and Related Phenomena. A hypothesized episode of accelerated crater erosion (AME 20); the Martian channels (AME1); terrestrial sedimentary rocks.

Examples

X1. Permanent feature. Valles Marineris. Abstract. "The floors of the Valles Marineris were mapped to establish the geographic distribution of the rhythmic layered deposits found in the lower elevations, and to establish the stratigraphy of all the geologic units within the canyons. The layered deposits form erosional remnants up to 5 km high, and once covered larger areas of the valley floors. Individual layers are 170-220 m thick, and can be traced laterally up to 50 km. The deposits erode with eolian fluting along steeper faces and mottling of exposed horizontal surfaces. A deep 'moat' formed by canyon wall collapse and minor erosion commonly separates the deposits from the walls. No landslides form in the deposits, even on the steepest surfaces. The canyon walls underwent extensive landsliding, and elsewhere exhibit spur-and-gully topography that is distinct from the eolian fluting. The material of the layered deposits is markedly different from that of the canyon walls. The initial tectonic grabens of the Valles Marineris were enlarged by collapse as the walls weathered to spur-and-gully topography. During the same period the layered deposits were emplaced, followed by further canyon wall collapse in some areas, and by an event that deeply eroded some of the deposits to nearly their present extent. The outflow channels associated with the Valles Marineris may date from this epoch of erosion. Landsliding predates the formation and deep erosion of the layered deposits, as evidenced by slides diverted by erosional remnants. Eolian erosion responsible for fluting could be continuing up to the present. The morphology and history of the sediments are consistent with deposition in standing bodies of water early in Martian history." (R1)

References

R1. Nedell, Susan S., and Squyres, Steven W.; "Geology of the Layered Deposits in the Valles Marineris, Mars," Eos, 65: 979, 1984. (X1)

AME20 Evidence for an Episode of Accelerated Crater Obliteration

Description. A study of some 50,000 Martian craters, taking into account their apparent age and degradation, reveals a brief (25 million years) episode of highly accelerated erosion about 200-450 million years ago. This event may have been synchronous with the formation of the Martian channels.

Background. This phenomenon of accelerated crater erosion seems to contradict the statement in AME18 that Martian topography is relatively uneroded. It seems to be a matter of degree.

Data Evaluation. The basic data are derived from the Mariner/Viking spacecraft photos, which are considered excellent. However, the determination of crater age and degree of obliteration requires human judgment and assumptions about the evolution of Mars and the whole solar system. Rating: 2.

Anomaly Evaluation. The favored implication of accelerated crater erosion is the brief presence of a considerable amount of water on Mars. There is no consensus concerning the origin of this water and what finally happened to it. However, most astronomers believe the water was part of Mars' original endowment and that it still survives just below the planet's surface. Rating: 2.

Possible Explanations. A fluvial period on Mars caused the accelerated erosion of the Martian craters as well as other water-shaped topographic features. Alternatively, an episode of powerful erosion by wind, dust, and volcanic ash is possible.

Similar and Related Phenomena. The Martian channels (AME1); the layered deposits (AME19); the apparent light erosion of ancient Martian features (AME18).

Examples

X1. Permanent condition. Martian surface. Abstract. "It is shown that Mars experienced a brief episode of an increased obliteration rate contemporaneous with the formation of the cratered plains units. The obliteration rate during this episode was at least an order of magnitude greater than the preexisting rate. The obliteration rates prior to and after the event were identical within a factor of 3. Regional variations in the observed densities of degraded craters on the cratered terrain are shown to result from differences in the intensity of the obliteration rate during the event. The identification and characterization of the event are accomplished by using crater diameter-frequency distributions, not only for the total crater population but also for four morphological classes. Distributions are selected from 38 regions within the Martian equatorial latitudes. Many of the water-formed features on the planet, such as the channels and polar laminated deposits, may have formed during the obliteration event." (R1) On reading this report, one gets the strong impression that Mars once experienced an intense episode of water erosion. (WRC)

References

R1. Jones, K. L.; "Evidence for an Episode of Crater Obliteration Intermediate in Martian History," Journal of Geophysical Research, 79:3917, 1974. (X1)

AME21 Pedestal Craters and Their Eroded Environs

Description. Craters that stand perched on elevated, roughly circular bases. The bases rise several tens of meters above the surrounding terrain and have volumes many times the volume of the crater itself. The pedestals, therefore, cannot be derived from the craters.

Data Evaluation. Mariner and Viking spacecraft photography. Rating: 1.

Anomaly Evaluation. An appealing explanation of the pedestal formation is given below; and, in itself, the pedestal crater is not particularly anomalous. Yet, the pedestal craters imply a prodigious amount of erosion of the surrounding terrain by water or some other agent. The fact that the agent is unknown and the disposition of all the eroded material undiscovered gives the total phenomenon a relatively high rating. Rating: 2.

Possible Explanations. The pedestal craters were originally made in a debris blanket which was subsequently eroded away, except for the pedestal supporting the crater, which was protected by a layer of ejecta from the crater.

Similar and Related Phenomena. Terrestrial erosion features protected by caps of resistant rock.

Examples

X1. Permanent features. "Modification of Martian craters also produces patterns not found on other planets, especially at high latitudes. Of particular interest are the so-called pedestal craters. These craters occur at the center of a roughly circular platform that stands several tens of meters above the surroundings. In most cases the platform cannot be formed of ejecta from the crater since the volume of the platform far exceeds that of the crater's bowl. Most planetary geologists now view the pedestal craters as indicating that thick blankets of debris used to lie on the surface. In the areas between craters the old debris has been removed, but the debris is retained around craters because of armoring of the surface by ejecta. The craters are thus left standing at the center of what remains of the old debris blanket. The pedestal craters are found mostly at high latitudes, both north and south, and are regarded as evidence of complex erosion and deposit histories at these latitudes, perhaps connected in some way with the long term climate changes." (R1)

References

R1. Carr, Michael H.; "The Surface of Mars: A Post-Viking View," Mercury, 12:2, January/February 1983. (X1)

AME22 Flow-Like Character of Crater Ejecta

Description. The curious flow-like appearance of the material surrounding Martian craters. Lunar crater ejecta, in contrast, appears to have been blasted into the space above the crater and flung outward. Martian crater ejecta seems to have flowed along the surface like "mud". This ejecta flows around obstacles, contrary to what one would expect from a debris cloud.

Data Evaluation. Mariner and Viking spacecraft photography. Rating: 1.

Anomaly Evaluation. Ostensibly a minor geological phenomenon, the peculiar character of Martian crater ejecta is fraught with radical interpretations, as detailed below. The possible flow of fluids out of the crater and perhaps even a thermal origin of the Martian craters are definitely contrary to the presently conceived history of the planet. Rating: 2.

Possible Explanations. The crater-forming projectile might have hit a water-saturated area, and the wet ejecta just flowed out. Unfortunately, the energy of the impacting projectile would probably have vaporized any water, militating aginst this theory. Another thought is that the crater weakened the crust, allowing subsurface fluids to flow out forming a mud flow. Finally, The Martian craters might be true mud pots on a colossal scale!

Similar and Related Phenomena. Pedestal craters (AME21) with their raised bases; terrestrial mud pots.

Examples

X1. Permanent features. "The material ejected from the Martian craters also has an unusual appearance. The ejecta appear to have been deposited by flow along the ground, rather than directly from being thrown out and freely falling as is generally the case on Earth and the Moon. On the Moon, most impact craters have hummocky ejecta close to the rim. Farther out the ejecta seem to fall in a linear pattern outward from the crater, a pattern which merges still farther outward with lines of secondary craters. In contrast, the ejecta around most Martian craters are disposed in thin sheets, each with a clearly delineated outer margin commonly marked by a low ridge. These ejecta look like mud flows and, indeed, where obstacles were in its path, the material clearly flowed around them." (R1)

References

R1. Carr, Michael H.; "The Surface of Mars: A Post-Viking View," Mercury, 12:2, January/February 1983. (X1)

AME23 The Tharsis Bulge

Description. A huge, 10-kilometer-high, 5,000-kilometer-diameter bulge in the Martian surface between 20-30° south and 10-85° west. The bulge is surrounded by volcanos, fractures, channels, and chaotic terrain.

Data Evaluation. Mariner and Viking spacecraft photography. Rating: 1.

Anomaly Evaluation. The origin of the Tharsis bulge is a mystery, although many scientists are inclined toward an internal origin, associated perhaps with some asymmetry in the formation of the planet's core. Rating: 2.

Possible Explanations. As mentioned above, some asymmetry during core formation. The bulge might be depositional; that is, an accumulation of ejecta and sediments.

Similar and Related Phenomena. The dichotomy of Martian topography (AME8).

Examples

X1. Permanent feature. "The 10-km high, 5,000-km diameter Tharsis bulge has clearly played a major role in the evolution of Mars. The planet's youngest and largest volcanoes lie on its flanks and fractures associated with the bulge affect about a third of the planet's surface. One question of interest to geologists is the extent to which the bulge is the result of accumulation of materials on the surface and the extent to which it results from uplift from below. The question has yet to be resolved. It appears, however, that the fractures and compressional ridges that abound around Tharsis are due to the presence of the bulge; they are not the result of stretching associated with its formation. No matter what its origin, the Tharsis region is clearly anomalous and various suggestions have been made for the primary cause. A plausible possibility is that the bulge is an indirect result of formation of the planet's core, which may have happened in an asymmetric way, causing anomalous chemistry and heat flow under Tharsis." (R1)

References

R1. Carr, Michael H.; "The Surface of Mars: A Post-Viking View," Mercury, 12:2, January/February 1983. (X1)

AMF LUMINOUS PHENOMENA ON MARS

Key to Phenomena

AMF0 Introduction

Mars, like our moon, displays transitory luminous phenomena. Transient Martian Phenomena (TMPs) are very rare compared to Transient Lunar Phenomena (TLPs), probably because Mars is so much farther away. Mars may be more active than we suspect. In any case, Mars and the moon are both considered to be relatively inactive in the geological sense, so any incandescent or self-luminous phenomenon bright enough to be seen from earth would be of great interest to planetologists.

Although occasional flares and flashes of light do appear on Mars, it is questionable whether they represent self-luminous events. The phenomena may be brighter than their surroundings only because they have higher reflectivity. If these events are not actually self-luminous, this section of the Catalog is unnecessary, and these phenomena should be placed in section AMW. However, self-luminous phenomena, such as volcanic eruptions, lightning, and earthquake lights cannot be ruled out completely at this time.

AMF1 Flares and Light Flashes

Description. Bright, white flares or flashes of light lasting from 5 seconds to several minutes. Short flashes are rare and may be associated with flickering around the polar caps. The longer-lived "flares" usually persist for about 5 minutes. Two separate phenomena may be involved here.

Background. Long-lived white clouds often appear on Mars (AMW1 and AMW3) and last for many hours. Circa 1900, considerable excitement was caused by Martian clouds, which some interpreted as signals from intelligent Martians. This was the era when the famous canals were also thought to be artificial.

Data Evaluation. Flickering, short-lived flashes are rare, while the longer so-called flares have been seen rather often, especially by Japanese astronomers. Brightenings lasting several minutes are backed by many sound observations, but systematic study is still lacking. Rating: 2.

Anomaly Evaluation. Several explanations for the short-lived flashes and flares have been profferred. It seems most likely, however, that the flares at least are just temporary clouds of vapor. There is nothing very anomalous here, assuming this interpretation is correct. Rating: 3.

Possible Explanations. Volcanic eruptions, which to be seen from earth would have to be immense; halo phenomena and glow discharges (X12) would also be far too weak for detection through terrestrial telescopes; the specular reflection of sunlight off ice or standing water would require immense, very flat, smooth areas. Probably the best theory is the sudden formation of large clouds of white vapor or fog from the Martian soil which quickly disperse.

Similar and Related Phenomena. Long-lived white clouds seen on Mars (AMW1 and AMW3); bright flashes seen on the moon (ALF3).

Examples

X1. December 1900. Beam of light. "Prof. Pickering makes the following statement relative to the light flash from Mars. 'Early in December we received from the Lowell Observatory in Arizona a telegram that a shaft of light had been seen to project from Mars (the Lowell Observatory makes a specialty of Mars) lasting seventy minutes. I wired these facts to Europe and sent out neostyle copies through this country. The observer there is a careful, reliable man and there is no reason to doubt that the light existed. It was given as from a well-known geographical point of Mars. That was all. Now the story has gone the world over. In Europe it is stated that I have been in communication with Mars, and all sorts of exaggerations have sprung up. Whatever the light was, we have no means of knowing. Whether it had intelligence of not, no one can say. It is absolutely inexplicable." (R1)

X2. May 30, 1937. Flashes. "On May 30, 1937, during a part of the interval between 4h 35m and 6h 0m G.C.T., while Mars was being observed through a 12-inch aluminized reflector, 250 magnification, with unusually good seeing, what appeared to be a series of bright flashes was seen, extending across the south polar cap about 1" north of the southern rim of the disk. They were irregularly intermittent and were estimated to be about one magnitude brighter than the rest of the cap. They were entirely unexpected, but when once seen they became the chief concern of the observation period.

Measured from the drawings and photographs the thickness of the south polar cap on the central meridian was about 1".87. The flashes extended almost midway between the northern edge of the cap and the southern rim of the disk. A line of tiny white spots seemed to extend across the cap, some of the spots coalescing to swell into a brilliant white spot which quickly became yellow, then red-yellow, the phenomenon passing from left to right across the polar cap." (R2)

X3. June 4, 1937. Martian "flares". "The classic instance of this phenomenon was seen on June 4, 1937, by Sizuo Mayeda in Japan. Observing the Alcyonius region of Mars with an 8-inch reflector, he saw an intense point of light suddenly appear near Sithonius Lacus. Far brighter than the planet's polar cap, it scintillated like a star and disappeared after about five minutes." (R5, R3)

X4. December 8, 1951. "At 21:00 UT I saw a sharp, bright, glaring spot suddenly appear on Tithonius Lacus. It was as brilliant as a 6th-magnitude star---decidedly brighter than the north polar cap---and shone with scintillation for about five minutes. Fading rapidly, by 21:05 it looked like a whitish cloudlet, as large as Tithonius Lacus. At 21:10 it was barely visible as a very faint and very large white spot, and by 21:40 this part of the Martian surface had returned to its normal state." (R3)

X5. July 1, 1954. A sudden brightening of Edom Promontorium, lasting 5 seconds. (R3, R4)

Bright flare on Mars, June 4, 1937. (X3)

X6. July 24, 1954. Another brightening of Edom Promontorium, but lasting about 58 seconds this time. (R4)

X7. November 5, 1958. A bright spot lasting 5 minutes observed neat Nix Tanaica by Sadao Murayama, at Tokyo. (R4)

X8. November 7, 1958. At the southwestern edge of Tithonius Lacus, a bright spot that faded after about 4 minutes. (R4)

X9. November 11, 1958. "At 0h 05m on November 11 (U.T. Nov. 10, 15h 05m), only 4 days after Tenabe's observation, Mr. Sanenobu Fukui at Kobe, using his 25-cm. reflector with the power 400X, also discovered a bright and white small spot near the N.E. portion of Solis Lacus. He recorded: 'I found this curious bright spot when I was studying the details of Solis Lacus, at about 15h 05m (U.T.). It was as bright as the North Polar Cap, but I could see it for only about 5 minutes. I computed its diameter at about 250 kms. from my drawing.'" (R4) Note that Solis Lacus may be the site of ice and water (AME16). (WRC)

X10. November 21, 1958. Abnormal brightening of the northern edge of Hellas and Edom Promontorium, lasting about 5 minutes. After fading, the spots flared up again. (R4)

X11. General observations. H. Heuseler, writing in Die Sterne listed 13 observations of Martian flares between 1937 and 1967. The Soviet astronomer V.D. Davydov has listed six other flares not mentioned by Heuseler. These are said to have occurred in 1896, 1900, 1903, 1911, 1924, and 1956. (R5)

X12. Possible explanations. Active volcanos (R5); halo phenomena involving clouds of ice crystals (R5); glow discharges caused by the frictional generation of electricity (R6). Actually, A.A. Mills (R6) does not suggest that glow discharges cause the visible flares, but it seems wise to at least list the possibility. (WRC)

References

R1. Scientific American, 84:179, 1901. (X1)

R2. Wilson, Latimer J.; "Apparent Flashes Seen on Mars," Popular Astronomy, 45: 430, 1937. (X2)

R3. Saheki, Tsuneo; "Martian Phenomena Suggesting Volcanic Activity," Sky and Telescope, 14:144, 1955. (X3-X5)

R4. Saheki, Tsuneo; "Some Important Martian Phenomena in 1958," Strolling Astronomer, 16:264, 1962. (X5-X10)

R5. "Bright Flares on Mars," Sky and Telescope, 39:83, 1970. (X3, X11, X12)

R6. Mills, A.A.; "Dust Clouds and Frictional Generation of Glow Discharges on Mars," Nature, 268:614, 1977. (X12)

AML THE CURIOUS SATELLITES OF MARS

Key to Phenomena

AML0 Introduction

Mars' satellites are so small that they weren't discovered until 1877. Nevertheless they have sparked two delightful debates down the years: (1) How did Swift describe the satellites so accurately in Gulliver's Travels in 1727? (2) Could Phobos be of artificial origin? These questions, especially the latter, are probably much ado about nothing, but still entertaining and instructive. The real scientific anomaly, the grooves of Phobos, was not dreamt of until the Viking spacecraft got close enough to get high-resolution photographs.

AML1 Possible Early Sightings of the Martian Satellites

Description. Observations of the Martian satellites before their official discovery by Hall in 1877.

Data Evaluation. Most of the data suggestive of pre-1877 observations of the two Martian satellites are merely vague literary references. Swift's description in Gulliver's Travels is more specific, but imprecise and of unknown origin. Truly scientific data are really nonexistent. Rating: 4.

Anomaly Evaluation. If the Martian satellites were actually discovered only a few years prior to 1877, it would simply mean that some other astronomer with a good eye and quality telescope had beat Hall to the punch. On the other hand, pre-Gulliver's Travels (1727) sightings of the satellites, given the state of telescope technology, would truly be anomalous. Our rating is based upon this remote possibility. Rating: 2.

Possible Explanations. Crude glass lenses were known in ancient times (M); but though the ancients were avid astronomers, no one has uncovered any evidence that they discovered the telescope. Much more reasonable is the theory that Swift combined mystical expectations (for the number of Martian satellites); simple luck (for their distances from Mars); and Kepler's laws (for their periods).

Similar and Related Phenomena. Early sightings of Martian craters (AMO2); early sightings of spoke-marks on Saturn's rings (ARL5); early sightings of Neptune's rings (ANL1).

Examples

X0. August 1877. On the 11th., Asaph Hall at the Naval Observatory, Washington, saw a small star near Mars but was not certain it was a satellite. On the 16th., he saw it again; and, later that night, found that the supposed star have moved with Mars and was therefore a Martian satellite. A second satellite was soon discovered. On August 18, 1877, Joseph Henry sent the following dispatch to European astronomers: "Two satellites of Mars have been discovered by Hall at Washington. First, elongation west, August 18th, eleven hours, Washington time. Distance, eighty seconds; period, thirty hours. Distance of second, fifty seconds." The distance given for the inner satellite turned out to be in error; but this was soon corrected. (R1)

X1. 1726. Publication of Gulliver's Travels. In the "Voyage to Laputa," Chapter III, we find: "They have likewise discovered two lesser stars, or satellites, which revolve about Mars; whereof the innermost is distant from the center of the primary planet exactly three of its diameters, and the outermost, five; the former revolves in the space of ten hours, and the latter in twenty-one and a half; so that the squares of their periodical times are very near in the same proportion with the cubes of their distance from the centre of Mars; which evidently shows them to be governed by the same law of gravitation that influences the other heavenly bodies." (R6)

Many have speculated as to how Swift arrived at the correct number of Martian satellites, and how he came so close to the proper orbits in the absence of any recorded observations of these satellites. It would take too much space to review such guesses. Generally, authors of these speculations credit pure luck/chance or educated guesses based on Swift's knowledge of Kepler's laws and the surmise (offered by Kepler) that since earth had one satellite and Jupiter four, Mars should have two. Such a geometric progression would demonstrate how orderly the heavens are. (R6-R8) Wilder explanations have included ESP, ancient astronauts, and ancient sightings with precocious technology. (R9)

X2. Circa 1737. Lecompte's book Empire of China, published in 1737, refers to satellites of Mars as actualities. Lecompte was a Jesuit missionary, who spent ten years in China at the end of the sixteenth century. So it is possible his information on Martian satellites actually predates Swift's publication of Gulliver's Travels. (R3)

X3. Circa 1744. A quarto pamphlet of ten leaves, published in 1746 by E. C. Kindermanns, relates how the author saw a single Martian satellite in 1744. The satellite is pictured as having an atmosphere and in size is 0.4 the diameter of Mars in size. (R4, R5)

X4. 1752. Voltaire, another satirist, in his Micromegas, third chapter, mentions that Mars has two moons---rather, his space travellers tell of seeing them! (R7)

X5. October 24, 1864, and January 3, 1865. After Hall's discovery, Lamey reported in Comptes Rendus that he had observed something resembling a ring of asteroids around Mars. Compiler's translation: "I find in my record of astronomical observations, under the dates of October 24, 1864 and January 3, 1865, the remark that red gleams were situated on each side of the disk of the planet Mars, corresponding nearly to the equatorial plane. These appearances and considerations of another kind reminded me of the gleam due to the existence of a ring of large asteroids encircling the planet, producing an analogy with the crepe ring of Saturn." (R2)

References

R1. Kirkwood, Daniel; "Mars and Its Satellites," Popular Science Monthly, 11:706, 1877. (X0)

R2. Lamey, Ch.; "Observations Tendant a

Faire Admettre l'Existence d'un Anneau d'Asteroides, autour de la Planete Mars," Comptes Rendus, 85:538, 1877. (X5)

R3. Symes, J.; "The Satellites of Mars," English Mechanic, 27:547, 1878. (X2)

R4. Copeland, Ralph; "On a Pretended Early Discovery of a Satellite of Mars," Astronomy and Astrophysics, 11:553, 1892. (X3)

R5. Copeland, Ralph; "On a Pretended Early Discovery of a Satellite of Mars," Royal Astronomical Society, Monthly Notices, 52:493, 1892. (X3)

R6. Lamont, Roscoe; "The Moons of Mars," Popular Astronomy, 33:496, 1925. (X1)

R7. Gould, S.H.; "Gulliver and the Moons of Mars," Observatory, 66:289, 1946. (X1, X4)

R8. Brinton, Henry C.; "Swift's Forecast of Mars' Satellites," Sky and Telescope, 15: 494, 1956. (X1)

R9. Moss, Ken D.; "Jonathan Swift and the Moons of Mars," Kronos, 8:17, 1983. (X1)

AML2 The Grooves of Phobos

Description. Long, parallel grooves extending most of the way around the surface of Phobos. Typically, the grooves are 100-200 meters wide and 20-30 meters deep. They are mostly lined with small, contiguous pits. The grooves are best developed near the large crater Stickney. Analysis suggests that there may be several distinct families of grooves, with each family on different sets of parallel planes and being of a different age.

Data Evaluation. Viking Orbiter photographs. Rating: 1.

Anomaly Evaluation. There are several potential explanations, most of them requiring no stretching of astronomical theory. But if Phobos is basically a complex, layered structure, it may once have been part of a larger layered structure. Such an implication increases the seriousness of the anomaly, for no one knows what this layered structure could have been. Rating: 2.

Possible Explanations. See X2 below for a list of the different suggestions.

Similar and Related Phenomena. Ganymede's terrain is also grooved.

Examples

X1. Permanent feature of Phobos. "As soon as the first high-resolution Viking photographs became available, a new and puzzling aspect of Phobos was revealed. Its surface is almost entirely covered by long nearly parallel grooves, typically 100 to 200 meters wide and 20 to 30 meters deep.

Since the Viking orbiters photographed almost all of Phobos, it was possible to map the distribution of grooves on its surface. As the map below shows, the grooves appear to be associated with the largest crater, Stickney, which is about 10 km in diameter. The grooves are widest and best developed near Stickney and may be surface manifestations of deep fractures caused by the severe impact that created the crater.

Under high resolution, the grooves appear not as simple cracks but as lines of more or less contiguous pits. (R10, similar descriptions, some with more detail, may be found in R1-R7, R9, R11)

Further studies of the grooves showed that they are more complex than first believed. "The complex system of linear fractures of Phobos, found by Viking Orbiter in 1976, can be classified morphologically according to their appearance as well as their global distribution on the surface of the moon into three types. The member of each set appears to lie along roughly parallel planes cutting through much of Phobos. We are convinced that the differences in morphology between the three kinds of grooves indicate the necessity for three different hypotheses for their cosmogony. The third kind of grooves (L-grooves) appear to form arcs of small circles normal to the Phobos-Mars direction. We suggest that they are the sur-

Photographs of Phobos taken by Viking Orbiter 1 on May 27, 1977. The right-hand photo has been computer-enhanced. (X1) (Courtesy NASA)

face manifestation of layering within Phobos." (R12)

"Present explanations of the grooves on Phobos, which include internal fracturing and secondary impacts from Stickney crater, are considered unsatisfactory by the author. Grooves can be separated into distinct families of different ages, whose members lie on parallel planes intersecting the satellite. This, and the tendency for grooves to consist of contiguous pits resembling secondary impact craters, suggests that grooves are of external origin, caused by the satellite encountering fields of parallel strips of small bodies in space. The most likely source of such ordered fields of small bodies is considered to be ejecta from large, basin-sized impacts on Mars. Such an origin for Phobos' grooves places constraints on the date of capture and the history of the axis of rotation." (R13)

X2. Theories of groove formation. The grooves were formed by ejecta from craters on the surface of Phobos (R1); from a highly ordered meteor swarm (R2), perhaps from debris expelled when the large Martian craters were created (R13); large-scale fracturing of Phobos associated with the formation of Stickney (R3, R6, R7, R9-R11); tidal stresses (R5-R9); Phobos is fundamentally a layered body (R6, R12)

References

R1. "Viking: Moons, Ice Caps and Magnetospheres," Science News, 110:212, 1976. (X1, X2)

R2. "Viking Looks at Phobos in Detail," New Scientist, 72:158, 1976. (X1, X2)

R3. "Chain Craters on Phobos," Astronomy, 5:55, January 1977. (X1, X2)

R4. "75 Miles from Phobos," New Scientist, 74:19, 1977. (X1)

R5. "Tidal Stresses Made Phobos Groovy," New Scientist, 74:394, 1977. (X1, X2)

R6. "Phobos Photographed by Mars Orbiter," New Scientist, 74:501, 1977. (X1, X2)

R7. "Phobos and Deimos: Similar and Yet..," Science News, 112:295, 1977. (X1, X2)

R8. Soter, Steven, and Harris, Alan; "Are Striations on Phobos Evidence for Tidal Stress?" Nature, 268:421, 1977. (X2)

R9. Thomas, P., et al; "Origin of the Grooves on Phobos," Nature, 273:282, 1978. (X1, X2)

R10. Veverka, Joseph, et al; "The Puzzling Moons of Mars," Sky and Telescope, 56: 186, 1978. (X1, X2)

R11. Thomas, Peter; "Surface Features of Phobos and Deimos," Icarus, 40:223, 1979. (X1, X2)

R12. Horvath, A., and Illes, E.; "On the Possibility of the Layered Structure of Phobos," Eos, 62:203, 1981. (X1)

R13. Murray, J. B.; "Grooved Terrains on

Planetary Satellites," Eos, 62:202, 1981.
(X1, X2)

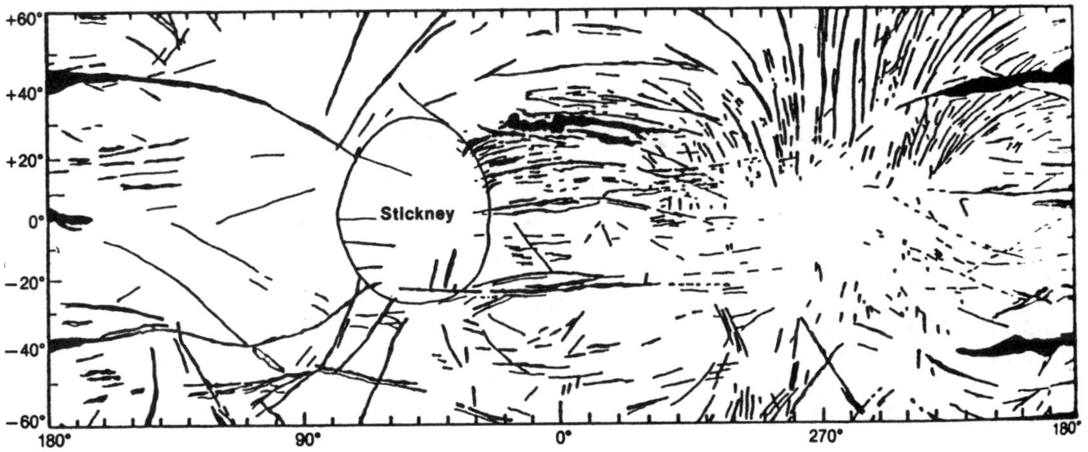

Map of the grooves of Phobos. (X1)

AML3 The Anomalous Acceleration of Phobos

Description. The small acceleration of Phobos' orbit amounting to about 3.3° per century.

Background. An important consideration in including this small effect is the suggestion in 1959, by the Soviet astronomer, I. Shklovsky, that the acceleration was so large that Phobos had to be hollow, given the calculated drag force and mass. He stated further that a hollow moon might well be of artificial origin! The popular press naturally made much of this at the time. The value of the acceleration later proved to be much smaller and the atmospheric model incorrect.

Data Evaluation. Recent measurements of the acceleration from the earth and spacecraft put it at 3.3° rather than the 19° estimated in 1964. A large change. Rating: 2.

Anomaly Evaluation. The drag of the Martian atmosphere causes Phobos to accelerate. No hint has been found that theory and experiment do not jibe. Rating: 4.

Possible Explanations. Atmospheric drag.

Similar and Related Phenomena. Terrestrial satellites in low orbit experience similar drag forces and accelerations.

Examples

X1. Permanent feature. In 1964 the acceleration of Phobos was estimated at 19° per century. (R1) On the basis of much better data, this figure was changed to 3.3° per century in 1979. (R2)

References

R1. "Mars' Inner Moon," Sky and Telescope, 28:138, 1964. (X1)

R2. "The Moons of Mars," Sky and Telescope, 57:11, 1979. (X1)

AMO MARTIAN TELESCOPIC ANOMALIES

Key to Phenomena

AMO0 Introduction

The major telescopic anomaly associated with Mars has to be the notorious canals. These lines, often thousands of miles long, form an intricate network, with many canals intersecting at dark spots called oases. Even though we know the canals are really not there, they can still be "seen" and have been captured forever in numerous photographs. Brushing them off as mere illusions is not the way to find the meaning of this important perceptual anomaly. The other phenomena placed in this section are of relatively minor importance: the springtime wave of darkening and the occasional appearance of intensely black spots. These are not well understood but hardly threaten our fundamental concepts about Mars. Finally, as with Mercury's craters and the Martian moons, we have claims of sightings of the Martian craters long before Mariner 4 sent back the first close-up photos of them.

AMO1 The Martian Canals

Description. Long, sharply defined, straight or slightly curved, dark lines observed on the surface of Mars. These lines or "canals" usually intersect at dark spots called oases and form an extensive, rather complex network on the lighter areas of Mars. Some of the canals on occasion appear to be double---the phenomenon of gemination. Individual canals may run for thousands of miles, with estimated widths of about 200 miles. The canal network has been captured, to a lesser degree, on photographs.

Background. The description just presented is the Lowellian view of the canal system, named after Percival Lowell, a champion of their artificiality. Many famous astronomers saw the same well-developed, clearly etched network of lines that Lowell did. Other astronomers, however, reported only a few smudgy lines; still others saw no lines at all. The physical reality of the Martian canals was long a subject of great contention. The close-up photographs of the Mariner and Viking spacecraft emphatically settled the question: they are not real. The astronomical anomaly has now become a psychological or perceptual anomaly.

Data Evaluation. Many capable, well-respected astronomers saw most of the Lowellian canal network; but others, just as reknowned, saw only wisps or nothing at all. Mariner and Viking photos show almost no correspondence with the complex canal maps. The data are very clear, although obviously conflicting, and available in abundance. Rating: 1.

Anomaly Evaluation. Science is based on the premise that all normal people see the same things. Evidently this is not the case with the Martian canals. Equally capable scientists saw radically different Martian surfaces through the telescope and even on the same photographs. Why did so many astronomers see the same canals in the same places when there is nothing objective there? Here we have a major perceptual anomaly. Rating: 1.

Possible Explanations. Hallucinations might be an answer; but why have so many hallucinated the same network?

Similar and Related Phenomena. Linear, canal-like features on Mercury (AHO4); the "spoke" system on Venus (AVO6); the naked-eye "canals" on the moon (ALO4).

Examples

X1. G. V. Schiaparelli's early observations. "These lines (the so-called canals) run from one to another of the dark spots of Mars, usually called seas, and form a very well-marked network over the bright part of the surface. Their arrangement seems constant and permanent (at least so far as can be judged by the observations of four and one-half years); but their appearance and the degree of their visibility is not always the same, depending on circumstances which we cannot at present discuss with full certainty. In 1879 many appeared which had not been seen in 1877 and in 1881-1882 all those which had been seen the first time were rediscovered, and other new lines as well. Their number could not be estimated as less than 60. Sometimes these lines or canals show themselves under the form of diffused and indistinct shading; at other times they appear as very definite and precise markings of uniform tone, as if they had been drawn with a pen. In most instances their curvature differs very little from a great circle, if indeed it does differ; some others however are much curved. The breadth of the finest can hardly be estimated at less than 2° (70 miles) of a great circle but in some cases it reached to about 4°. As to the length, that of the shortest is certainly less than 10°, others extend to 70° and 80°. The color is sometimes as dark as in the seas of Mars, but often it is brighter. Each canal terminates at its two extremities either in a sea or in another canal. I know of no instance where one end remains isolated in the midst of one of the bright areas of the surface without resting on lines and dark spaces.

Now in many of these lines it has fallen to me to observe the curious and unexpected circumstance of a doubling or reduplication; this happens in the following manner: To the right or left of a pre-existing line, which suffers no change from its previous direction or position, another line appears, nearly equal to the first and parallel to it; in some instances a slight difference of appearance being visible and sometimes also a slight divergence of direction. The distances between the pairs of lines formed in this manner varies from 6° to 12° of a great circle." (R1 and R5)

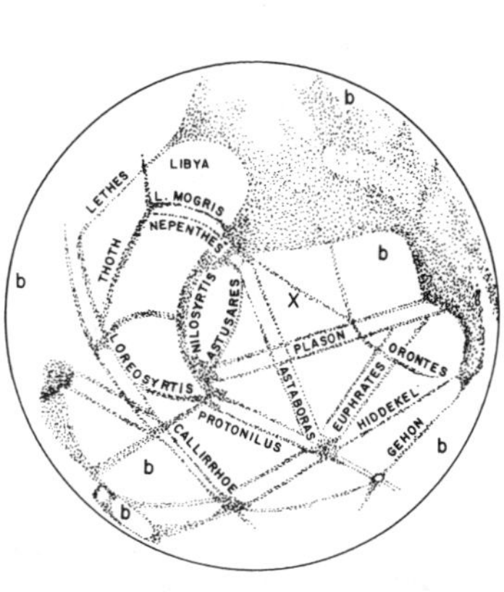

Schiaparelli's June 1888 drawing of Mars, made using the 18-inch refractor at Milan. (X1)

X2. Percival Lowell. A recapitulation of his observations (or beliefs). "(1)Mars is covered with a network of lines of individually uniform width, of exceeding tenuity, and of great length. These are the Martian canals. (2) These lines are 'joys of geometric beauty; they look to have been laid down by rule and compass.' (3)'Interdependence, not independence, marks the attitude of the canals. Each not only proceeds with absolute directness from one point to another, but at its terminals it meets canals which have come there with like forthrightness from other far places upon the planet.' (4) The total number of canals is great (being eventually represented as several hundred). (5) At times, one observes a doubling of certain of the canals. What has appeared as a single line appears suddenly as a pair of parallel lines ('gemination'). (6) The course of the canals always or almost always lies along one of the 'great circles' of the planet, i.e. taking the shortest route between two points on a sphere. (7) Small, dark, circular areas are usually to be seen at the crossing or juncture of canals; these are termed 'oases'. (8) In general, the canals were seen to darken in the spring and fade in the late summer, but 'each canal had its own times and seasons for showing or remaining hid. What dated the one left the other unaffected.'" (R45)

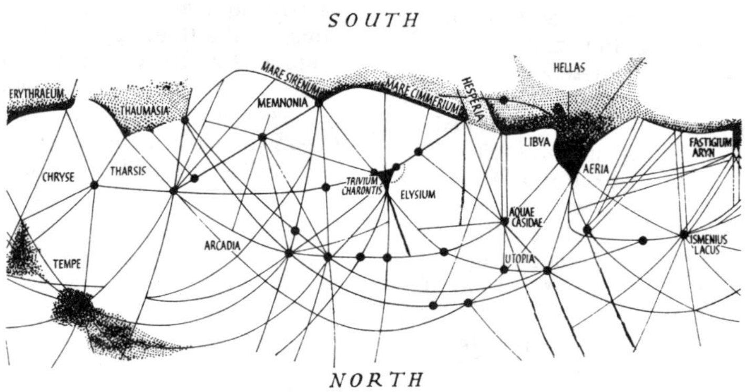

A Mercator Projection of Mars drawn by Lowell, showing the major canals, as he saw them. The black dots are oases. (X2)

X3. E.M. Antoniadi. Antoniadi was an early and very vigorous opponent of the canal idea. Many astronomers agreed with him. His views were summarized in a 1933 paper in Nature. "It is at Flagstaff that the planets Mercury, Venus, Jupiter and Saturn, as well as the Jovian satellites, have been seen covered with linear markings, which were afterwards pronounced unreal by their most prolific discoverer. It is further imprudent for a believer in linear canals on the planet Mars to mention the name of Prof. G.W. Ritchey, since that great astronomer repeatedly asserted to me that, having carefully scrutinised Mars with the fine 60-inch reflector of Mount Wilson, constructed by himself, he never saw a single straight line, or canal, on the planet, though details were observed beyond the range of the most powerful refractors in existence. My late-lamented mentor and friend, E.W. Maunder, remarked in 1894 that the so-called canals of Mars often disobey the laws of perspective; and I have demonstrated in my book the veracity of this statement, by showing that canals considered to run along arcs of great circles appear straight, not only at the center of the disc, but also at 30° and more from the centre---a fact which, of course, establishes their illusive character. I have also shown that the canals, after defying perspective, further defy diffraction, since they vanish in a large telescope, whereas a real, dark, planetary line, such as Cassini's division of Saturn's ring, becomes naturally darker and broader with an increase of aperture. Lastly, I pointed out in 1900 that, in the 33-inch refractor I was using, no straight lines could be detected on Mars, while delicate detail, far beyond the reach of the small telescopes with which the illusory canals had been discovered, was held quite steadily under fine

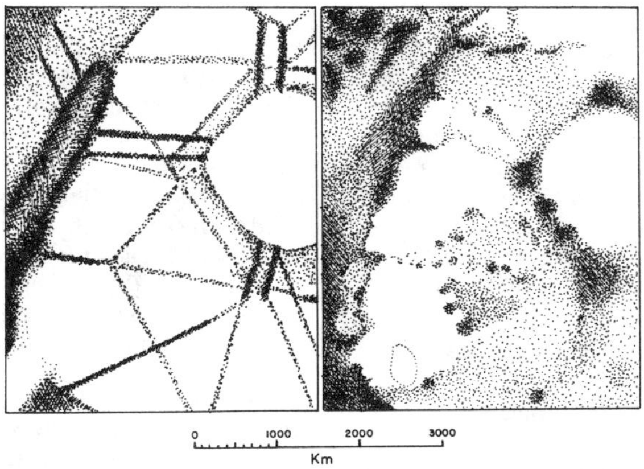

Two drawings of the Elysium region of Mars. Left: by Schiaparelli, from his 1877-1890 observations. Right: by Antoniadi, from his 1911, 1924, and 1926 observations. (X1, X3)

seeing. Such objections, whether considered severally or collectively, are, of course, fatal to the alleged reality of the linear markings. The planet is far too distant, from 35 to 250 millions of miles, to show canals. The truth is that a tired eye is liable to glimpse straight dark lines on the so-called continental regions of that neighbor world when viewed through an ordinary telescope; but these optical illusions vanish in a powerful instrument, which reveals broad, complex, dusky streaks, forming no network on the planet. The surface features of Mars are thus shown to be infinitely irregular and natural, like those of all the other bodies of the solar system." (R38, R22, R26, R28)

X4. Many astronomers nevertheless were certain they observed the famous canals essentially as Lowell saw them; others saw a vague pattern of poorly defined streaks.

The experience of J.H. Worthington at Flagstaff is instructive. "When I first looked at Mars at Flagstaff (September 27, 1909), I saw with great difficulty three streaks, presumably canals. The seeing was bad, and the general faintness of the planet's markings at that time is admitted by all. I continued to observe Mars on every possible night (which was nearly every night) until October 25, and as my eye became accustomed to the work I saw more and more. The canals were seen repeatedly better---this with the 24-inch refractor generally stopped down to about 18 inches. I found that with more than 20 inches the air was nearly always too unsteady, and with less than 15 inches too much separating power was lost. The canals were seen best with a power of 300 diameters. Clearer they became each night until, on October 25, the seeing being the best I ever experienced, the canals came out with amazing clearness and steadiness, sharp and clean, like telegraph wires against the sky, the oases also being exquisitely defined. Whereas on previous nights the canals could be held only by short glimpses of perhaps half a second at a time, they were now steadily visible for three or four seconds together, when a short flicker would sweep over them; during the lucid intervals the limb also of the planet was perfectly steady; as I have never seen it before or since. Of the objective existence of these markings in the image at the focus of the telescope there could be no manner of doubt, and Lowell's representations of them are nearer the actual appearance than any I have seen, though even in his drawings the lines seem hardly fine enough. The effect produced on my mind by this remarkable definition, which lasted for upwards of one and a half hours (from about 8.30 until after 10 p.m.), was staggering and ineffaceable. Soon after ten the definition went to pieces. It may be relevant to mention that a few evenings previously I had obtained a fair and convincing view of the canals with the 40-inch reflector (full aperture and a power of about 700), when they had appeared hazy and broader, but the image had been very unsteady, and only obtained in very short flashes; but nothing that I had hitherto seen had prepared me for the astonishing steadiness and <u>fine-</u>

ness of the details visible on this superb night..... In the face of all this positive evidence, and in the absence of any evidence that the observing conditions at Meudon, just outside Paris, ever approach these best conditions at Flagstaff, I find it impossible myself to attach any serious weight to the ingenious and plausible contentions of M. Antoniadi, which seem to have been much too hastily accepted in this country." (R29)

In 1947, E. Pettit summarized the observational evidence. "Few places are known in the world where the whole pattern of canals has been seen, although there are many places where the canals have been seen occasionally a few at a time. It was nine years after Schiaparelli saw them in 1877 before anyone else saw them. Perrotin in Vienna saw them on April 15, 1886, and in this country H. C. Wilson (late editor of Popular Astronomy) saw three of them at Cincinnati on March 6 of the same year. Every astronomer of that time attempted to see canals, but only a few succeeded. Young, Swift, Hall, and Hale, for example, could make nothing of them; W. H. Pickering saw them with difficulty at Harvard, and Keeler had the same experience at Allegheny. The best results at the opposition of 1892 were obtained at the Lick Observatory with the 36-inch by Holden, Campbell, Schaeberle, and Hussey. Barnard is popularly credited with not having seen the canals, although his drawings clearly show them. He observed with the 12-inch and, to a lesser extent, with the 36-inch telescopes. Speaking of a small spot near the Solis Lacus he says 'It is connected with the great sea south by a slender threadlike line. There is a small canal running north from Solis Lacus to a diffused dusky spot which does not appear on Schiaparelli's chart.' I think his chief concern was that he never saw the whole pattern at one time. Lowell was probably the first in this country to see the canal pattern, when he began work in 1894. Philip Fox told me thirty years ago that he saw the canals at Flagstaff 'stand out like an etching, and this was the experience of others who worked there. It is only with the most favorable atmospheric conditions when the whole pattern appears that we may hope to photograph canals as we can see them, and the exposures must be so short that large reflectors will be required." (R40)

As the following testimony of R. J. Trumpler demonstrates, some astronomers saw only vague approximations of the famous Lowellian network. "At the close approaches of 1924 and 1926 I undertook a careful study of the surface of Mars by photographic and direct visual observation with the 36-inch refractor of the Lick Observatory. These observations confirm the existence of a general network of dark lines and spots on Mars. They contradict the view of former observers that these lines are narrow, uniform, straight canals arranged in regular geometrical figures. This network was found to cover the whole surface of Mars, not only the yellow-orange areas thought to be deserts, to which the canals have formerly as a rule been confined, and it exhibits, after all, not much regularity, the marks being densely crowded in some regions and sparsely distributed in others. The lines themselves are not of uniform canal-like character, but show a wide variety of formations from faint narrow lines to diffuse darkbands, several hundred miles wide. Even the same line may change in width, intensity and definition along its course and a few cases were noted in which such lines break off or run out before reaching an end point. No impression of artificiality is conveyed by this network, but it seems perfectly feasible to interpret it as a natural topographic feature of the planet's surface." (R36)

Almost all the references at the end of this phenomenon profile contain affirmations of the canals' objective existence.

X5. Photographic evidence of the canals. Lowell and his colleagues at Flagstaff took many pictures of the Martian canals. As with telescopic observation, some easily see canals in the photos, others only a few broad streaks. Some comments on the photos follow:

"They show the snow-caps and the darker and lighter shading; but with normal sight I can see a number of most beautiful lines---the canals---and with the aid of a very low-power handglass, not strong enough to show the fibre of the paper, I can count thirty-six of these perfectly straight lines---one of them double. They come out from the snow-cap, they pass through the darker shading, they join and cross and pass over the planet's Equator, and meet others radiating from the opposite Pole. And all these most delicate lines on a disc 4 millimetres in diameter! Certainly a triumph in astronomical photography. Anyone who sees these photos must be at once convinced of the reality and the doubling of the canals. I count thirty-six canals, and I am not an expert. Professor

Lowell sees fifty-six on the photographs." (R21)

After Lowell exhibited his photographs at England's Royal Institution. "For myself, looking at the photographs of Mars thrown by the lantern on the screen, I could certainly see 'canals'; that is to say, dark, linear markings such as one sees in drawings of the network on the Martian surface, but not so many in number. This was not an optical illusion on my part, although I at first thought it might have been, for other persons saw exactly the same thing, and these markings are seen quite plainly on the original plates." (R27)

"I consider the photographs taken by Dr. Lowell and his co-workers one of the great proofs that what he sees, as he sees it, is correct." (R33)

E. M. Antoniadi stated that Lowell's photographs showed only a few broad, diffused streaks which looked nothing like a network. (R28)

Additional references describing the canal photos and reactions to them are R14-R18, and R39.

X6. Analyses of the Martian canal network, assuming the Lowellian vision. W. A. Webb compared the network to terrestrial networks of various kinds. "...we see that the communication networks, as exemplified by the railroads of Iowa and Ohio, are strikingly similar in their analysis to the Martian canal network. We see that Trumpler, in recording only the darkest canals, produced a map of Mars whose network characteristics resemble those of rural Iowa, while Lowell, in recording all that he could surely see, whether faint or well defined, produced a network whose characteristics are most like those of industrial Ohio. Although the Martian network appears to be clearly of the communication type, I do not wish to imply that it has been proved to have a design requiring intelligence equal to man's for its creation. Additional evidence giving more certain criteria for patterns indicating high intelligence is required before such a hypothesis can be proposed. (R42, R44)

Wells, on the other hand, analyzed the Martian canals as if they were natural lineaments like those mapped on the moon (ALO4). "Wells demonstrates that the canals can be mapped into two distinct but overlapping grid systems---one diagonal (trending NW-SE, NE-SW) and the other meridional-latitudinal (trending N-S, E-W). Furthermore, because Mars, unlike the Moon, presents more than one face to the Earth, the grid pattern can be observed throughout the whole 360° of longitude. One feature of the diagonal grid system is that the elements are not generally precisely 'perpendicular', Wells has thus been able to plot contour maps of the Martian surface bounding the three areas within which the northern apex angles of the diamond patterns were, respectively, greater than 90°, 80°-90°, and less than 80°. The isogonal maps produced were symmetrical about the almost antipodal meridians 290° and 125°." (R48)

X7. Ground-truth: how the Lowellian vision compares with close-up mapping by spacecraft.

After the Mariner 9 mission. "The Lowellian canal network has been compared with the results of Mariner 9 photography of Mars. A small number of canals may correspond to rift valleys, ridge systems, crater chains, and linear surface albedo markings. But the vast bulk of classical canals correspond neither to topographic nor to albedo features, and appear to have no relation to the real Martian surface." (R51, R49, R50)

After the Viking missions. P. Moore has compared a modern map of Mars, compiled at Flagstaff and based on spacecraft photography, with the 1909 canal network drawn on the Flammarion/Antoniadi map. Moore's conclusion: "There is no correlation at all, at least with any physical irregularity---one cannot be sure about the albedo features, which are probably variable to some extent.

Let me give a few examples. First, there is the 'focal point' of the darkish patch which has always been known as the Trivium Charontis. From it, Flammarion and Antoniadi (hereafter abbreviated to F. and A.) recorded several radiating canals; the Styx, Erebus, Orcus, Tartarus, Laestrygon, and the double Cerberus. There is a darkish patch extending vaguely in the direction of the Cerberus, but the others are conspicuous only by their absence. Yet F. and A. showed them plainly; and in his later book The Planet Mars (which I have translated, and which was published in English in 1975) Antoniadi described the Laestrygon as 'a beautiful filamentary streak...I saw it as a black line for an eighth of a second in 1896 on a 6".7 disk. With the 33-inch Meudon refractor in 1909 I was able to see it as a series of filaments and small lakes'. The Styx was 'a broad, knotty streak', and so on..... Antoniadi himself saw several double canals---such as the Jamuna, again in the cratered area north of Aurorae Sinus and not so far from the site of Viking 1. On the site---nothing at all; merely a medley of craters and fine features, not even roughly aligned. Then there is Hellas, regarded formerly as a plateau and now known to be a basin. Making an X inside it Schiaparelli recorded two fine canals, the Alpheus and the Peneus, 'visible at first glance' and also seen later by Antoniadi and many others. They are not there..... I do not think that I need labour the point further. I have done my best to plot the Lowell-type canal network accurately upon the Mariner chart, and the lack of any correlation seems to be conclusive. There is a lesson to be learned; there are so many observers who have drawn the Gehon, Hiddekel, Araxes and other canals with the aid of small telescopes---no doubt with absolute honesty, but subconsciously influenced by the fact that these canals were regarded as fully authenticated, so that failure to see them would indicate a lack of perception. I feel, though, that Antoniadi was being rather harsh on Lowell. He was correct in saying that the Lowell canals were illusory; but we now know that the same applies to the canals shown by Schiaparelli (for whom Antoniadi had great admiration) and also to those recorded by numerous other observers. Lowell was not alone in erring. At least we have solved the problem once and for all. Nobody has ever seen a canal on Mars; there are no canals to see." (R52) The astronomical problem has been solved but not the psychological one. (WRC)

X8. Related lines and geometric figures on Mars. Since the genesis of this class of markings is probably the same as that of the canals, it is logical to include them here.

"It is a rather curious coincidence that at each of the recent very near approaches of the earth to Mars, strikingly regular, although only temporary, geometrical figures should have appeared upon its surface. The well-known cross, centered in the approximate circle of Hellas, appeared to Schiaparelli in 1879. (Figure 1) It may have appeared in the unusually close apparition of 1877, but in that year he saw only the circle and the single vertical canal. The diameter of the circle is 900 miles. The cross has of late years been replaced by an irregular curved structure. The next very close approach of our planet occurred in

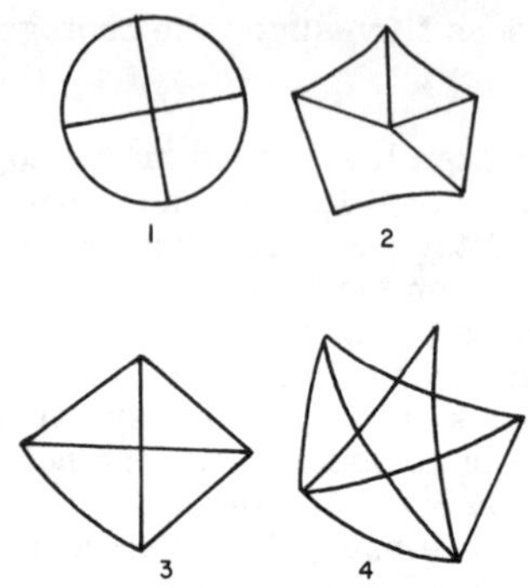

Other regular geometric figures seen by some on Mars. See text for details. (X8)

1892. In that year a regular pentagon with central radiating canals was seen in Arequipa. (Figure 2) One of the canals was missing, but it may have been too delicate for our 13-inch telescope, or it may have been covered by a temporary Martian cloud when we made our drawing. The center of the pentagon was located at Ascracus Lake, as it is now called, although then unnamed, and the diameter of the figure was 800 miles."

Pickering continued with descriptions of other geometric figures and, at the end, wondered if these regular figures might be attempts of intelligent life on Mars to communicate with earth! (R35)

X9. Possible explanations. Intelligent life on Mars created the canals. (R1, R20, R27, R47, R32, R35, R45) Lines of sand dunes, the aeolian hypothesis. (R46, R47, R51) Systems of ridges. (R47) Natural fractures of the surface. (R47, R51) Crater rays. (R51) Accidental alignments of natural features. (R51) Natural Martian rivers. (R1) Rifts in glaciers. (R1) For the gemination phenomenon only; purely optical effects. (R7-R9) Optical illusion wherein random markings on the Martian surface are interpreted as systems of lines---a type of psychological explanation. (R10, R11, R25, R31, R50) Purely psychological origin, as based on a psychologist's evaluation of Lowell's life:

"Summary. The thesis of the present paper is that Lowell's energetic investigations of Mars, a certain proportion of his 'findings,' and a large proportion of his conclusions were heavily influenced by unconscious forces, taking the final form of incompletely sublimated voyeuristic impulses. These impulses were a product of Lowell's unresolved oedipal conflicts." (R45)

References

R1. Wilson, H. C.; "Mars and His Canals," Sidereal Messenger, 8:13, 1889. (X1, X9)
R2. Hussey, William J.; "The Lines on Mars," Science, 20:235, 1892. (X4)
R3. Maunder, E. Walter; "The Canals of Mars," Knowledge, 17:249, 1894. (X4)
R4. "Observations of Mars," Nature, 51: 40, 1894. (X4)
R5. "Schiaparelli on Mars," Nature, 51:87, 1894. (X1)
R6. Brenner, Leo; "Chart of Mars," English Mechanic, 61:171, 1895. (X4)
R7. Antoniadi, E. M.; "On the Optical Origins of Martian Gemination," English Mechanic, 67:500, 1898. (X9)
R8. Phillips, Theodore E. R.; "The Doubling of the Canals of Mars," British Astronomical Association, Journal, 8:236, 1898. (X9)
R9. Antoniadi, E. M.; "Martian Gemination," English Mechanic, 67:288, 1898. (X9)
R10. "The 'Canals' of Mars, An Optical Illusion?" English Mechanic, 77:407, 1903. (X9)
R11. "The Canals on Mars," Nature, 68: 461, 1903. (X9)
R12. "Variations of the Martian Canals," Nature, 69:496, 1904. (X4)
R13. "Visibility of the Martian Canals," Nature, 70:416, 1904. (X4)
R14. "Photographs of the Martian Canals," Nature, 72:302, 1905. (X5)
R15. "Canals of Mars Photographed," Scientific American, 93:107, 1905. (X5)
R16. Butler, Charles P.; "The Canals of Mars Photographed," Knowledge, 2:204, 1905. (X5)
R17. Lowell, Percival; "First Photographs of the Canals of Mars," Royal Society, Proceedings, 77:132, 1906. (X5)
R18. "Photographs of the Canals on Mars," Knowledge, 3:369, 1906. (X5)
R19. "The Riddle of Mars," Scientific American, 97:25, 1907. (X4)
R20. "The Evidence of Life on Mars," Scientific American, 97:287, 1907. (X9)
R21. Cane, Frank Edward; "The Lowell Photographs of Mars," English Mechanic, 86:149, 1907. (X5)
R22. Antoniadi, E. M.; "On Some Objections to the Reality of Prof. Lowell's Canal System of Mars," British Astronomical Association, Journal, 20:194, 1910. (X3)
R23. Lowell, Percival; "The New Canals of Mars," Nature, 82:489, 1910. (X4)
R24. "New Canals and Lakes on Mars," Nature, 84:20, 1910. (X4)
R25. Dawson, Alfred; "The Mars Photographs of Professor Percival Lowell," English Mechanic, 91:102, 1910. (X9)
R26. Antoniadi, E. M.; "Mars," English Mechanic, 91:124, 1910. (X3)
R27. Hollis, H. P.; "Professor Lowell's Lectures," English Mechanic, 91:265, 1910. (X5, X9)
R28. Antoniadi, E. M.; "The 'Canals' of Mars," English Mechanic, 91:311, 1910. (X3, X5)
R29. Worthington, James H.; "Markings of Mars," Nature, 85:40, 1910. (X4)
R30. Antoniadi, E. M.; "Observations of Mars," Nature, 85:305, 1911. (X4)
R31. Watson, H.; "Notes on Mars," English Mechanic, 96:59, 1912. (X9)
R32. Pickering, William H.; "The Canals of Mars," English Mechanic, 102:222, 1915. (X9)
R33. Hamilton, George Hall; "The Canals of Mars Seen Photographically," Observatory, 39:363, 1916. (X5)
R34. Wicks, Mark; "Mars," English Mechanic, 121:147, 1925. (X4)
R35. Pickering, William H.; "Curious Geometrical Figures Appearing upon Mars," Scientific American, 134:57, 1926. (X8, X9)
R36. Trumpler, Robert J.; "Mars' Canals Not Man-Made," Science Newsletter, 12: 99, 1927. (X4)
R37. Atkins, E. A. L.; "Mars," English Mechanics, 1:278, 1927. (X4)
R38. Antoniadi, E. M.; "The Minor Details of the Planet Mars," Nature, 131:802, 1933. (X3)
R39. "The Surface of Mars," Nature, 147: 30, 1941. (X5)
R40. Pettit, Edison; "The Canals of Mars," Astronomical Society of the Pacific, Publications, 59:5, 1947. (X4)
R41. "Martian 'Canals' Seen," Science News Letter, 66:77, 1954. (X4)
R42. Webb, Wells Alan; "Analysis of the Martian Canal Network," Astronomical Society of the Pacific, Publications, 67:283, 1955. (X6)
R43. Richardson, Robert S.; "Strange Blue Streaks on Mars," Science Digest, 42:20, August 1957. (X4)
R44. Webb, Wells A.; "On the Rejection of the Martian Canal Hypothesis," Scientific Monthly, 85:23, 1957. (X4, X6)
R45. Hofling, Charles K.; "Percival Lowell and the Canals of Mars," British Journal of Medical Psychology, 37:33, 1964. (X2, X9)
R46. Gifford, F. A., Jr.; "The Martian Canals According to a Purely Aeolian Hypothesis," Icarus, 3:130, 1964. (X9)
R47. Sagan, Carl, and Pollack, James B.; "On the Nature of the Canals of Mars,"

Nature, 212:117, 1966. (X9)
R48. "Martian Canal Patterns," Nature, 222: 818, 1969. (X6)
R49. Hartmann, William K.; "Geological Observations of Martian Arroyos," Journal of Geophysical Research, 79:3951, 1974. (X7)
R50. Hughes, David W.; "Martian Canals after Mariner 9," Nature, 258:288, 1975. (X7, X9)
R51. Sagan, Carl, and Fox, Paul; "The Canals of Mars: An Assessment after Mariner 9," Icarus, 25:602, 1975. (X7, X9)
R52. Moore, Patrick; "Requiem for the Canals," British Astronomical Association, Journal, 87:589, 1977. (X7)
R53. Lowell, Percival; "On the Suggested Movement of the Canals of Mars," Popular Astronomy, 23:478, 1915. (X4)

AMO2 Pre-Mariner Observations of Martian Craters

Description. The observation of craters on the Martian surface through the telescope prior to the close-up photographs made by Mariner 4 in 1965.

Background. The Mariner photos of a heavily cratered Martian surface came as a great surprise to most astronomers. Despite many thorough studies of the planet with the best telescopes, the open literature recorded no craters.

Data Evaluation. The only written evidence seems to be a personal letter from John Mellish in which he states that both he and E. E. Barnard had indeed seen Martian craters. No observatory records or drawings have ever been found. Rating: 3.

Anomaly Evaluation. In the absence of drawings it is impossible to tell whether Mellish or Barnard saw the Martian craters or whether, like the canals, what they saw was illusory. Actually, Mellish's letter, quoted in X1 below, portrays a planet rather different from that which exists; i.e., oases as water-filled craters! In any case, the Martian craters are often very large; and it is not impossible that they might have been glimpsed in moments of good seeing, although it would have been surprising in 1935. Rating: 4.

Possible Explanations. None necessary.

Similar and Related Phenomena. Pre-spacecraft observations of the craters on Mercury (AHO0)

Examples

X1. Permanent features. A quotation from John Mellish in a letter to Walter Leight, dated January 18, 1935. "There is something wonderful about Mars. It is not flat, but has many craters and cracks. I saw a lot of the craters and mountains one morning with the 40" (Yerkes refractor) and could hardly believe my eyes. That was after sunrise and Mars was high in a splendid sky. I used a power of 750 and after seeing all the wonders, I went to Barnard and showed him my drawings, and told him what I had seen. I had never heard of any such thing having been seen, and he laughed and told me he would show me his drawings made at Lick in 1892-93. He showed me the most wonderful drawings that were ever made of Mars. The mountain ranges and peaks and craters and other things, both dark and light, that no one knows what they were. I was thunderstruck and asked him why he had never published these. He (Barnard) said, no one would believe him and (others) would only make fun of it. Lowell's oases are crater pits with water in them, and there are hundreds of brilliant mountains shining in the sunlight. Barnard took whole nights to draw Mars and would study an interesting section from early in the evening when it was coming on the disk until morning, when it was leaving. He made the drawings four or five inches (in) diameter and it is a shame that those were not published." (R1) Mellish's letter was also published in the February 1975 issue of the Observer, a bulletin of the Lehigh Valley Amateur Astronomical Society. (WRC)

Brief mentions of the Mellish/Barnard observations are to be found in R2 and R3, where it is noted that no drawings to support the claims have ever been found.

References

R1. Gordon, Rodger W.; "Mellish and Barnard---They Did See Martian Craters!" Strolling Astronomer, 25:196, 1975. (X1)
R2. Young, Andrew T.; "Mercury's Craters from Earth," Icarus, 34:208, 1978. (X1)
R3. Baum, Richard; "Historical Sighting of the Craters of Mercury," Strolling Astronomer, 28:17, 1979. (X1)

AMO3 The Springtime Wave of Darkness

Description. The seasonal darkening of the Martian hemispheres. In the Martian spring, the appropriate polar cap recedes and a wave of darkening sweeps equatorward. The darkened areas turn from brown to blue-green. In the fall the dark areas retreat poleward.

Background. The springtime wave of darkening was seized upon by Lowell and others as a certain sign that water released from the melting polar caps was reviving dormant vegetation. The canals served as conduits for this water in this view.

Data Evaluation. The springtime wave of darkening is a well-established phenomenon of the Martian cycle of seasons. Rating: 1.

Anomaly Evaluation. Mars, by virtue of the inclination of its axis of rotation, the presence of an atmosphere, and a palpable inventory of water, should show some seasonal changes by analogy with the earth. Several reasonable, season-sensitive processes are listed below. Any one of these, or some combination, could well explain the wave of darkening. The problem is that no one knows which are the most important, or even if there are other processes at work. In sum, the darkening phenomenon does not seem very mysterious. Rating: 3.

Possible Explanations. (1) Seasonal dust storms; (2) seasonal melting or sublimation of frost covering dark areas; (3) seasonal changes in atmospheric water vapor and its chemical reactions with surface materials; (4) seasonal biological activity.

Similar and Related Phenomena. The Martian canals which also darken in the spring (AMO1); possible biological activity on Mars (AME14); anomalous frost deposits (AME10); spectroscopic evidence of vegetation (AME17); seasonal changes on earth, as they might be seen from Mars.

Examples

X1. General observations."Seasonal Changes. Not only the polar caps, but also the dark surface markings, including the canals, exhibit conspicuous seasonal changes, which have been carefully studied for many years at the Lowell Observatory, and elsewhere. The darkening, or the appearance, of the dark markings occurs in both hemispheres during the Martian seasons corresponding to our late spring. They become more intense in early summer, remain near maximum intensity for 50 days or more, and begin to fade as the fall season approaches. The fading, which extends into the winter, is in color as well as intensity; the blue-green of the spring and summer turns to chocolate brown. The dates of the changes in the intensity and color of the dark markings are such as would be expected, if these changes are produced by the growth and decline of vegetation. This view was expressed by Lowell and by E. C. Slipher, and many astronomers concur in it." (R1) If the Martian canals have no objective existence, as stated in AMO1, why do their intensities change with the seasons? (WRC)

X2. A detailed analysis of the darkening with the season. "Abstract. The seasonal darkening of the dusky areas of Mars starts with maximum thickness of the winter polar clouds. The regional brightness of the polar caps is connected with the profile of the relief. The average intensity of the dark areas increases from the poles towards the equator; the amplitude of the darkening waves decreases from the poles towards the equator. The combined action of the two darkening waves shows that the action of the darkening generating element is constant for all areographic latitudes during the Martian year. The distribution of the total intensity of the dark areas, the sizes and frequency in areographic latitude of dark blocks or nuclei composing the dark areas of the planet, depend on the duration of the action of the darkening generating element." (R2)

X3. Statistical analysis of the darkening phenomenon. "Abstract. The progressive springtime darkening of the dark areas of Mars is discussed in terms of two models ---one in which the darkening is due to biological activity in response to the increased temperature and humidity, the other in which fine dust is windblown off dark Martian highlands in spring. The observational data on darkening are subjected to a statistical-significance analysis. Although there are dark areas that are exceptions, a very significant correlation emerges between latitude and time of maximum darkening. Other significant correlations are found between brightness of a given dark area and time, between minimum and maximum contrast of the dark areas during the seasonal changes, and between minimum contrast and adjacency to bright areas---correlations expected on the dust models. The present data do not permit a choice between the biological and the windblown-dust models." (R3)

In a later study, Sagan and Pollack opted for the windblown dust model. (R4)

References

R1. Baker, Robert H.; "Mars and Its Satellites," Astronomy, New York, 1938, p. 190. (X1)

R2. Focas, J. H.; "Seasonal Evolution of the Fine Structure of the Dark Areas of Mars," Planetary and Space Science, 9:371, 1962. (X2)

R3. Pollack, James B., et al; "A Statistical Analysis of the Martian Wave of Darkening and Related Phenomena," Planetary and Space Science, 15:817, 1967. (X3)

R4. Sagan, Carl, and Pollack, James B.; "Windblown Dust on Mars," Nature, 223: 791, 1969. (X3)

AMO4 Transitory Dark Spots

Description. Small, intensely dark spots appearing temporarily on the Martian surface. In general, the spots seem to persist for an hour or more. Although the spots are small in the telescopic image, they must be 100 miles and more in size to be seen from earth.

Data Evaluation. A few amateur astronomers have reported these spots. None of the famous Mars observers of the past (Schiaparelli, Lowell, Antoniadi, etc.) have mentioned them in the literature examined so far. Rating: 2.

Anomaly Evaluation. As with the springtime wave of darkening, there is no shortage of possible explanations. In such situations, the phenomenon is probably not very anomalous, even though its exact cause cannot be ascertained. Rating: 3.

Possible Explanations. Dark areas are temporarily stripped of lighter-colored dust by winds; surface areas become temporarily wet due to an upwelling of subsurface water; small volcanos emit clouds of dark dust and ash; unnoticed clouds cast shadows.

Similar and Related Phenomena. The Martian springtime wave of darkening (AMO3).

Examples

X1. October 27, 1924. Dark wedge at Martian terminator. "A little after 8h, while I was studying the disk, I suddenly saw---at least it impressed me as being sudden---a very definite break in the terminator, a jagged saw-toothed cut, in the vicinity of Hellas, just appearing from the unilluminated portion of the disk. At that time it may well have been over that region preceding Hellas, since the terminator side of the planet in that latitude was fairly bright, lacking in markings of note and desert-like in colour. This dark wedge was so definite and black ---as dark as the night sky in the field--- that my first thought was that it must be some optical trouble in the telescope." The marking lasted 1 hour 40 minutes. (R1)

X2. May 1, 1952. This and X3-X5 from reports of James C. Bartlett. "Near to the southern terminus of Mare Tyrrhenum I was surprised to find a very small, intensely dark spot looking much like a satellite shadow, though perhaps not quite as black." Bartlett was referring to the black spots cast upon Jupiter by its Galilean satellites. (R2)

X3. July 3, 1952. A minute black spot on the northern edge of the Hellespontus. (R2)

X4. July 7, 1954. A small, round, intensely black spot, not more than 2" in diameter. Located at Castorius Lacus. (R2)

X5. September 23, 1954. A dark black spot near Ascuris Lacus. (R2)

References

R1. Hamilton, G. H.; "A Dark Entering Wedge at the Terminator of Mars," Popular Astronomy, 33:287, 1925. (X1)

R2. Avigliano, D. P.; "Mars, 1954---Unusual Observations," Strolling Astronomer, 10:26, 1956. (X2-X5)

AMW ATMOSPHERIC PHENOMENA ON MARS

Key to Phenomena

AMW0 Introduction

Martian air is thin. The atmospheric pressure at the surface is less than 0.1% that on earth. Carbon dioxide makes up 95.32% of the planet's atmosphere, with the rest being mainly nitrogen and argon. Water vapor is present in only minute amounts. Still, the potential is there for weather and even climate, because Mars spins once in just over 24 hours and its axis of rotation is inclined by 25°. These variations stir up the Martian atmosphere.

Clouds frequently appear on the face of Mars; they are hardly anomalous. Somewhat more interesting are moving dark lines in the region of the Tharsis ridge, which seem to be atmospheric bores akin the the earth's Morning Glory bores. The most spectacular weather phenomenon of all is the dust storm---not the common small ones but those that encompass the entire planet every few years. How do these immense wind storms get started and what turns them off?

An atmospheric puzzle of long standing is the so-called 'blue clearing'. Usually the Martian surface details do not show up in photographs taken in blue light; but this is not always the case. For a few days, at seemingly random intervals, the blue veil is pulled away as if by magic, and the surface can be photographed at the shorter wavelengths. This is a perplexing phenomenon involving dust in the atmosphere, properties of surface materials, and the angles of illumination and viewing.

AMW1 Bright Spots during Planet-Wide Dust Storms

Description. Bright regions photographed by orbiting spacecraft during planet-wide dust storms when all surface detail is obscured. These bright spots coincide with the locations of craters.

Data Evaluation. Mariner 9 photographs. Rating: 1.

Anomaly Evaluation. Since the bright spots are very likely due to aerodynamic effects in the vicinities of the craters, this is not an important anomaly. Rating: 3.

Possible Explanations. The multiple scattering of light by dust clouds entrained in some craters during dust storms.

Similar and Related Phenomena. None.

Examples

X1. Dust storm phenomena. Abstract. "Mariner 9 photographs of the southern hemisphere of Mars taken during the 1971 planet-wide dust storm display circular bright spots at a time when all near-surface features were totally obscured. Correlating the positions and diameters of these spots with topography shows that they correspond to craters. About half of all the large craters in the study area were brightened. The associated craters are large and flat-floored, have significant rim uplift, and contain dark splotches on their floors. The depth/diameter relationship of the bright spot craters is comparable to that of a planet-wide sample. Depth may not be important in selectively brightening certain craters. The visibility of bright spots in A-camera photographs is strongly dependent on the wavelength of the filter used during exposure. It is proposed that bright spots result from the multiple scattering of incident light in dust clouds entrained within craters during dust storms. The appearance of the dust clouds is a function of the availability of a dust supply and, perhaps, air turbulence generated by winds flowing over upraised rims and rough crater floors. Bright spots persist during the final stage of the planet-wide dust storm. If bright spots are dust clouds, this persistence demonstrates that crater interiors are the last regions of clearing." (R1)

References

R1. D'Alli, Richard E.; "The Significance of Bright Spots Observed during the 1971 Martian Dust Storm," Icarus, 31:146, 1977. (X1)

AMW2 Moving Dark Lines

Description. Long, dark lines that progress across the face of Mars during morning hours in the late spring and early summer. The phenomenon seems confined to a small region near the Tharsis bulge. The lines may be as long as 1,000 miles and exhibit kinks and other topographically-induced variations. Often a system of ten or so parallel lines will develop.

Data Evaluation. Viking Orbiter photographs. Rating: 1.

Anomaly Evaluation. The close resemblance of these lines to bores in the terrestrial atmosphere relegates them to the more-curious-than-anomalous file. Rating: 3.

Possible Explanations. Thermally induced progressive waves or atmospheric bores.

Similar and Related Phenomena. The terrestrial Morning Glory phenomenon (GWC12); internal waves in the earth's oceans (GHW2).

Examples

X1. Frequent occurrence. During late spring and early summer, in the region of the Tharsis ridge, the Viking Orbiters have photographed long, dark, moving lines that stretch for about 1,000 miles. The wave-like lines are seen in the Martian morning and, as the morning progresses, the line usually becomes a system of up to ten or so evenly spaced waves 10-20 kilometers apart. Mountains clearly deflect and disturb the moving lines, indicating that the phenomenon is confined to the lower atmosphere. Kinks and bifurcations are often noted. The lines may be a daily occurrence. (R1, R2)

References

R1. "Rare Martian Weather Wave---with a Kink," Science News, 118:7, 1980. (X1)
R2. Hunt, Garry E., et al; "Daily and Seasonal Viking Observations of Martian Bore Wave Systems," Nature, 293:630, 1981. (X1)

AMW3 Vertical Cloud Columns

Description. Rapidly changing, vertical cloud columns.

Background. Large, white clouds appear frequently in the Martian atmosphere and are not considered anomalous on a planet with polar ice and probable subterranean water reservoirs.

Data Evaluation. The total evidence consists of two Viking Orbiter photographs taken a few seconds apart that show an obvious cloud. The identification of this cloud as a very unusual vertical column depends upon one's interpretation of a fuzzy shadow cast upon the ground by the cloud. The data are sparse and controversial. Rating: 3.

Anomaly Evaluation. Mars today is not considered to be very active thermally, although many ancient volcanos are obvious. Any evidence of steam or volcanic activity would be of great interest. Nevertheless, a bit of residual thermal activity would not be highly anomalous. Rating: 3.

Possible Explanations. Vented steam associated with subterranean heat sources; ditto for water geysers and vented low-temperature water vapor, which would quickly condense in the cold atmosphere; volcanic dust plumes constitute another possibility. A meteor impact might also release a plume of water vapor. Whirlwinds or dustdevils.

Similar and Related Phenomena. Bright, white spots or flares seen on Mars (AMF1); "normal" Martian clouds; transient points of light on the moon (ALF3).

Examples

X1. August 1, 1978. "Leonard Martin of Lowell Observatory claims to have found evidence that Mars may not, in fact, be dead. In a pair of close-up pictures of the Martian surface, taken by Viking Orbiter 1, a lone whitish cloud, casting a slender, faint shadow on the landscape, appears near the base of a suspected volcano. According to Martin, the unusual shape of this shadow and the rapid cloud changes over the 4 1/2-second interval between the two pictures strongly suggest that it was explosively expelled from the surface as a geyser or steam vent.

The mystery cloud was located at 16°S, 79°W---just north of Solis Lacus. In a photomosaic of the general region, the cloud is visible near the base of a mountain capped by a six-mile-diameter crater. Martin believes that the crater probably had a volcanic origin, rather than being an impact feature. If so, this would help support the presence of a nearby steam vent." The semitransparent shadow reveals that the cloud is a vertical column extending to an altitude of almost half a mile, with a base nearly 1800 feet across. The shadow changed markedly in the 4 1/2 seconds. Other scientists feel that the cloud's shadow was too fuzzy to make any conclusions are the cloud's shape or altitude. (R1)

References

R1. Villard, Ray; "Martian Steam Vent Suspected," Star & Sky, 3:6, February 1981. (X1)

AMW4 Planet-Wide Dust Storms

Description. Dust storms that, every few years, grow to envelope the entire planet, blotting out surface detail for months.

Data Evaluation. Many such dust storms have been observed from earth and from spacecraft. Rating: 1.

Anomaly Evaluation. Although a potential bootstrap mechanism is proposed below, no one understands why a few dust storms trigger planet-wide storms while the majority of local storms do not. Rating: 3.

Possible Explanations. One of the frequent local dust storms somehow triggers a bootstrap effect, which probably involves the enhanced absorption of solar radiation during the day and more efficient radiative cooling at night. The consequence would be wider temperature variations, which could feed the storm.

Similar and Related Phenomena. Terrestrial hurricanes also seem to need some sort of triggering mechanism in order to develop into full-fledged storms.

Examples

X1. General observations. "Among the most striking seasonal phenomena on Mars are the global dust storms. Local dust storms are common in the southern hemisphere during its summer. Occasionally one of these storms grows rapidly and, within a few days, engulfs the whole planet, after which the dust takes three to four months to settle out of the atmosphere. The precise cause of the storms is not known but it is possible that dust raised into the atmosphere by local winds increases absorption of sunlight within the atmosphere during the day and increases the efficiency of heat radiation at night. As a result much wider daily variations of temperatures occur at high elevations in the atmosphere than if the atmosphere were clear. These temperature differences between the night and day sides of the planet cause large winds which flow from one hemisphere to the other and raise more dust so that the process feeds on itself. Dust raising ceases when the atmosphere is so loaded with dust that its temperature is constant, irrespective of elevation." (R3) Much more detail and a theoretical model may be found in R2.

X2. Circa 1845. Except for a broad belt around the equator, the usually ruddy face of Mars turned to a brilliant white. (R1) It is uncertain just what this phenomenon could have been---dust or exceptional cloud cover and/or frost or ice. This seems the best place to file it for this edition. (WRC)

References

R1. "Extraordinary Appearance of the Planet Mars," Eclectic Magazine, 6:568, 1845. (X2)

R2. Zurek, Richard W.; "Martian Great Dust Storms: An Update," Icarus, 50: 288, 1982. (X1)

R3. Carr, Michael H.; "The Surface of Mars: A Post-Viking View," Mercury, 12:2, January/February 1983. (X1)

AMW5 Isotopic Anomalies in the Martian Atmosphere

Description. Large excesses or deficiencies in the abundances of isotopes in the Martian atmosphere when compared to the abundances predicted by the prevailing models of solar system genesis and the evolution of Mars' atmosphere.

Data Evaluation. Viking spacecraft compositional analyses. Although these data are considered to be of high quality, additional experiments on future spacecraft would be desirable, considering the complexities and remoteness of the experiments. Rating: 2.

Anomaly Evaluation. The solar system argon anomalies were of great concern initially, as described in more detail in AVW4, for it was thought that all the terrestrial planets would possess much the same noble gas inventories. The model of solar system formation has now been modified to account for these anomalies. By delaying the final accretion of Mars, additional outgassing can occur, and the seriousness of the argon anomaly for Mars has been reduced. Rating: 3.

Possible Explanations. The Martian argon deficiency is now attributed to enhanced outgassing that took place from the planetesimals that later accreted to Mars. The nitrogen-15 excess is thought to be due to recent cosmic ray interaction with oxygen-16.

Similar and Related Phenomena. The excess argon in the atmosphere of Venus (AVW4).

Examples

X1. Nitrogen-15. This isotope is present in excess in the Martian atmosphere, according to Viking spacecraft analyses. Nitrogen-15 does not have a long-lived radioactive parent, so it must have been produced more recently, probably by cosmic rays interacting with oxygen-16. By "excess", scientists mean that there is far more nitrogen-15 present than can be accounted for by the best models of Martian atmospheric evolution. (R1)

X2. The noble gases. "An acceptable model for the terrestrial planets must account for: (1) an abundance of primordial neon and argon in Venus' atmosphere ~70 times larger than that for earth; (2) an abundance of primordial neon and argon in Mars' atmosphere ~ 180 times less than that for earth; (3) ratios of primordial neon to argon similar for all three planets; (4) occurrence of the planetary pattern for Ne, Ar, Kr and Xe in meteorites and in the atmospheres of Mars and earth, with a departure for Kr on Venus; (5) abundance ratios $^{129}Xe/^{132}Xe$ and $^{40}Ar/^{36}Ar$ for Mars higher than for earth; (6)a ratio $^{20}Ne/^{22}Ne$ for Venus similar to solar wind, though higher than for earth; and (7) a ratio $^{36}Ar/N$ for Venus higher than for earth by a factor of 20. The problem may be summarized as follows. Venus has an unexpectably high concentration of ^{36}Ar, a ratio $Ne/^{36}Ar$ similar to earth, Mars and the planetary component of meteorites, but a distinctly different value for the ratio $^{20}Ne/^{22}Ne$. The concentration of ^{36}Ar in Mars' atmosphere is low and cannot be attributed simply to inefficient degassing." The low concentration of ^{36}Ar in the Martian atmosphere may be a consequence of the enhanced degassing from planetesimals with radii between 5 and 100 kilometers early in solar system history, before they accreted to Mars. (R2)

References

R1. Yanagita, Shohei, and Imamura, Mineo; "Excess ^{15}N in the Martian Atmosphere and Cosmic Rays in the Early Solar System," Nature, 274:234, 1978. (X1)

R2. McElroy, Michael B., and Prather, Michael J.; "Noble Gases in the Terrestrial Planets," Nature, 293:535, 1981. (X2)

AMW6 The Blue Clearings

Description. The sudden, short-lived appearance of Martian surface details on photographs taken in blue light through terrestrial telescopes. Ordinarily, surface details cannot be seen in blue light, even though they are readily visible at other visible wavelengths. Blue clearings last only a few days and may occur at any time. Blue clearing may prevail for close-up spacecraft even though it does not for terrestrial observers.

Data Evaluation. Scores of blue clearings have been recorded in the literature, but usually only at oppositions. Recently, the phenomenon has been studied at all elongations, at all wavelengths, and close-up from spacecraft. Rating: 1.

Anomaly Evaluation. The current explanation of the blue clearings, as described below, appears to account for most aspects of the phenomenon, although it is rather complex and a few doubts remain. Rating: 3.

Possible Explanations. Early theories of planet-wide ice crystal and/or dust layers have been discarded. Present preference is for a combination of effects: (1) a thin dust haze over bright areas of Mars; (2) the varying contrast of the Martian surface with changing angles of illumination and viewing.

Similar and Related Phenomena. None.

Examples

X1. General description. "It has been known for many years that photographs of Mars taken in blue light do not usually show those surface features of the planet which can be photographed in yellow or red light. The accepted interpretation of this observation is that a cloud or haze layer exists in the Martian atmosphere and that this haze scatters short wave lengths of light more strongly than long wave lengths. Thus the surface detail is more obscured the shorter the wave length. This explanation received strong substantiation when Slipher (1937) found that there were occasional periods during which the surface markings could be photographed in blue light. Only a cloud layer of some sort could exhibit such variability in transparency. As a result, the phenomenon has come to be called the 'blue haze.'

Over the years since this first observation of 'blue clearing,' a number of other cases of temporary dissipation have been recorded. The characteristics of these clearings have been: (1) the clearing reveals large portions of the surface; (2) the clearing usually occurs for several days near the opposition date and can be detected at most oppositions; (3) it takes only a few nights for the haze to dissipate; and (4) it takes only a few nights for the haze to re-establish itself." (R3) See X2 for further comments on the correlation of clearings with oppositions. (WRC)

The conclusions of a more recent study are as follows: "(1) The observed contrast changes which have traditionally been called 'blue clearings' are due to variations in the bright areas, not the dark ones; (2) Syrtis Major does not appear to undergo limb darkening in red light, while surrounding bright areas do; (3) The classical 'blue clearing' phenomenon is not limited to blue light; it is a contrast enhancement which occurs at all visible wavelengths (from 4000 to 6200 A, at least; (4) The contrast of Syrtis Major with its surroundings showed a definite dependence on phase angle in 1969; (5) A variable diurnal variation in the visibility of Syrtis Major occurs." (R15)

Moroz has presented an overview of the whole blue haze phenomenon. "In the spectral region $\lambda < 4000$ A the contrast between the details of the Martian surface is considerably less than in the region $\lambda < 5500$ A. Photographs from the earth do not give the possibility, as a rule, of distinguishing dark and bright regions in blue light. They picture Mars as a disk with almost constant brightness. Polar caps are distinguished by a little bright spot and sometimes blue clouds become brighter on the edge. However on rare days dark regions suddenly begin to be just about visible on the blue photographs. These appearances are named as 'blue clearings'.

It was supposed for a long time that it is the so-named 'blue haze' in the atmosphere of Mars which creates the large absorption and decreasing contrasts. It was also supposed that in the period of clearings the optical density of the blue haze reduces for an unknown reason and leads to an increase in contrasts. Serious doubts about the correctness of these ideas arose earlier but the strongest blow was struck by the phototelevision pictures obtained by Mariner-6. Color dependence of contrast between dark and bright regions, as had been proved by Sharonov earlier (1961), appeared to be the property of Martian rocks. The optical thickness of the Martian atmosphere appeared to be insignificant in the blue spectral region and incapable of having a noticeable influence on the contrasts.

A basically new explanation of 'blue clearings' was advanced independently by Boyce and Thompson (1972) and Prokofyeva (1973). They explain blue clearings by the appearance of a thin dust layer in the atmosphere predominantly above the bright regions. The relatively small life-time of the appearance (several days) may be in agreement with particles being of small size if we propose that the blue clearing appears not because of the subsidence of the particles but because of their transfer to the dark regions." (R16)

X2. Correlation of blue clearings with oppositions. It was remarked early in the history of the blue clearing phenomenon that the blue clearings seemed to be connected in some way with inferior oppositions; that is, they were noticed primarily when Mars was closest to earth. (R1) Historical records soon showed, however, that blue clearings did not occur at all oppositions. It was also pointed out that most observatories did not systematically observe Mars except during opposition, so that a strong bias in observations was bound to exist. (R5) Later studies of Mars at times well before and after opposition revealed blue clearings as early as 175 days before opposition and 349 days after opposition. (R13) In essence, oppositions seemed to have little to do with the onset of blue clearings. (WRC)

X3. Correlation of blue clearings with geomagnetic activity. Bigg has proposed that the earth is shielded to some extent from the solar wind during new moons and the inferior oppositions of Mercury and Venus. (GEM2-X5) He also provides some evidence that the earth may likewise shield Mars at opposition, for the blue clearings, as discussed in X2, do seem correlated with oppositions, too. (R3, R4) But in X2 it is demonstrated that blue clearings do not seem to have any cause-and-effect relationship to oppositions. (WRC)

X4. Blue clearing observed by spacecraft but not from earth. When Mariners 6 and 7 flew past Mars, crater details were photothrough blue filters, even though similar details could not be seen from earth; i.e., blue clearing existed for the spacecraft but not terrestrial telescopes. (R11) Considerable debate then erupted involving the filter ranges of the spacecraft cameras, exactly which wavelengths are involved in blue clearing, contrast effects, etc. (R10, R12) The upshot seems to have been that the old 'blue haze' theory was no longer viable; there was a more complex phenomenon at hand. (WRC)

X5. Possible explanations. Several theories have been proposed down the years: ice crystals in the Martian atmosphere (R2, R4, R9); a dust haze (R8), possibly of meteoric origin; and, widely accepted today, a thin layer of dust above the bright regions. (R15, R16)

References

R1. Martz, E.P., Jr.; "Variation in Atmospheric Transparency of Mars in 1938," Astronomical Society of the Pacific, Publications, 66:45, 1954. (X2)

R2. "Blue Haze on Mars Believed Ice Crystals," Science News Letter, 74:265, 1958. (X5)

R3. Hess, Seymour L.; "Blue Haze and the Vertical Structure of the Martian Atmosphere," Astrophysical Journal, 127:743, 1958. (X1, X5)

R4. Urey, Harold C.; "The Blue Haze of Mars," Astrophysical Journal, 128:736, 1958. (X5)

R5. Richardson, Robert S., and Roques, Paul E.; "An Example of the Blue Clearing Observed 74 Days before Opposition," Astronomical Society of the Pacific, Publications, 71:321, 1959. (X2)

R6. Bigg, E.K.; "Lunar and Planetary Influences on Geomagnetic Disturbances," Journal of Geophysical Research, 68:4099, 1963. (X3)

R7. Atkinson, Gerald; "Planetary Effects on Magnetic Activity," American Geophysical Union Transactions, 45:630, 1964. (X3)

R8. Palm, A., and Basu, B.; "The Blue Haze of Mars," Icarus, 4:111, 1965. (X5)

R9. Firsoff, V.A.; "On the Nature of the 'Violet Layer' of Mars," Observatory, 88:223, 1968. (X5)

R10. Opik, E.J.; "Mars: The Changing Picture," Irish Astronomical Journal, 9:136, 1969. (X4)

R11. Leighton, Robert B.; "The Surface of Mars," Scientific American, 222:27, May 1970. (X4)

R12. Hess, Seymour L., et al; "Blue Haze and Mariner 6 Pictures of Mars," Sci-

ence, 176:906, 1970. (X4)
R13. Capen, Charles F.; "Martian Blue-Clearing during 1967 Apparition," Icarus, 12:118, 1970. (X2)
R14. Well, E.H., and Hale, D.P.; "Martian Blue Haze Clearings and Flash Phenomena," Nature, 232:324, 1971. (X5)
R15. Boyce, Peter B., and Thompson, Don T.; "A New Look as the Martian 'Violet Haze' Problem I. Syrtis Major-Arabia, 1969," Icarus, 16:291, 1972. (X5)
R16. Moroz, V.I.; "The Atmosphere of Mars," Space Science Reviews," 19:763, 1976. (X1, X5)

AN INTRODUCTION TO NEPTUNE

Key to Categories

ANB PROBLEMS WITH NEPTUNE'S ORBIT
ANF NEPTUNE'S INTRINSIC RADIATION
ANL RING AND SATELLITE IRREGULARITIES
ANO TELESCOPIC ANOMALIES

Neptune is the outermost of the so-called "giant" planets, although it is only one-third the diameter of Jupiter and is considerably more dense. Its two known satellites are in obvious disarray and seem to indicate past catastrophism at the edge of the solar system. Neptune may be ringed like the other three giant planets, but if the ring is really there, it seems to be incomplete or distorted according to present data. The reflectivity or albedo of Neptune is very high and varies substantially. Just why, no one knows. Like some of the other planets, Neptune emits more energy than it gets from the sun. Perhaps it is still losing primordial heat or possesses heat-producing radioisotopes. Neptune's anomalies are still few in number and, as in the case of Uranus, may be expected to multiply when spacecraft finally give us a good look at the planet.

ANB PROBLEMS WITH NEPTUNE'S ORBIT

Key to Phenomena

ANB0 Introduction

None of the outer planets follows the dictates of celestial mechanics to the letter. Neptune, in particular, strays far from its predicted orbit. There must, say many astronomers, be a tenth planet that gravitationally pulls Neptune from its assigned path. For a long time, Pluto was thought to be the culprit, but now its mass has been found to be far too small to have the observed effect on Neptune.

ANB1 The Large Residual in Neptune's Orbit

Description. The long-standing, large discrepancy between the observed and predicted positions of Neptune.

Data Evaluation. Data from two centuries of astronomical observations (including prediscovery sightings of Neptune) are available. Rating: 1.

Anomaly Evaluation. It has proven impossible to account theoretically for the observed positions of Neptune using currently accepted laws and celestial mechanics and the effects of the nine recognized planets of the solar system. Predictions of Neptune's future positions inevitably fail in just a few years. Historically, such anomalies have led to the discoveries of new planets and, in the case of Mercury, have given support to new physical theories (Relativity). In Neptune's case, a new trans-Plutonian planet seems the most likely outcome. Rating: 2.

Possible Explanations. A tenth planet gravitationally perturbs Neptune's orbit. Actually, several small planets or even distributed mass could do the job.

Similar and Related Phenomena. Planet X (AXO1); the advance of Mercury's perihelion (AHB1).

Examples

X1. Anomalies observed in Neptune's orbit. "A new computation of the orbit of Neptune has revealed an unexpected discrepancy between predicted and observed positions. According to Dr Dennis Rawlins of Baltimore, an alien perturbation may be responsible for the failure of observations made in 1795 to fit modern predictions. Digital computing techniques have enabled us to gain considerable knowledge about the detailed motion of the planets. In the past few years mathematicians have successfully accounted for several alleged divergences of theory and observation and, as a result, sizes and distances in the solar system are known with great precision. Only Neptune is refusing to fit this neat picture. Although Neptune was not discovered until 1846, prediscovery records can be traced in star catalogs back to 1795, and it is these observations that are causing the trouble. What Dr Rawlins finds is that the position predicted for 1795 is seven seconds of arc (equivalent to 100 000 miles) away from the observed position. It is impossible to attribute a difference as large as this---described by this astronomer as the Great Unexplained Residual---to instrumental effects, and so another explanation of the discord must be sought. It is certainly possible that a small undiscovered planetary body is responsible." (R2, R1)

The situation worsened when Pluto was found to have a smaller mass than previously assigned. Astronomers "...now know that the best calculated orbits for all of the outer planets---Jupiter, Saturn, Uranus, Neptune, Pluto---fail to agree with observations to some degree. The new, lower estimate of Pluto's mass in the calculations complicates matters ever further. (P. K.) Seidelmann and his group have prepared a series of charts comparing the observations and the predicted positions of both Uranus and Neptune. In just a 10-year period, since the early 1970s, the charts show a noticeable diversion from the straight line that indicates Neptune's expected position. 'It's running off the fit already,' says Seidelmann. And, he adds, looking at a plot of Uranus's right ascension, 'there seems to be something funny here too.' Pluto also appears to be showing such residuals, or deviations, although the 51-year duration of observations is too short to make much of that. A variety of such plots all show the same kinds of discrepancies, 'Historically, everybody who has been able to fit the observations of Neptune in the past has failed to predict its positions in the future,' says Seidelmann." (R5)

X2. Explanations of Neptune's residual. A tenth planet beyond Pluto is usually blamed for the discrepancies in Neptune's orbit, and those in the orbits of the other outer planets as well. For example, T. C. Van Flandern calculates that a planet with two to five times the mass of the earth, located 50 to 100 astronomical units from the sun, could account for Neptune's orbital irregularities. (R4) The entire subject of Planet X, the oft-suggested planet beyond Pluto is covered in category AX, in another volume of this Catalog. (WRC)

References

R1. Rawlins, Dennis; "The Great Unexplained Residual in the Orbit of Neptune," Astronomical Journal, 75:856, 1970. (X1)

R2. "News from the Outposts of the Solar System," New Scientist, 48:165, 1970. (X1)

R3. Rawlins, D., and Hammerton, M.; "Is There a Tenth Planet in the Solar System," Nature, 240:457, 1972. (X2)

R4. "What's Bothering Neptune?" Science News, 119:68, 1981. (X1, X2)

R5. Frazier, Kendrick; "A Planet beyond Pluto," Mosaic, 12:27, September/October 1981. (X1, X2)

ANF NEPTUNE'S INTRINSIC RADIATION

Key to Phenomena

ANF0 Introduction

Neptune's distance precludes any detection of any small-scale and transient sources of radiant energy. However, measurements of millimeter radio emissions from the planet-as-a-whole are feasible; and they suggest that Neptune is a net energy producer.

ANF1 Measurements of Intrinsic Energy from Neptune

Description. The emission of more radiant energy from Neptune than the planet intercepts from the sun.

Data Evaluation. Terrestrial measurements of the radio brightness of Neptune in the millimeter region of the spectrum. The planet's overall energy balance is inferred from these limited measurements. Rating: 2.

Anomaly Evaluation. Neptune is much more dense than Uranus (2.27 vs. 1.56), so that a heat-producing inventory of radioisotopes is more reasonable. Actually, we know so little about the planet that only educated guesses are possible. Rating: 3.

Possible Explanations. An internal energy source, such as radioactive isotopes is possible. Also, Neptune could still be radiating primordial heat acquired during the accretion phase.

Similar and Related Phenomena. Neptune's brightness changes in the visible range (ANO1); the intrinsic radiation from Venus (AVF1), Jupiter (AJF1), and Saturn (ARF2).

Examples

X1. Radio astronomy measurements. "Drs Ivan Pauliny-Toth and Ken Kellermann of the US National Radio Astronomy Observatory, Greenbank, West Virginia, have completed their measurements of the microwave spectrum of Neptune and Uranus. At millimetre wavelengths the radiation from these planets is entirely thermal in origin, so that surface temperatures can be estimated by measuring the intensity of the emissions. The brightness temperatures found at 9.5 and 3.5 mm by Pauliny-Toth and Kellermann are considerably in excess of the temperature of a planet heated only by the Sun's rays For instance, Neptune has a brightness temperature of 88 K at 3.5 mm, which is twice the value expected from the solar flux. Clearly Neptune and Uranus, in common with Jupiter, have substantial internal heat sources. It is widely believed that the fission of atomic nuclei in radioactive elements deep inside the planets causes the temperature excess." (R1)

References

R1. "News from the Outposts of the Solar System," New Scientist, 48:165, 1970. (X1)

ANL RING AND SATELLITE IRREGULARITIES

Key to Phenomena

ANL0 Introduction

Neptune possesses at least two satellites; and these are very poorly behaved as satellites go! In fact, Triton's retrograde orbit and Nereid's very high eccentricity have impelled astronomers to hypothesize that a celestial visitor once radically perturbed the Neptunian system.

Does Neptune, like Uranus, Saturn, and Jupiter, also have a ring? If it did not, Neptune would be the only giant planet without one. Except for some suspicious 1846-1847 sightings of a Neptunian ring, astronomers have generally supposed that Neptune is ringless. However, the modern discovery of a Uranian ring through the photometric observation of stellar occultations has encouraged some researchers to try their luck with Neptune. The results have been pro and con. The jury is still out.

ANL1 Disarray among Neptune's Moons

Description. Triton's retrograde orbit; Nereid's highly eccentric orbit.

Data Evaluation. Triton and Nereid have been observed with precision for many years. Rating: 1.

Anomaly Evaluation. No one believes that Triton and Nereid formed where they now orbit. It is much less anomalous if they are assumed to have been displaced by a celestial interloper; and, given the modern tolerance for catastrophic scenarios, a Neptunian interloper would not be especially anomalous. Rating: 3.

Possible Explanations. A large object invaded the Neptunian system, perturbing the original, well-behaved orbits of Triton and Nereid.

Similar and Related Phenomena. Pluto is often considered to be an escaped satellite of Neptune and, therefore, another sign of past catastrophism.

Examples

X1. Triton. "Triton is the only large, relatively close satellite in the solar system having a retrograde orbit and so highly inclined (160 degrees) to its planet's equator." (R1)

X2. Nereid. "Nereid has twice the eccentricity of any other satellite or planet in the solar system." (R1)

References

R1. Frazier, Kendrick; "A Planet beyond Pluto," Mosaic, 12:27, September/October 1981. (X1, X2)

ANL2 Neptune's Elusive Ring and Its Possible Incompleteness

Description. The apparent presence around Neptune of a broken or twisted ring of matter.

Data Evaluation. The 1846-1847 visual sightings of a Neptunian ring are generally discounted. The modern photometric observations made during stellar occultations have been both positive and negative. Unfortunately, the positive observations imply difficult-to-believe ring geometries. More and better data are required here. Rating: 3.

Anomaly Evaluation. The anomaly here is not the presence of a ring (three other planets have them), but rather in the apparently strange shape of the Neptunian ring. An incomplete ring, if verified, would be unique in the solar system. A ring of matter should spread out circumferentially and complete itself. A shepherd satellite might help keep a gap open, but no such satellite can be seen from earth. A judgment of anomalousness is difficult under such circumstances. Assuming no shepherd satellite and the reality of an incomplete ring: Rating: 2.

Possible Explanations. An incomplete ring might be maintained by a shepherd satellite. The 1984 observations that suggest an incomplete ring might be explained if the star's passage was tangent to the ring rather than intersecting. The past visual observations of Neptune's ring could have been accurate if planetary rings are highly dynamic (as Saturn's now seem to be) and on occasion expand greatly, rendering them highly visible temporarily. Since the rings of Jupiter, Uranus, and now Neptune were all reportedly seen before modern "discovery", it seems wise to at least consider this possibility.

Similar and Related Phenomena. The discovery and elucidation of the Uranian ring system (AUL1).

Examples

X1. 1846-1847 observations. "Several European astronomers have pronounced in favor of the existence of a ring around the planet Neptune. Mr. Lassell, of Liverpool, observing with his Newtonian reflector, of two-feet aperture, first announced its existence, in October, 1846; and in January last, Prof. Challis, of Cambridge, using the large Northumberland reflector, was disposed to believe Mr. L.'s assertion. The ratio of the diameter of the ring to that of the planet is about that of 3 to 2. The angle made by the axis of the ring with a parallel of declination, in S. preceding or N. following quarter, is about 65°. Other observers, however, with equal means, cannot detect any such ring." (R1)

R. Baum has written an excellent histori-

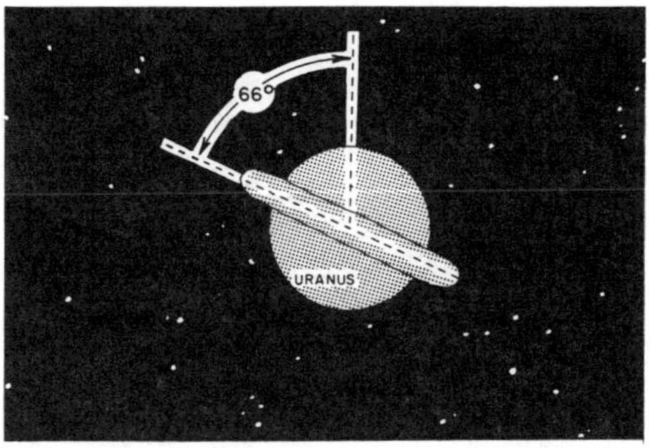

The ring of Neptune as drawn by Challis in 1847. (X1)

cal review of the supposed early discoveries of Neptune's ring. He summarized the evidence thusly: "In the winter of 1846-7, Neptune was found to be accompanied by a thin, flat, elliptical form of high eccentricity, which when most clearly seen gave the undoubted impression of a ring obliquely inclined to the line of vision. This feature was neither clear nor obvious, but tenuous and faint, and required reasonable conditions for its detection. It was only definitely seen in large apertures." (R3) Baum mentioned that the Northumberland telescope employed by Challis was actually a refractor, contrary to the first quotation above. Baum's general conclusion was that the ring seen by W. Lassell and J. Challis was illusory.

Two more recent reviews of the early reports by Hoyt and Hetherington come to similar conclusions. (R4, R5)

X2. A 140-year interim. "During the latter part of the 19th century and the early 20th century, extensive measurements of the diameter of Neptune and of the position of Triton were made by several astronomers with large refractors of optical quality much superior to Lassell's reflector. The fact that this repeated scrutiny did not reveal a ring demonstrates clearly that Neptune does not have one fitting the description given by Lassell and Challis." (R4)

X3. 1968 occultation measurements. Observers in New Zealand recorded a dip in the light curve of a star before and after the star was occulted by Neptune. These dips were interpreted as being the consequence of a ring of matter with an inner radius of about 28,600 kilometers and an outer radius of 32,900 kilometers. (R8-R10)

X4. 1981 occultation measurements. The careful observation of two stellar occultations by Neptune showed no hint of the existence of any ring. (R6, R7)

X5. 1983 occultation measurements. On June 15, several groups, from Tasmania to Taiwan, photometrically observed the passage of a star past Neptune. Again, no hint of a ring was detected. (R11)

X6. 1984 occultation measurements. When the star SAO 186001 brushed by Neptune on July 22, a number of astronomers were watching it carefully to see if its light would be diminished by an encircling, Saturn-like ring of matter. (The ring of Uranus was discovered in a similar fashion.) Observers at the European Southern Observatory, in Chile, and the Cerra Tololo Observatory, also in Chile some 90 kilometers away, detected a 1-second, 35% reduction of the star's light at the same instant. These data suggest the presence of "something" 10-20 kilometers wide---hardly an undiscovered satellite, but very possibly a ring. But given the geometry shown in the illustration, there should have been two occultations, but only the one of the right was registered. Conclusion: Neptune may have a partial ring or, alternatively, a grotequely twisted one. (R12-R14) The existence of a partial ring is the real anomaly here. (WRC)

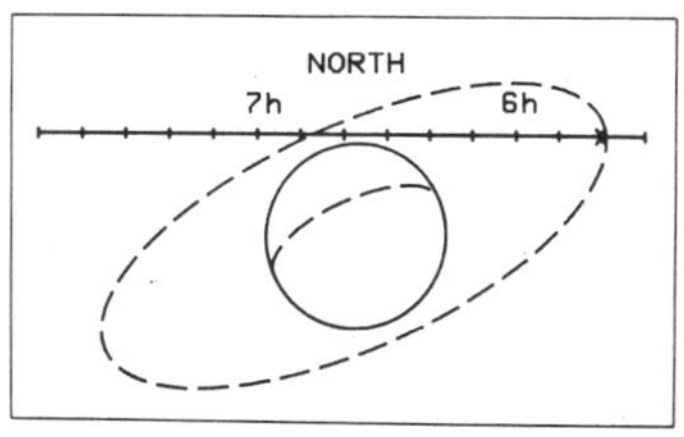

July 22, 1984. The star SAO 186001 brushed past Neptune on the trajectory shown. An occultation was detected at the right-hand intersection but not at the left. (X6)

References

R1. "Neptune, Its Supposed Ring and Satellite," American Journal of Science, 2: 4:287, 1847. (X1)

R2. Challis, J.; Astronomische Nachrichten, 25:231, 1847. (X1)

R3. Baum, Richard; "Neptune 1846-1847," The Planets, New York, 1973, p. 120. (X1)

R4. Hoyt, William G.; "Reflections Concerning Neptune's 'Ring'," Sky and Telescope, 35:284, 1978. (X1, X2)
R5. Hetherington, Norriss S.; "Neptune's Supposed Ring," British Astronomical Association, Journal, 90:20, 1979. (X1)
R6. Elliot, J. L., et al; "No Evidence of Rings around Neptune," Nature, 294:526, 1981. (X4)
R7. Kerr, Richard A.; "Neptune's Rings Fading," Science, 213:1236, 1981. (X4)
R8. Waldrop, M. Mitchell; "Neptune: A Ring at Last," Science, 217:143, 1982. (X3)
R9. "Neptune's Ring," Astronomy, 10:62, September 1982. (X3)
R10. Hecht, Jeff; "Does Neptune Have Rings Too?" New Scientist, 94:760, 1982. (X3)
R11. Kerr, Richard A.; "Neptune Ring Fades Again," Science, 222:311, 1983. (X5)
R12. Eberhart, J.; "Signs of a Puzzling Ring around Neptune," Science News, 127:37, 1985. (X6)
R13. "Neptune Reveals Its Ring," New Scientist, 5, January 17, 1985. (X6)
R14. Kerr, Richard A.; "What's Going On at Neptune?" Science, 227:734, 1985. (X6)

ANO TELESCOPIC ANOMALIES

Key to Phenomena

ANO0 Introduction

As in the case of Uranus, Neptune is too distant for terrestrial observers to see much planetary detail, although a few cloud-like structures have been reported. The only visual anomaly that has been found is the planet's varying brightness.

ANO1 Neptune's Variable Brightness

Description. Changes in the visible brightness by as much as 30% (up to 400% in the infrared). Both correlation and anticorrelation with solar activity have been claimed.

Data Evaluation. The existence of both correlation and anticorrelation with solar activity casts suspicion on the quality of the data. Until this conflict is resolved, the data rating will have to be low. Rating: 3.

Anomaly Evaluation. At Neptune's distance the sun's rays are so weak that any small variation in them due to changes in solar activity would appear to be completely inadequate to cause significant albedo changes. Thus, the brightness changes, if truly sun-controlled, are rather mysterious. Rating: 2.

Possible Explanations. Solar radiation and perhaps the solar wind act as triggers for chemical and/or meteorological processes which, in turn, change Neptune's albedo.

Similar and Related Phenomena. The varying brightness of the Saturnian moon Titan (ARL8) and Uranus (AUO1).

Examples

X1. Correlation of Neptune's brightness with the solar cycle. According to an analysis of luminosity data by Balasubrahmanyan and Venkatesan, the brightness of Neptune changes in step with the sunspot cycle, being brightest when solar activity is high and vice versa. (R1) See AUO1-X1 for a quotation from this source. (WRC)

On the other hand Lockwood and Thompson have found a significant anticorrelation between solar activity and Neptune's brightness! (R3) See ARL8-X1 for a pertinent quotation from this source. (WRC)

X2. Infrared brightness. In 1976, Richard Joyce, of Kitt Peak, and colleagues found that the infrared brightness of Neptune had increased by a factor of four between April 1975 and March 1976. (R2)

References

R1. "Why the Solar Cycle Alters the Planets' Brightness?" New Scientist, 52:146, 1971. (X1)

R2. "Neptune Brightens Up His Image," New Scientist, 73:393, 1977. (X2)

R3. Lockwood, G. W., and Thompson, D. T.; "A Relationship between Solar Activity and Planetary Albedos," Nature, 280:43, 1979. (X1)

AP INTRODUCTION TO PLUTO

Tiny Pluto swings around the sun at the fringes of the solar system. Little is really known about the planet. When it was discovered in 1930, its mass was thought to be about the same as that of the earth. More recent estimates make it less than 1% the mass of earth. It has become quite apparent that this small chunk of matter could not be the perturber of Uranus and Neptune that astronomers were looking for when they found Pluto. It is probably only an escaped moon of Neptune. The fact that Pluto's orbit intersects that of Neptune and the observation that Neptune's moons are askew strengthen the suspicion that some celestial event tore Pluto loose from Neptune long ago.

No Plutonian anomalies worth recording here have been uncovered. True, the planet does rotate a bit slowly---it has a 6.39-day period of rotation---and its recently discovered satellite, Charon, is one-tenth of Pluto's mass, an unusually large fraction for a solar system moon. These attributes of the Plutonian system do not seem to warrant more than this brief mention. Once we can see Pluto close-up via spacecraft cameras, many anomalies will surely appear. The whole history of astronomy, from the invention of the telescope to the Voyager flights, assure us that better instruments will conquer the distances that conceal anomalies from us.

AR INTRODUCTION TO SATURN

Key to Categories

ARF SATURN'S INTRINSIC RADIATION
ARL SATELLITE AND RING ANOMALIES

In a good telescope, Saturn is a spectacular sight. The rings, of course, make all the difference. Although other planets have ring systems, Saturn's are bright and vivid, even in small telescopes. The three major rings are thousands of miles in width, but their thicknesses seem to be measured only in feet. Such obvious features are duly registered in all the textbooks. But the idiosyncracies of the rings---those known before the Voyager encounters with Saturn---are never mentioned. Saturn has a long telescopic history of bright spots appearing on the rings, along with out-of-perspective shadows of the planet and changes in ring size and brightness. The Voyager missions lengthened the list of anomalies considerably. These spacecraft sent back photos showing that the major Saturnian rings are actually made up of thousands of small ringlets, some of which are eccentric, others twisted, and many spiral. The material of the rings seems to be predominantly water ice, but where all this ice came from and when are not known. It is not beyond the bounds of possibility that Saturn's rings are young in terms of solar system age.

Saturn's strongly banded sphere displays few anomalies of consequence. Astronomers see Saturn's surface as highly convective, with turbulence and transient phenomena likely. Temporary white spots, for example, occasionally manifest themselves on the surface. These are not considered anomalous enough to include here. The precise nature of Saturn from cloud top to core is not certain, but it is probably much like Jupiter; that is, mostly hydrogen and helium. The planet does emit considerable infrared energy, indicating that also like Jupiter it is a generator of energy in quantities over and above that received from the sun. Of all Saturn's emissions, the intense bursts of radio energy (the Saturnian Electrostatic Discharges or SEDs) evoke the most scientific excitement. The SED bursts peak about every 10 hours, and astronomers are not sure whether they emanate from the planet's atmosphere or rings.

ARF SATURN'S INTRINSIC RADIATION

Key to Phenomena

ARF0 Introduction

Like Jupiter, Saturn emits more energy than the sun delivers to it. Whereas Jupiter's intrinsic radiation has not been explained to everyone's satisfaction (it's probably due to gravitational contraction of the planet), the theoretical model of Saturn is qualitatively and quantitatively very successful. Consequently, Saturn's energy production is no longer considered anomalous.

The radio energy emitted by Saturn, however, is still high on the list of anomalies. Specifically, it is Saturn's electrostatic discharges that continue to puzzle planetary scientists. Do these bursts of energy, which resemble those from lightning, come from colossal storms in the planet's atmosphere or are Saturn's rings arcing over like a bad commutator?

ARF1 The Saturn Electrostatic Discharges (SEDs)

Description. Bursts of radio frequency energy resembling terrestrial lightning discharges that peak every 10 hours, 10 minutes.

Data Evaluation. Radio physics experiments aboard Voyagers 1 and 2. Rating: 1.

Anomaly Evaluation. Both explanations noted below imply bizarre situations to say the least: (1) Immense electrical discharges across the ring structures; (2) An apparently stationary electrical storm of huge proportions in Saturn's atmosphere. While these phenomena are remarkable in size and character, they do not seem to challenge electrical theory too much. The real problems are the source of electrical energy in the rings, or the origin and nature of the postulated atmospheric storm. Rating: 2.

Possible Explanations. Electrical discharges in the B ring, possibly stimulated by a ring gap or object orbiting at the 10 hour, 10 minute period; a large electrical storm in Saturn's atmosphere rotating with the atmosphere at the measured SED period.

Similar and Related Phenomena. The spokes on Saturn's rings (ARL5), which also peak at the SED period.

Examples

X1. General description of the Saturn Electrostatic Discharges (SEDs): "David Evans, Joseph Romig, and James Warwick of Radiophysics, Inc., of Boulder reported that SED appeared as 30- to 250-millisecond bursts over the radio-frequency range of 20 kilohertz to 40 megahertz scanned by Voyager's radio astronomy instrument. The mean power in a 50-millisecond burst was about 1 billion watts, peak power being at least 10 billion watts. (This is the latest of several sets of updated figures.) The power of a terrestrial lightning stroke of radio frequencies typically reaches about 200,000 watts. The number of SED bursts detected by Voyager tended to peak every 10 hours and 10 ± 5 minutes. The approximate 10-hour periodicity of SED suggests two possible locations for its source. One is within the equatorial region of Saturn's atmosphere, where the speed of the wind added to the rotation of the planet would carry a lightning storm around the planet every 10 hours. The second is within the B ring at a distance of 1.81 Saturn radii. The Radiophysics group prefers the ring source because radio-frequency signals generated by atmospheric lightning would probably not make it through the ionosphere. Even if they did, the frequency distribution of lightning-generated signals is never as flat as that of SED, they note." (R4, R1)

Other investigators have correlated the SED activity with both the rotation of Saturn's magnetic field and the frequency of spokes appearing on the rings. All three phenomena have almost exactly the same period from maximum to maximum. (R1)

X2. Possible explanations. A large electrical storm in Saturn's atmosphere. (R5-R9) An object or gap in Saturn's B-ring. (R2, R4, R8)

References

R1. Kerr, Richard A.; "Spokes, SKR, and SED: A Connection?" Science, 218:276, 1982. (X1)

R2. Evans, D.R., et al; "The Source of Saturn Electrostatic Discharges," Nature, 299:236, 1982. (X2)

R3. Hoagland, Richard C.; "The Phantom of the Rings," Science Digest, 90:80, July 1982. (X1, X2)

R4. Kerr, Richard A.; "Lightning on Saturn or Ring Discharges," Science, 216:1211, 1982. (X1, X2)

R5. Kaiser, M. L., et al; "Atmospheric Storm Explanation of Saturnian Electrostatic Discharges," Nature, 303:50, 1983. (X2)

R6. Burns, Joseph A., et al; "Saturn's Electrostatic Discharges: Could Lightning Be the Cause?" Icarus, 54:280, 1983. (X2)

R7. Burns, Joseph A., et al; "Saturn's Electrostatic Discharges (SED): Exotic Ring Phenomena or Just Lightning?" Eos, 63: 156, 1982. (X2)

R8. Evans, D.R., et al; "Saturn's Electrostatic Discharges: Properties and Theoretical Considerations," Icarus, 54:267, 1983. (X1, X2)

R9. Eberhart, J.; "Saturn's Belt of Lightning: 40,000 Miles of Zap," Science News, 123:292, 1983. (X2)

ARF2 Measurements of Saturn's Intrinsic Energy

Description. The radiation by Saturn of about twice as much energy as it receives from the sun.

Data Evaluation. Modern infrared measurements of Saturn from high altitude aircraft. Rating: 2.

Anomaly Evaluation. Since Voyager measurements confirm that Saturn's atmosphere is depleted in helium by just the right amount to account for the energy radiated by the planet, as discussed below, this phenomenon is considered satisfactorily explained. Rating: 4.

Possible Explanations. The gravitational energy of atmospheric helium is converted into heat as it sinks through the hydrogen layers towards Saturn's center.

Similar and Related Phenomena. Jupiter's intrinsic energy (AJF1), which is thought to be the consequence of compression of the planet-as-a-whole by gravity.

Examples

X1. Experimental results. "Abstract. The total power emitted by Jupiter and Saturn has been measured by observing the planets from a jet aircraft at 15-km altitude with a telescope system open from 1.5 to 350μ. The two planets were found to radiate 2.7 and 2.4 times the amount of power they receive from the Sun, respectively. These new results put observational restraints on models for the internal structures and atmospheres of the two planets." (R1)

X2. Possible explanations. Whereas the excess energy radiated by Jupiter is thought to be due to gravitational compression of the entire planet, in the case of Saturn the favored explanation is that atmospheric helium settles down through the planet's hydrogen toward the center of the planet, releasing its gravitational energy as heat. If this were the case Saturn's atmosphere should be depleted in helium when compared to Jupiter's. The Voyager measurements of helium abundances indicate that there is just enough depletion to account for the measured excess heat radiation. (R2-R5) Phase-change energy sources have also been suggested. (R6)

References

R1. Aumann, H. H., et al; "The Internal Powers and Effective Temperatures of Jupiter and Saturn," Astrophysical Journal, 157:L69, 1969. (X1)
R2. Kerr, Richard A.; "Voyager 1 at Saturn," Science, 210:1107, 1980. (X2)
R3. "Saturn's Mysteries Revealed---But Not Much," New Scientist, 90:70, 1981. (X2)
R4. "Sorting Out the Saturn System," Science News, 119:212, 1981. (X2)
R5. "Puzzling Over Saturn's Internal Heat," Eos, 62:538, 1981. (X2)
R6. Franck, S.; "On the Luminosity of Saturn," The Moon and the Planets, 25:131, 1981. (X2)

ARF3 Sudden, Temporary Cessation of Radio Emissions

Description. A mysterious, 4-day cutoff of Saturnian radio frequency emissions.

Data Evaluation. Detected by two separate radio physics experiments on one of the Voyager spacecraft. Rating: 2.

Anomaly Evaluation. Mysterious temporary failures of spacecraft equipment have occurred before and are obviously not anomalous in the scientific sense. Setting aside this possibility, it is virtually inconceivable that Saturn could "turn off" for 4 days; however, the existence of a cloud of plasma blocking radio signals is not unreasonable. Based on the latter suggestion: Rating: 2.

Possible Explanations. Temporary experiment failure; the interposition of an absorbing or reflecting medium between the spacecraft and Saturn.

Similar and Related Phenomena. Polar radio blackouts on earth (GER7) and other radio outages (GER10).

Examples

X1. August 1981. From a report on Voyager discoveries: "The radio emissions, for example are so powerful that the Voyagers detected them from more than 600 million kilometers away. (Jupiter's still stronger signals have been known on earth since the early days of radio astronomy.) But as Voyager 2 was pulling away from Saturn late last month, heading for its 1986 rendezvous with Uranus, a totally unexpected thing happened. 'For some reason,' says Frederick L. Scarf of TRW, 'Saturn turned off for four days.'" The same dropout occurred on two different radio physics experiments. (R1)

References

R1. "Saturn: Scanning the Unseen Scene," Science News, 120:182, 1981. (X1)

ARL SATELLITE AND RING ANOMALIES

Key to Phenomena

ARL0 Introduction

Saturn's several moons and incredibly complex and beautiful rings are really much more interesting than the huge planet itself.

The moons, though much smaller than Jupiter's Galilean satellites, pose an equally impressive array of problems: Why does Hyperion rotate chaotically? Why is the leading face of Iapetus so dark? How did such small moons acquire enough energy to display extensive surface evolution? Saturn's moons are so different from one another that some scientists are asking whether they actually formed where we now see them, or whether they were captured from some other part of the solar system.

But the glorious rings of Saturn steal the show for sheer complexity, dynamicism, and anomalousness. First and foremost, the theory of gravitational resonances customarily employed to explain the presence of ring gaps now seems woefully inadequate. There are tens of thousands of ringlets and accompanying gaps; there are spirals, braided rings, travelling wavelets, stacked layers of ring material, and many other enigmatic features. Resonances are too few and weak to do the job by far. Next, strange, shadowy spoke marks form and dissolve continuously on the rings, taunting us into invoking electrostatic levitation and other ideas that would have been thoroughly ridiculed before the Voyager flights. Even before Voyager, several telescopic ring anomalies were all but ignored: the "knots" sometimes seen when the rings are seen edgewise; bright spots on the rings; and out-of-perspective shadows of the planet on the rings. Finally, where did the rings come from? When did they form? Several ring features suggest that the rings are actually recent acquisitions that are still evolving---and rapidly at that---when one considers the age of the solar system.

ARL1 Pre-Spacecraft Observations of Extra Rings

Description. Telescopic observations of additional rings interior and exterior to the bright A, B, and C rings. These additional rings are difficult to see and may not be permanent.

Background. The dusky of D ring (old nomenclature).of Saturn was seen as early as 1907 and perhaps even earlier. Likened to the Loch Ness Monster, some observers have seen it, but most cannot. (R3) After the highly respected E. E. Barnard looked carefully for this ring in 1908 to no avail, despite good eyes and excellent telescope, the dusky ring was not seriously looked for again until it resurfaced in the 1950s. In addition to several modern sightings, there is also photographic evidence. The Pioneer and Voyager spacecraft have confirmed additional interior and exterior rings, so that the telescopic observers did have a physical basis for what they saw, although we cannot really say with certainty that they saw the same structures the spacecraft photographed.

Data Evaluation. The several telescopic observations of the dusky or D ring are bolstered by photographic evidence. The same is true for an inner ring between the C ring and the atmosphere of Saturn. These additional rings are obviously very difficult to see. Only a handful have seen the dusky ring, and those who have seen it do not see it all the time. Rating: 2.

Anomaly Evaluation. Since the Pioneer and Voyager spacecraft actually found new interior and exterior rings, the earlier sightings are not overly remarkable; that is, they are more like the early sightings of Martian craters (really there) rather than the Martian canals (not really there). The real anomalies associated with these additional rings are their apparent variable nature and the inability of expert astronomers with top-notch equipment to see them at all (a psychological anomaly). Rating: 2.

Possible Explanations. The additional rings may vary in brightness due to the influx and loss of material.

Similar and Related Phenomena. Pre-Space Age observations of Martian craters (AMO2) and those of Mercury (AHO0); the canals of Mars (AHO1); the Venusian spoke system (AVO6).

Examples

X1. An elusive, "dusky" ring exterior to the A ring. Called the D ring in the older literature, modern nomenclature has marked it as the F ring, while labelling a new ring interior to the C ring as the D ring!

"The first intimation of a dark annular appendage exterior to the bright rings of the Saturnian system was obtained in 1907 by the French astronomer M. G. Fournier. Using the 28-cm refractor of the Jarry-Desloges Observatory on Mount Revard (altitude 1550 m) in Savoy, he made his first observations of the 'new ring' on 1907 September 5d 10h 15m. Subsequently it was again recorded by him on September 7d 10h 25m when under very fine seeing conditions." Although reported in several scientific publications, this early sighting received little attention. "In September, 1908, M. E. Schaer of the Geneva Observatory, completed a Cassegrain reflector with which, during the autumn and winter months of that year, he obtained a series of observations on the globe and rings of Saturn. Thus it was that, while engaged in this pur-

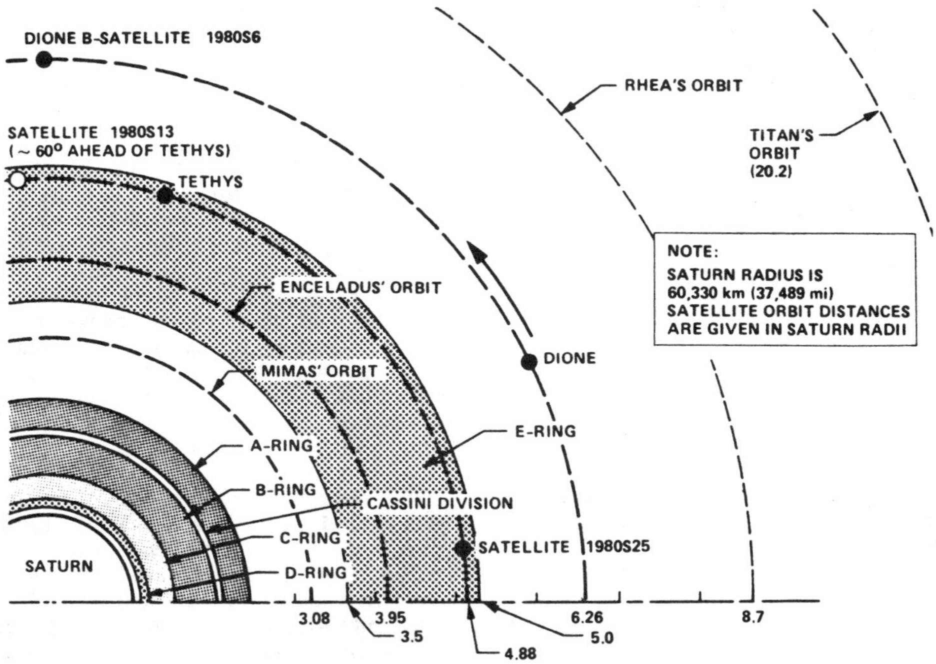

General nomenclature now applied to the various features of Saturn's ring system. (Adapted from Morrison, R5)

suit, the Swiss astronomer detected what appeared, to him at least, to be a new feature ---namely a dark ring surrounding the known bright rings. This was during the early part of October and, judging by the telegram sent to the Astronomische Nachrichten, it would seem that the earlier discovery by Fournier had either been overlooked or was not known to the Swiss observer. Following the receipt of an announcement from the Geneva Observatory, on 1908 October 10 the rings were examined at Greenwich with the 28-inch refractor by three observers, namely Bowyer, Lewis, and Eddington. In view of their partial confirmation of the exterior dark ring these observations are here considered in detail." .....(Details omitted)....."Probably the last observations of the exterior ring were made in 1909 January by Schaer, who found the feature decidedly more easily detected than previously."

In contrast to these positive sightings, E. E. Barnard, using the 40-inch refractor at Yerkes Observatory in 1908, could find no trace of the exterior dusky ring.

In the 1950s, however, R. M. Baum again spied the dusky ring. "The first observation was made on 1952 April 1d 20h 00m U. T., when with an eyepiece of X216, with seeing at 4 on a scale of 10, considerable detail was noticed. At 20h 40m on this evening seeing rose to 6 and, with this happy conspiration of atmospheric circumstances, traces of some dusky nebulous matter in the form of an additional ring were recorded beyond Ring A, which the writer could not recall having observed in 1951. The appearance was so unexpected that strong doubts as to its objective character were entertained, resulting in its being dismissed as some form of illusion though consideration of the steel engraving aspect of the global details suggested the improbability of such views. However, all doubts were finally dispelled later the same evening at 21h 15m when the dusky ring was again strongly noticed as a pale nebulous feature extending beyond the bounds of the bright ring system." Further sightings of this dusky ring came later in 1951 and 1952. American observers T. R. Cave, Jr., and T. Cragg, of Mount Wilson, also saw this feature in 1952-1953. (R1, R2)

In 1966-1967, some photographic evidence of the dusky ring was obtained by W. A. Feibelman. "On six nights between October 27, 1966, and January 16, 1967, about fifty photographs of Saturn of considerable length were taken with the 30 in. refractor at the Allegheny Observatory of the University of Pittsburgh. On all moderately long (5-10 min) and very long (30 min) exposures on backed Kodak '103a-0' without filter between October 27 and December 12, 1966, a very thin extension of the nearly edge-on ring system can be seen. The thin line extends to more than twice the known ring diameter and has an approximate equivalent brightness of $m_{pg} \approx 15$/sq. sec of arc or fainter. When two or three exposures are viewed in superposition the thin line is clearly seen when in registration. After December 12 the thin line became more difficult to detect and by January 16 no trace of it could be found. Printed reproduction of the thin line would be very difficult, but two of many microdensitometer

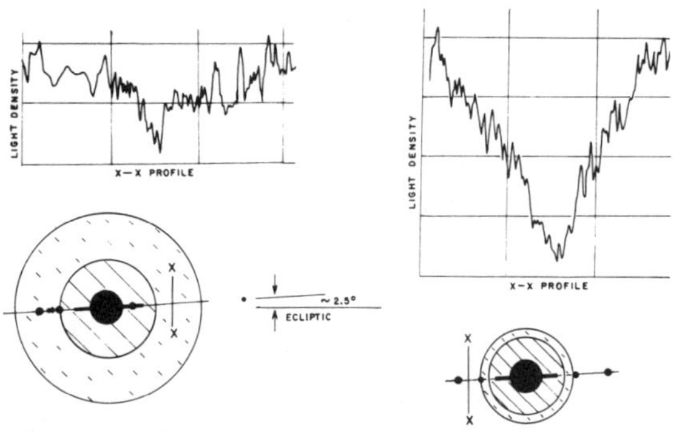

Microdensitometer traces of Saturn's rings made 1966-1967. (X1)

traces are presented in Figs. 1a and 2a. These two tracings were for November 14 and December 12, 1966, respectively, and Figs. 1b and 2b show the locations of the scans X-X. The positions of the satellites are indicated and were obtained from short exposures.....The inner cross-hatched circle indicates the extent of the greatly over-exposed image of Saturn, while the outer cross-hatched area represents the photographic halo. The over-exposed image causes a general density gradient on the densitometer traces which accounts for the sloping background level, but a noticeable increase in density at the position of the thin line is consistently evident on some thirty scans at different positions. A tentative interpretation of the thin line is that it represents a tenuous, outer ring seen only briefly when the ring plane is nearly edge-on to our line of sight." (R3)

Spacecraft photographs of the Saturnian ring system reveal two rings exterior to the A ring. "Far beyond the main rings are two additional rings of apparently different character, the G Ring at 2.8 and the E Ring, stretching from 3.5 to 5.0 Saturn radii. Both are tenuous, and they appear to lack the fine-scale structure of the main rings. Apparently the processes that produce intricate structure are confined to regions nearer the planet. The E Ring is apparently associated in some way with Enceladus." (R5) It is impossible to tell whether the modern E ring is the same as the older, very elusive D ring. (WRC)

X2. A faint ring interior to the C ring now called the D ring. A report from 1970. "From the Pic du Midi Observatory in France comes news that Dr. Pierre Guerin has discovered a fourth, inner, ring round Saturn (Sky and Telescope, vol. 40, p. 88). The fourth ring is only 30,000 miles above the surface and is extremely faint; its existence, however, is confirmed by independent photographs taken in New Mexico." (R4)

The Voyager spacecraft detected a ring interior to the C ring. "Between the classically known rings and the top of Saturn's atmosphere are a number of very thin ringlets discovered by Voyager. Collectively called the D Ring, these ringlets appear to be as intricate in structure as the more dense regions farther out." (R5)

References

R1. Baum, R. M.; "On the Observations of the Reported Dusky Ring outside the Bright Rings of the Planet Saturn," British Astronomical Association, Journal, 64:192, 1954. (X1)

R2. Cragg, Thomas A.; "A New Ring around Saturn?" Strolling Astronomer, 8:22, 1954. (X1)

R3. Feibelman, W. A.; "Concerning the 'D' Ring of Saturn," Nature, 214:793, 1967. (X1)

R4. "News from the Outposts of the Solar System," New Scientist, 48:165, 1970. (X2)

R5. Morrison, David; "The New Saturn System," Mercury, 10:162, 1981. (X1, X2)

ARL2 Knots in Saturn's Rings When Viewed Edgewise

Description. Bulges, "condensations", and "knots" seen in the plane of Saturn's rings when seen edgewise. The knots measure several hundred kilometers in diameter and are not always observed.

Data Evaluation. A handful of good telescopic observations have been uncovered, including one photographic example. The Pioneer and Voyager photos do not seem to show any knots. Rating: 1.

Anomaly Evaluation. From the observations at hand, it is impossible to tell whether the knots are accumulations of ring material or merely waves in the rings or something else entirely. In any case, the knots represent unexplained, probably transitory perturbations of ring structure---one of many "ring problems". Rating: 2.

Possible Explanations. Temporary concentrations of ring material; waves in the rings, which appear as knots when viewed edgewise.

Similar and Related Phenomena. White spots on Saturn's rings seen broadside (ARL3); waves in the rings (ARL14).

Examples

X1. September 16, 1877. Saturn's rings were viewed edgewise. Each ansa contained a knot. (R1)

X2. June 5, 1891. Rings seen edgewise. "The following ansa presented, in a marked form, the beaded appearance. The preceding ansa showed the same aspect, but less decidedly." While classified as an example of "knots", the article's drawings (not reproduced) revealed several wave-like structures. (R2)

X3. Frequently during 1907. Seen by Campbell, at Lick; Barnard, at Yerkes; and Lowell, at Flagstaff. General description: "At the present time, a month after the earth has passed through the plane of the rings, they are seen not quite edgewise, and appear in a powerful telescope as a thin line of light. A splendid opportunity is thus afforded of discovering whether the rings are everywhere of the same thickness. As knots are seen along the thin thread of light, we must perforce conclude that the rings are not quite flat, but that they have condensations here and there, and that the particles that make them up crowd together more at some places than they do at others. This is indeed no new discovery. Nearly half a century ago Otto Struve suspected that he saw bumps on the ring system, and the observations of 1907 are but a confirmation, using better telescopes of what Struve saw." (R3) The 1907 episode of knots brought forth several other reports. (R4-R6) In R5, Lowell introduces the term "tores", but everyone persisted in calling them knots.

X4. During 1951. Baum, using his 3-inch refractor, saw luminous condensations on the rings, which were set equidistantly on each side of the globe of Saturn. (R7) Baum did not specify whether there were more than two knots. (WRC)

X5. November 27, 1966. N. Sekiguchi, of the Tokyo Astronomical Observatory, was able to photograph knots in Saturn's rings, when viewed edgewise. In these photographs, there are seen four knots of matter on each side of Saturn itself, at exactly the distance of Cassini's division. (R8)

X6. General observation. "In so far as this pertains to the Cassini Division, it leads to the interesting speculation that the gap actually marks the location of another Saturnian satellite, too small to be seen as a separate body in the midst of the bright rings (except perhaps when the rings are edge on, when 'ring knots' of a few hundred kilometres' diameter are often seen)." (R9)

References

R1. Ingall, Herbert; "Saturn," English Mechanic, 26:67, 1877. (X1)

R2. Waugh, W.R.; "Saturn," English Mechanic, 53:360, 1891. (X2)

R3. Mitchell, S.A.; "'Knots' in the Rings of Saturn," Scientific American, 97:376. 1907. (X3)

R4. Butler, Charles P.; "Bright Knots on Saturn's Rings," Knowledge, 5:13, 1908. (X3)

R5. Lowell, Percival; "The Tores of Saturn," English Mechanic, 87:320, 1908. (X3)

R6. Bolton, Scriven; "Saturn and His Rings," English Mechanic, 86:586, 1908. (X3)

R7. Baum, R. M.; "On the Observations of the Reported Dusky Ring outside the Bright Rings of the Planet Saturn," British Astronomical Association, Journal, 64:192, 1954. (X4)

R8. "There Is Something between Saturn's Rings," New Scientist, 41:644, 1969. (X5)

R9. Van Flandern, T. C.; "New Saturnian Satellites?" Observatory, 99:8, 1979. (X6)

ARL3 Bright Spot on Saturn's Rings

Description. A bright white spot, semicircular or oval in shape, seen occasionally on Saturn's rings adjacent to the shadow of the planet cast on the rings. The spot does not rotate with the rings. Its brightness often exceeds that of all other portions of the Saturn system. The spot has been reported to pulsate in brightness.

Data Evaluation. The 1889 and 1939-1941 apparitions of the spot were well-reported, but the other appearances were noticed by only a few astronomers. Nevertheless, there is ample testimony to the objective existence of the spot. Rating: 1.

Anomaly Evaluation. If the spot is a consequence, as suggested below, of real changes in the albedo of the rings or the emission of intrinsic radiation, we have a fascinating anomaly with no ready explanation. Rating: 2.

Possible Explanations. Some have considered the spot to be merely a contrast phenomenon; i.e., an irradiation effect. Since the spot always occurs next to the planet's shadow on the rings, it could be an albedo change associated with the eclipse of the rings. Given the intense relative brightness of the spot, it might be a region of electrical activity associated with Saturn's electrostatic discharges and self-luminous.

Similar and Related Phenomena. Knots on Saturn's rings (ARL2); Saturn's electrostatic dischanges (ARF1), which are ring-related; the post-eclipse brightening of Jupiter's moon, Io (AJX6).

Examples

X1. 1840-1842. The Italian astronomer De Vico records a "fixed bright point attached to the open ansa," which he saw on several consecutive nights in 1840 and once more in 1842. (R7)

X2. 1889. "The remarkable white spot, first seen by Dr. Terby on March 6, has been so conspicuous during the past week that it has doubtless been observed by a large number of your astronomical readers. I had a fine view of it on April 18, and again to-night. On both occasions it occupied the same position as shown in Dr. Terby's sketch. That portion of it on ring B appeared to be much brighter than that on ring A. Not the least remarkable feature in connection with it was its peculiarly pure white colour, which contrasted strongly with the other lines associated with the planet. Its following edge was clearly convex, and it appeared to be oval in shape. I could trace it from the inner edge of ring B, to a little beyond the in-

Bright spot occasionally observed on Saturn's ring adjacent to the planet's shadow.

ner edge of A. Cassini's division, which was beautifully defined on both evenings, could not be seen through the spot." (R1)

From Lewis Swift, at Warner Observatory. "The white spot on Saturn's ring, recently announced by Terby of Belgium, was observed at this Observatory on the evening of March 14th both by Professor Brooks (who was my guest that night) and myself. In consequence, however, of its faintness, and of the bright moonlight in which it was viewed, it was a difficult object; but as we both saw it in the same position and of the same size and shape, there could be no doubt in the mind of either of us that we had seen the 'spot,' which appeared as a narrow band extending across both outer rings, its western boundary being in contact with the black notch termed the shadow of the ball on the ring. As we found the spot in the same place as at discovery, it cannot belong to the ring itself, as the latter revolves, and, so doing, would cause the spot to be seen on all parts of the ring. We are led, therefore, to believe that that the phenomenon must be produced by reflected sunlight from the globe of the planet, though in just what manner produced we are not able to determine." (R3)

W.R. Brooks, Smith Observatory, discovered that the white spot pulsates in brightness at irregular intervals of a few minutes. (R4)

A report from Taylor, who was using a 5-foot reflector. "Bad weather prevented anything further being done until Monday, March 25, when the sky was very clear and the definition excellent. Saturn bore the highest powers, details of the ring and ball being very prominent; but no 'region blanche' on the ring could be made out. The shadow of the planet on the ring was not perfectly regular, but had a curious notched outline, extending further from the planet on the middle ring than on the outer one. The gauze ring was very brilliant; the satellites were all visible; but nothing corresponding to Terby's announcement was seen." (R2)

In April, W.R. Brooks was still finding the spot on <u>every</u> attempt. He remarked that the spot seemed to be growing smaller and that it seemed to be <u>intrinsically</u> brighter than the rest of the ring. (R6) See R5 and R7 for additional descriptions of the 1889 spot.

X3. 1919. Observations of a bright spot next to the black shadow of Saturn on the rings were made in March and November. (R8, R9)

X4. 1939-1941. H.M. Johnson noted the spot first in 1939. After reviewing the many observations from 1889, he began to study the phenomenon systematically. "Reinvigorated by these early observations, I began to observe Saturn more often, making special notes on the spot 20 times between 1940 October 13 and 1941 March 29, U.T. At the same time several additional observations were forwarded to me by Messrs. W.H. Haas of Ohio, F.R. Vaughn, jun., of Iowa, and D.P. Barcroft of California. We attempted to gain impressions of the spot's size, shape, brightness and whiteness, etc., and variations in these qualities; some conflict resulted. To me, using the 8-inch reflector and 213X, the spot was always present in all grades of seeing; the ratio of the area of the spot to that of the shadow averaged near 1/2 during the period of observation, varying unprogressively between 1/3 and 1, being largest perhaps but not most conspicuous in poorest seeing though even then never exceeding unity; the spot was somewhat diffuse though occasionally was 'sharply bounded,' the contour being 'irregularly spindle-shaped'; the spot was usually considered to be the whitest part of the whole Saturnian system, 'being whiter than the outer half of ring B or the equatorial zone on the planet itself.'" The observations of Haas and Vaughn supported those of Johnson (above) except that they thought the spot was larger than Johnson estimated. (R10)

<u>References</u>

R1. Elger, Thos. Gwyn; "Bright Spot on the Rings of Saturn," <u>English Mechanic</u>, 49: 195, 1889. (X2)

R2. "White Spot on Saturn's Rings," <u>Observatory</u>, 12:195, 1889. (X2)

R3. Swift, Lewis; "The White Spot on Saturn's Ring," <u>Sidereal Messenger</u>, 8:189, 1889. (X2)

R4. Brooks, William R.; "Changes on Saturn," <u>Scientific American</u>, 60:197, 1889. (X2)

R5. Brooks, William R.; "Variability of White Region on Saturn's Ring," <u>English Mechanic</u>, 49:110, 1889. (X2)

R6. Brooks, William R.; "White Region on Saturn's Ring," <u>Sidereal Messenger</u>, 8: 233, 1889. (X2)

R7. Brooks, W.R.; "The White Region on Saturn's Ring," <u>Observatory</u>, 12:262, 1889. (X1, X2)

R8. Sargent, Frank; "White Spot on Saturn,"

English Mechanic, 109:235, 1919. (X3)
R9. Best, C.; "Bright Region on Saturn's Ring," English Mechanic, 112:191, 1920. (X3)
R10. Johnson, Hugh M.; "The White Spot on Saturn's Rings," British Astronomical Association, Journal, 51:309, 1941. (X4)

ARL4 Anomalous Shadows of the Planet on the Rings

Description. (1) Distinct notches or peaks in the shadow of the planet on the rings; (2) Wrong curvature of the shadow, considering the rules of perspective; (3) Notched shadows on both sides of the globe of Saturn. The three categories are not completely independent, because the notched shadows are often curved the wrong way, too. Anomalous shadows are rare. They may change from day to day.

Data Evaluation. Many anomalous shadows are described in the older literature. If they occur today, no one writes about them! To some extent at least, the prominence and degree of shadow distortion depends upon the observer; a fact suggesting a psychological factor. Rating: 2.

Anomaly Evaluation. If these remarkable shadows are only subjective phenomena, the anomaly rating must be low. But objective effects may be involved, because even under the same lighting conditions the same observer does not always see anomalous shadows. These effects are presumed here to be albedo changes and/or changes in the rings themselves. Rating: 3.

Possible Explanations. Contrast effects come to mind first here; and some psychological experiments seem to support this idea. (R11) The rings themselves may exhibit changes in reflectivity and physical geometry; both of which may in turn be related to a change in the level of solar intensity. Since the occasional bright white spot on the rings is inevitably associated with the planet's shadow (ARL3), the anomalous shadows may be simply normal shadows with encroaching bright spots.

Similar and Related Phenomena. The white spot that sometimes appears on Saturn's rings (ARL3); the post-eclipse brightening of Io (AJX6).

Examples

X1. Notched shadows. "Whilst observing with a friend, on the night of the 12th. inst., we turned the 6 in. refractor on to Saturn. we noticed that the outline of the shadow of the planet on the ring was decidedly 'notched,' or, as described by Webb, 'curved the wrong way for its perspective.' It appeared to follow close to the lines of Cassini's division and then suddenly curve inward on touching ring B, then continued its line of perspective on that ring to and through ring C." (R2)

H. Watson examined many drawings of Saturn's shadow as thrown on the rings, and found that many showed a notched or peaked appearance, but others did not. (R3)

Notched shadow of Saturn cast upon the rings. (X1)

W. F. A. Ellison expressed well the consternation of astronomers over the presence of a notch and, often, incorrect curvature of the shadow. "The peak on the shadow is a puzzle to me---not how it can exist as a depression in a surface will cause a deviation in the outline of a shadow thrown obliquely on the surface, but how we can see it, being almost exactly in a straight line between Saturn and the Sun, which casts the shadow. One would have supposed that, by all rules of geometrical optics, the form of the shadow, to an observer so situated, could not be affected by the form of the surface on which it is cast. We see no corners on the edge of the earth's shadow in a lunar eclipse, in spite of the irregular surface which receives it. Yet the peaks are unmistakably there in the case of Saturn." (R7)

X2. Shadows of Saturn with the "wrong" curvature. Note that the notched shadows of X1 usually show the wrong curvature on one or both sides of the notch.

"On the evening of April 25th, 1889, at about 8:30 P. M., I was examining Saturn with a power of about 180 on a 4 1/8-inch achromatic by Brashear, when, much to my surprise, I found the shadow of the globe on the rings curved the wrong way, i. e. from the globe, as shown in the following drawing. Thinking my eyes might be deceiving me I called my wife, and without telling her what I had seen, requested her to describe the shape of the shadow. She described the shadow as having its right hand edge curved away from the planet. I wrote to Professor Comstock of the Washburn Observatory about it, and was informed by him that while my observation of Saturn was unusual, it was far from being unprecedented; that the same appearance was observed in 1875 with the 26-inch achromatic at Washington, and that Webb, in 'Celestial Objects for Common Telescopes,' says: 'The outline of this shadow has often been found curved the wrong way for its perspective.' Professor Comstock also adds, 'I do not know that any satisfactory explanation for this anomaly has ever been given.'" (R5)

Shadow of Saturn on the rings has the wrong curvature. (X2)

1904. Italy. "During a series of observations of Saturn made at Aosta (Italy) in October, November, and December, 1904, Signors M. Amann and Cl. Rozet observed a secondary shadow, other than that of the planet, projected on to the illuminated surface of the rings. First seen on October 20, this shadow was thinner and much less accentuated than that of the planet, whilst its curvature was in the opposite sense to that of the latter body. From October 20 to November 15, despite the fact that numerous opportunities of observing it occurred, the shadow was not seen, but from the latter date until the end of December it was shown on twenty-six drawings of the system. On seven drawings made between December 22 and 27, the shadow appeared bifurcated where it traversed the inner ring, and on November 28 and 29 a third line of shadow, narrower and feebler than the preceding and much further from the planet, was seen." (R6) Other shadows with the wrong curvature are described in R1 and R4.

X3. Notched shadow on both sides of the planet. When the rings of Saturn are tilted toward the earth, but with both ansa on a horizontal plane, notched shadows are on rare occasions seen on both sides of Saturn, giving the planet an "eared" look. The shadows are not always of the same size. (R8-R10) Given the distance of the sun, only a single shadow should be cast. (WRC)

References

R1. Ryle, R. J.; "Saturn and His Shadow," English Mechanic, 43:102, 1886. (X2)

R2. Terby, F., and Clapham, T. R.; "Saturn," English Mechanic, 45:79, 1887. (X1, X2)

R3. Watson, H.; "Saturn," English Mechanic, 45:97, 1887. (X1, X2)

R4. Elger, Thos. Gwyn; "Bright Spot on Rings of Saturn," English Mechanic, 49:195, 1889. (X2)

R5. Jenks, Aldro; "On the Reversed Curvature of the Shadow on Saturn's Rings," Sidereal Messenger, 9:255, 1890. (X2)

R6. "Secondary Shadow on Saturn's Rings," Nature, 71:401, 1905. (X2)

R7. Ellison, Wm. F. A.; "The Shadow of Saturn," English Mechanic, 98:425, 1913. (X1)

R8. Watson, H.; "Shadows of Saturn," English Mechanic, 99:73, 1914. (X3)
R9. Watson, H.; "Study of Saturn's Shadows," English Mechanic, 99:187, 1914. (X3)
R10. Hepburn, Patrick H.; "Saturn---Shadow of Ball," English Mechanic, 99:254, 1914. (X3)
R11. Bartrum, C.O.; "A Note on the Appearance of Saturn at Opposition," British Astronomical Association, Journal, 24:359, 1914. (X1-X3)

ARL5 Dark Spokes on the Rings of Saturn

Description. Dark, radial, wedge-shaped spokes that continuously appear and disappear in the B ring. Spoke length is typically 10,000 kilometers, with a base width of about 2,000 kilometers. Spokes originate most frequently on the morning ansa, and may develop over a length of several thousand kilometers in a period of only several minutes. See X1 below for more details.

Data Evaluation. Many Voyager spacecraft photographs. Rating: 1.

Anomaly Evaluation. Originally, the spokes baffled everyone, but now a rather successful group of theories depending upon the levitation of dust particles seems to explain most features of the spokes. Rating: 3.

Possible Explanations. Electrostatic levitation of dust particles above the normal plane of the rings in the presence of plasma clouds is the currently favored theory. The electrostatic alignment of ring particles has been suggested and found wanting. (R13, R12)

Similar and Related Phenomena. Saturn's electrostatic discharges (ARF1), which are concentrated in the same region where spoke production is highest.

Examples

X0. Historical note. Hints of the Saturnian spokes appear in older books. For example, tears or gores in the rings were mentioned in the 1908 book The Heavens and Their Story, by A. Maunder and W. Maunder. (R9)

X1. General description based on the Voyager encounters with Saturn. "Among the most interesting variable ring features are the narrow radial markings or spokes in the B ring. These features were first observed as dark wedge-shaped projections rotating in the outer half of the ring, predominantly on the morning ansa (the part of the ring just emerged from Saturn's shadow). At high resolution, spokes were easily observed on both ansae and across most of the B ring. Spokes have not been seen to cross the Cassini division and seldom extend farther in than about 105,000 km from the center of Saturn. This point appears to be a sharp cutoff for most of the spokes and is a bright region (in backscattered light) in the B ring. Characteristic lengths and widths of these features are 10,000 and 2,000 km, respectively. What were discovered as dark markings during inbound passage (backscattered light) became bright markings during outbound passage

Voyager photograph of spokes on Saturn's rings. (X1) (Courtesy NASA)

(forward-scattered light). Viewed against the planet's disk, the spokes show no optical density variations, and no spokes are visible from the unilluminated side of the rings. In the highest resolution images of the spokes taken in forward-scattered light, the B ring resolves into radially aligned areas of enhanced brightness among ringlets. This is especially true in the outer optically thin region of the ring. Only in the optically thick regions do the gaps appear to be filled by spoke material. These observations indicate that small particles constitute a large fraction of the B ring and are more visible, perhaps because they are elevated above the ring plane, within the spokes." (R6)

From the Voyager 2 summary report in Science, more details: "As in the Voyager 1 data, the Voyager 2 images reveal the spokes more easily on the morning ansa (just rotated out of Saturn's shadow). Some spokes can be tracked as they rotate through 360° or more; however, it is not clear whether we really see the identical spoke pattern or a new one that it 'reprinted' on top of the old one. If spokes are created as radial features, then mapping their orbital motion backward in time should allow a determination of the point of origin. Spokes mapped backward in this way were found not to have a common orbital longitude of creation. Places of origin also appear unrelated to the shadow of the planet on the rings, although a few spokes appear to have formed within the shadow. After Voyager 2 passed through the ring plane, the image shown in Fig. 44 (not reproduced) was taken of the unilluminated side of the rings, showing spokes for the first time with this viewing perspective. In this image the B ring is illuminated predominantly by sunlight scattered off Saturn's atmosphere and appears brightest in backscattered light. Bright spokes are visible in the region with Saturnshine phase angles of about 80° to 120°. The fact that these features are seen only in this region of favorable Saturn illumination indicates that they are dark-side phenomena and not bright-side spokes shining through optically thin parts of the B ring. The shapes seen on the dark side indicate a morphological behavior similar to that observed on the illuminated side, with spokes tilting away from the radial direction in the sense of the Keplerian motion of ring particles. Furthermore, dark-side spokes are not tilted enough to be created on the illuminated side and then passed through the ring plane a quarter of an orbit later. The conclusions that these features may be created on the dark side of the B ring while others form in Saturn's shadow have implications for the mechanism of spoke formation, casting some doubt on the idea that charging of small ring particles by photoionization alone is responsible for levitating them out of the ring plane." (R14)

Rapid formation of spokes. "The famous dark spokes on the B ring remain as perplexing as ever. Voyager 2 showed that their edges are sharp to a resolution of 60 kilometers. The spacecraft also made a time-lapse movie (which has been dubbed the Saturn 500) following a group of spokes around the ring. (The ring particles are in orbit around Saturn and carry the spokes along with them.) The movie shows that as the older spokes fade, a fresh spoke forms, leaping radially across 20,000 kilometers of ring surface within about 12 minutes. 'Our impression is that the spokes are somehow imprinted on the rings by some action of Saturn's magnetic field,' says (Bradford A.) Smith. 'But the fact that they deposit so quickly surprises me.'" (R7)

Some other interesting features of the spokes: "Several narrow spokes were observed during formation along radial lines in the sunlit portion of the ring. The formation time is typically ~ 5 min for a 6000-km-long spoke. The rate of spoke formation is highest at the morning ansa outside Saturn's shadow. Several spokes have been found where one edge revolves with Keplerian speed whereas the other edge stays radial. Recurrent spoke patterns have been observed at the period of Saturn's rotation. From edge-on views of the ring system, an upper limit for the height of spokes of 80 km is derived." (R17) In a statistical study of spoke activity, a period of 621 minutes was found, which is consistent with the 639.4-minute rotation period of Saturn's magnetic field. Maximum spoke activity coincides with the morning ansa, which is also the location of the souce of the electrostatic discharges (ARF1). (R16) Note that the occasional white spot and anomalous shadows also occur at the junction between sunlight and shade. (WRC)

X2. Possible mechanisms of formation. C.K. Goertz and G.E. Morfill suggest the levitation of electrically charged dust particles above the plane of the rings in the presence of a plasma cloud. (R17, R19) Most theories, in fact rely upon electrostatic levitation of dust. (R1, R5, R11, R18) Some rely on photoionization and other processes. J. F. Carbary et al proposed the electrostatic alignment of ice particles in the rings (R13), but A. J. Weinheimer and A. A. Few, Jr., do not believe the forces are strong enough to cause the postulated alignment. (R12)

References

R1. "Voyager 1's Saturn: Moonwatch," Science News, 118:325, 1980. (X2)
R2. "Saturn: More and More...," Science News, 118:307, 1980. (X2)
R3. Kerr, Richard A.; "Voyager 1 at Saturn," Science, 210:1107, 1980. (X2)
R4. "Voyager Discovers Spokes in Saturn's Rings," New Scientist, 88:276, 1980. (X1)
R5. Morrison, David; "The New Saturn System," Mercury, 10:162, 1981. (X1, X2)
R6. Smith, Bradford A.; "Encounter with Saturn: Voyager 1 Imaging Science Results," Science, 212:163, 1981. (X1)
R7. Waldrop, M. Michell; "The Puzzle That Is Saturn," Science, 213:1347, 1981. (X1)
R8. "Sorting Out the Saturn System," Science News, 119:212, 1981. (X1, X2)
R9. Young, Janet E.; "Saturn's Rings," British Astronomical Association, Journal, 91:407, 1981. (X0)
R10. Kerr, Richard A.; "Spokes, SKR, and SED: A Connection?" Science, 218:276, 1982. (X1)
R11. Hill, Jay Roderick, and Mendis, D. A.; "The Dynamical Evolution of the Saturnian Ring Spokes," Journal of Geophysical Research, 87A:7413, 1982. (X2)
R12. Weinheimer, Andrew J., and Few, Arthur A., Jr.; "The Spokes in Saturn's Rings: A Critical Evaluation of Possible Electrical Processes," Geophysical Research Letters, 9:1139, 1982. (X2)
R13. Carbary, J. F., et al; "The Spokes in Saturn's Rings: A New Approach," Geophysical Research Letters, 9:420, 1982. (X2)
R14. Smith, Bradford A., et al; "Voyager 2 Encounter with the Saturnian System," Science, 215:499, 1982. (X1)
R15. Cuzzi, J. N.; "Mysteries of the Ringed Planets," Nature, 300:485, 1982. (X2)
R16. Porco, C. C., and Danielson, G. E.; "The Periodic Variation of Spokes in Saturn's Rings," Eos, 63:156, 1982. (X1)
R17. Grun, Eberhard, et al; "The Evolution of Spokes in Saturn's B Ring," Icarus, 54:227, 1983. (X1, X2)
R18. Smoluchowski, R.; "Formation of Fine Dust on Saturn's Rings as Suggested by the Presence of Spokes," Icarus, 54:263, 1983. (X2)
R19. Eplee, Robert E., Jr., and Smith, Bradford A.; "Spokes in Saturn's Rings: Dynamical and Reflectance Properties," Icarus, 59:188, 1984. (X1, X2)

ARL6 Kinked and Inhomogeneous Rings

Description. Rings of particles orbiting Saturn that display kinks rather than smooth ellipses and the presence of lumps or discontinuities. The F ring, just outside the A ring are kinked and lumpy. In the Encke Division, kinked and discontinuous strands are evident. These phenomena may be time-variable.

Data Evaluation. Pioneer and Voyager spacecraft photography. Rating: 1.

Anomaly Evaluation. The comfortable "gravitational" explanation of the kinks and clumping, due supposedly to the presence of shepherd satellites, seems to be undercut by the kinked strands in the Encke Division, where no satellites are in evidence. Until this paradox is resolved, kinkiness and clumpiness will have a moderate anomaly rating. Rating: 2.

Possible Explanations. The gravitational effects of the two F-ring shepherd satellites promise to explain some of the F-ring features---at least theoretical analysis and computer-modelling show how kinks may be formed.

Similar and Related Phenomena. Knots on Saturn's main rings, as seen through terrestrial telescopes (ARL2).

Examples

X1. General description. The F ring. "Like a thin pencil line circumscribing the main rings, the F Ring orbits Saturn at a distance of 2.33 Saturn radii, or 140,600 km. Discovered by Pioneer Saturn, the F Ring, which is slightly eccentric, is gravitationally confined between two shepherd satellites, 1980-S26 and 1980S27. Each of these satellites is also in a slightly eccentric orbit, and at times the inner shepherd can approach within grazing distance of the ring. The total width of the F Ring is about 100 km. Within this span are several perhaps discontinuous strands each less than 10 km across; Voyager 2 photographed as many as five in one image. Still greater complexity was revealed by the photopolarimeter trace, which resolved the brightness of these strands into many additional components. At moderate resolutions the F Ring appears clumpy, and at the highest resolution, Voyager photographs show kinks and apparent intertwining of strands. The kinks seen by Voyager 1 were spaced about 7,000 km apart, ten times the scale of the kinky ring in the Keeler Gap." (R7)

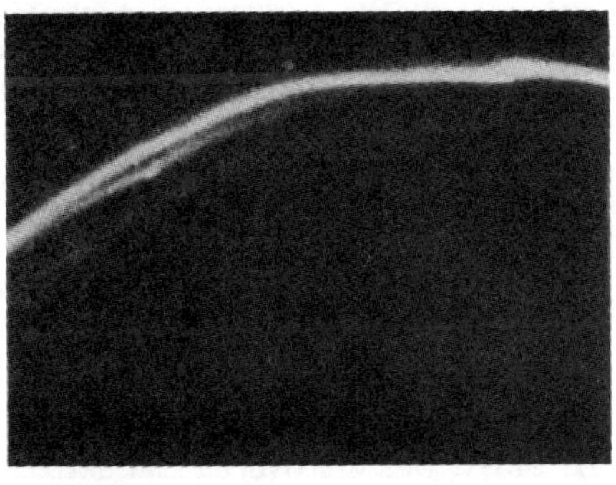

The braided F ring. (X1) (Courtesy NASA)

Another view of the F ring. "The rings of Saturn can only be described as maddening. Consider the F ring. During Voyager 1's approach to Saturn it was just visible in the images as a pencil-thin line just outside the broad, bright rings known from the earth. Then Voyager 1 took close-ups; in violation of all commonsense celestial mechanics, the F ring was kinked, clumpy and appeared to have three braided strands. In the 9 months between Voyager 1 and Voyager 2 theorists expended considerable effort trying to understand the F ring. Electromagnetic effects on very fine dust particles, perhaps? Wave-like resonances with the little shepherd satellites just inside and outside the ring? Nothing was proved, but several ideas looked promising. Then came Voyager 2. No braids in the F ring. No kinks. Five strands. And, as if to mock the theorists further, there was the 'Encke doodle,' a faint thin strand wandering through the Encke division in the A ring, with a well developed set of kinks and not a shepherd moon in sight." (R9)

X2. Kinky ringlets in the Encke Division. (See also the last few lines of X1 above.) "The Keeler Gap, also sometimes called the Encke Division, contains a pair of extraordinary ringlets that are both discontinuous and kinky. A few other ringlets are clumpy, but these are the only ones known that apparently consist only of short arcs. The kinks, of unknown origin, are spaced a few hundred kilometers apart. The high-resolution profile of the Keeler Gap obtained by the photopolarimeter showed more than a dozen very faint features less than a kilometer wide, as well and one of the two kinky rings. Moonlets within the gap have been suggested as the cause of all this structure, but none were seen by the Voyager cameras." (R7)

X3. Possible origins of the kinks. As mentioned in X1 and X2, the gravitational influences of the two shepherd satellites are generally considered to be the cause of the kinking and clumping of the F ring. Indeed, theoretical analysis and computer modelling of the region have yielded promising results. The interweaving or braiding of the F ring strands, however, seems more difficult to explain. (R3, R10) Also to be explained: (1) Why didn't the Voyager 2 pictures show kinking; and (2) What causes the kinking in the Encke Division, where no shepherd satellites are in evidence? (WRC)

References

R1. "Voyager 1's Saturn: Moonwatch," Science News, 118:325, 1980. (X1)

R2. Smoluchowski, R.; "The F-Ring of Saturn," Geophysical Research Letters, 8:623, 1981. (X1)

R3. Dermott, Stanley F.; "The 'Braided' F-Ring of Sature," Nature, 290:454, 1981. (X3)

R4. "Pow-Wow of the Planet People," Science News, 120:278, 1981. (X1)

R5. "Saturn's Mysteries Revealed---But Not Much," New Scientist, 90:70, 1981. (X1)

R6. "New Theory on Braided Ring," Astronomy, 9:63, August 1981. (X3)

R7. Morrison, David; "The New Saturn Sys-

tem," Mercury, 10:162, 1981. (X1-X3)
R8. Smith, Bradford A., et al; "Encounter with Saturn: Voyager 1 Imaging Science Results," Science, 212:163, 1981. (X1)
R9. Waldrop, M. Mitchell; "The Puzzle That Is Saturn," Science, 213:1347, 1981. (X1-X3)
R10. Kerr, Richard A.; "The F Ring Becomes a Little Less Baffling," Science, 218:276, 1982. (X3)
R11. Smith, Bradford A., et al; "Voyager 2 Encounter with the Saturnian System," Science, 215:499, 1982. (X1)

ARL7 Dark-Sided Iapetus

Description. The large, dark red area on the leading face of Saturn's moon, Iapetus. The dark area "appears" to be superficial dust swept up from orbit preferentially by the leading face.

Data Evaluation. A long history of telescopic observation, plus Voyager spacecraft photography. Rating: 1.

Anomaly Evaluation. Although the popular hypothesis that Iapetus sweeps up dark dust on its leading face while in orbit is attractive, one must wonder why other Saturnian moons do not show this preferential darkening---or, for that matter, other moons around other planets, where similar situations might occur. None of the explanations below and in X3 is completely convincing. Rating: 3.

Possible Explanations. The sweeping up of dark chondritic material in orbit by Iapetus's leading face. Alternatively, the bombardment may remove superficial ice or bright material to expose dark material underneath. See X3 for further discussion.

Similar and Related Phenomena. None.

Examples

X1. Telescopic observations. "The remarkable brightness variation of Saturn's satellite Iapetus was discovered by J.D. Cassini over three centuries ago. More recent observations by visual observers have revealed that the satellite's brightness varies smoothly over a range of about two magnitudes, reaching maximum brightness near greatest western elongation and minimum brightness near greatest eastern elongation. Comparison of the more accurate visual light curves suggests that the amplitude of brightness variation may be variable from one apparition to the next." The range of Iapetus' brightness variation greatly exceeds that of any other satellite. (R2)

X2. Spacecraft observations. "Voyager 1 observations of Iapetus confirmed ground-based observations (going back to the discovery of the satellite by J.D. Cassini in 1671) which show that the hemisphere leading in orbital motion about Saturn is about an order of magnitude darker than the trailing hemisphere. Although Voyager 1 images did not have sufficient resolution to show much topographic detail, a dark ring about 400 km in diameter centered near 5°N and 245°W was discovered that extends from the dark hemisphere into the bright hemisphere. A number of craters, just at the limit of resolution, were also seen. Surface resolution was improved from about 50 km/lp in the Voyager 1 images to about 20 km/lp in the Voyager 2 images. The highest resolution observations were obtained for the bright, trailing hemisphere at high northern latitudes. The Voyager 2 images reveal large numbers of craters and clarify the character of the boundary of the dark region, particularly where it extends into the trailing hemisphere..... The principal dark region is seen to be centered almost perfectly in the leading hemisphere. Ground-based telescope observations of the leading hemisphere must include bright regions near the poles; hence no ground-based observations of Iapetus are completely restricted to the dark material. Circular dark patches, probably dark-floored craters, occur in the bright region near the boundary and deep inside the trailing hemisphere. There are also some craters (for instance, near 200°W) that exhibit low-albedo patches suggestive of dark areas on their floors, perhaps parts of their walls facing the dark region. The contact relations in the trailing hemisphere strongly suggest that the dark material is superimposed on the bright, densely cratered terrain and is therefore younger than the cratered terrain. Although craters are clearly seen at the boundaries of the dark region, there is no hint, even in images in which many individual pictures have been summed to increase the signal-to-noise ratio, of bright-floored or bright-rimmed craters within the dark region that are well away from the boundary. Either the dark material is very thick or it is replenished at such a rate that fresh craters which penetrate the dark layer are quickly covered again." (R7)

X3. Possible explanations. The fact that the intensely dark (actually dark red) region of Iapetus is centered nicely on the leading face of the spheroid has encouraged deposition theories. The leading face of Iapetus should sweep up dust located in the moon's orbit much faster than any other portion of the satellite. One possible source of dark dust is Saturn's moon Phoebe, which orbits farther out than Iapetus. Perhaps chondritic material is continually breaking off from Phoebe and being collected by Iapetus as it falls in toward Saturn. Alternately, the impinging particles could hit Iapetus with so much energy that its normal icy surface is melted and vaporized, leaving behind high concentrations of dust. Or the postulated bombardment may blast away a thin veneer of bright material on the leading face to expose underlying dark material. Extrusion of dark material from the interior of Iapetus is not considered a viable explanation because it seems so unlikely that the dark area would occur precisely in the center of the leading face. (R8)

References

R1. Zellner, Ben; "On the Nature of Iapetus," Astrophysical Journal, 174:L107, 1972. (X1)

R2. Millis, R. L.; "UBV Photometry of Iapetus," Icarus, 18:247, 1973. (X1)

R3. "Voyager 1's Saturn: Moonwatch," Science News, 118:325, 1980. (X2)

R4. Smith, Bradford A., et al; "Encounter with Saturn: Voyager 1 Imaging Results," Science, 212:163, 1981. (X2)

R5. Cruikshank, D. P., et al; "The Dark Side of Iapetus," Icarus, 53:90, 1983. (X2, X3)

R6. Waldrop, M. Mitchell; "Saturn Redux: The Voyager 2 Mission," Science, 213:1236, 1981. (X3)

R7. Smith, Bradford A., et al; "Voyager 2 Encounter with the Saturnian System," Science, 215:499, 1982. (X2)

R8. Maran, Stephen P.; "Iapetus: Saturn's Mysterious Moon," Natural History, 93:92, October 1984. (C2, X3)

R9. Squyres, Steven W., et al; "Voyager Photometry of Iapetus," Icarus, 59:426, 1984. (X2)

ARL8 Titan's Variable Brightness

Description. The long-term fluctuation in the brightness of Titan, Saturn's largest satellite. The changes amount to several per cent, and may be correlated with changes in solar activity.

Data Evaluation. Several terrestrial studies using filter photometry have documented substantial brightness changes for Titan. Although one experiment found Titan's brightness to be anticorrelated with solar activity, no confirmation has been found in the literature examined so far. Voyager photos reveal substantial north-south asymmetry in the moon's brightness. Rating: 2.

Anomaly Evaluation. If Titan's brightness variations are simply seasonal changes in cloud cover, the anomaly is not very significant, although the precise mechanism of albedo change with solar heating has not been elucidated. The mystery deepens if the variations are tied to the solar cycle, for there is no clear-cut way for changes in solar activity to force albedo changes at such great distances. Rating: 3.

Possible Explanations. Since Titan does possess a well-tilted spin axis, changes in seasonal heating would be significant and could alter cloud cover and, in consequence, albedo. Solar wind changes might chemically or electrically stimulate brightness changes.

Similar and Related Phenomena. The planet Neptune displays similar brightness changes (ANO1); terrestrial weather changes correlated with solar activity (GWS).

Examples

X1. Telescopic observations. In the early 1970s, several astronomers suggested that the brightness of Titan, Saturn's largest moon, might vary over long periods of time. In 1974, Noland et al confirmed such variation using filter photometry. (R1) G. W. Lockwood reported similar results in 1975. (R2) In 1979, Lockwood published a paper in Nature connected Titan's variability with the solar cycle: "We have observed a significant anti-correlation between solar activity and the brightnesses of two Solar System objects. Both the planet Neptune and Saturn's satellite Titan increased in brightness by several per cent between 1972 and 1976 and subsequently became fainter by comparable amounts. This period corresponds to the decline of solar activity at the end of solar cycle 20 (1972-1976), followed by the rapid increase of activity at the beginning of cycle 21. Solar minimum and the maximum observed brightness of Titan and Neptune both occurred in 1976. ...We hypothesize that what we have observed are changes in planetary albedos induced by solar activity. Such changes may have an important bearing on the energy balances of the outer planets and their satellites." (R3)

X2. Voyager spacecraft observations. Abstract. "Voyager 1 images of Titan, when normalized to remove limb darkening, reveal an axially symmetric brightness pattern with significant north-south asymmetry. This interhemispheric contrast seems to be a response to seasonal solar heating variations resulting from Titan's inclined spin axis. The contrast significantly lags the solar forcing, indicating that its production involves the atmosphere well below the unit optical depth level. The contrast has a significant effect on Titan's disk-integrated brightness as seen from earth, and probably accounts for most of the observed long term variation, with solar UV variations accounting for the remainder." (R5)

X3. Possible explanations. Noland mentioned five possible mechanisms for Titan's brightness variations: (1) Fluctuations in the solar constant (unlikely); (2) Random fluctuations in cloud cover; (3) Seasonal (29.5-year) fluctuations in cloud cover; (4) Fluctuations in cloud cover related to the solar cycle of about 11 years; and (5) Changes in aspect. (R1)

References

R1. Noland, M., et al; "New Evidence for the Variability of Titan," Astrophysical Journal, 194:L157, 1974. (X1, X3)

R2. "Titan: Locked and Varied," Science News, 107:137, 1975. (X1)

R3. Lockwood, G. W., and Thompson, D. T.; "A Relationship between Solar Activity and Planetary Albedos," Nature, 280:43, 1979. (X1)

R4. "Sun Games with Neptune and Titan," Cycles, 31:214, 1980. (X1)

R5. Sromovsky, Lawrence A., et al; "Implications of Titan's North-South Brightness Asymmetry," Nature, 292:698, 1981. (X2, X3)

ARL9 Changes in Saturn's Rings Observed within Historical Times

Description. Secular and transient changes in the sizes, brightnesses, and colors of Saturn's rings. Necessarily included by such a definition are variations in the gaps between the rings.

Data Evaluation. The claims for telescopically observed ring changes are all quite old and, in addition, rather vague. Trouvelot's observations of the 1880s are the most detailed, but even here there is a frustrating lack of distance measurements and other specifics. Rating: 3.

Anomaly Evaluation. Given the dynamic nature of Saturn's rings and the varying forces applied to them, changes are to be expected. Such variations seem well within the capacity of current theories to explain. Astronomers have been surprised at the complexity of the rings and their obviously dynamic nature; but this is hardly anomalous. Rating: 3.

Possible Explanations. Saturn's rings are dynamic and sensitive to the changing gravitational and electromagnetic environments created by the accompanying satellites, Saturn's magnetic field, and the solar wind.

Similar and Related Phenomena. The apparent changes in the F ring between Voyagers 1 and 2 (ARL6); knots in the rings (ARL2); bright spots on the rings (ARL3); the ring spokes (ARL5); the possible youth of the rings (ARL16).

Examples

X1. Widening of the rings. "Mr. William B. Taylor recalls attention to the announcement made by Otto Struve in 1851, that the observations of two hundred years showed the rings of Saturn to be widening, and the inner edge of the inner bright ring to be approaching the body of the planet. Later observations tend in the same direction; and, though there may have been unintentional exaggeration in Struve's numerical results, there seems little reason to question the general fact." (R1)

X2. Trouvelot's study of changes in Saturn's ring. "I will commence first with those occurring on the exterior ring A, and of which M. Perrotin has lately interested the Academy. It is very plain that great changes are produced on this ring and on the ring B near it. Encke's division does not exist, or if it does, it nearly coincides with Cassini's division. It is certain that I see nothing in its place with an eight-inch glass; but on the contrary I see a division larger and clearer than it was and much nearer the division Cassini.

On the 15th of February of this year (1884) I found for the first time that the division Encke had changed place. At the same time I found also that the Zone A, situated between this division and that of Cassini, was whiter and more luminous than I had apparently ever seen it. These changes must have been recent, for the 11th and 12th of February having observed Saturn, I took notice that this division of Encke was perfectly visible, and on the drawing that should accompany this note, is in its usual place.

On the evening of the 15th I also discovered these changes on the ring B. We know the ring is divided into three parallel zones. Since I have observed Saturn these zones have always appeared pretty nearly the same size. The internal zone, which is near the nebulous ring, is the deepest; and the external one, which forms the internal border of Cassini, is more brilliant. During this evening I found this last zone was narrower than the eastern limb and its size was diminished by less than one-half. It was also brighter, and could be distinguished from the grayish intermediate zone. I observed the same phenomenon the 20th of February, but this time on the western limb; whereas on the opposite limb it was difficult to recognize it.

In 1882 I observed the same changes very apparent on the internal zone of the ring B which touched the nebulous ring and which was so deep it could hardly be distinguished. The nebulous ring showed the same remarkable changes. It was easily seen at the east and with difficulty distinguished at the west. The brilliant and narrow zone which is now on the ring A, between the division of Cassini and the new division of Encke was shown with variations of brightness on the limbs, so much that it was brilliant on the one and hardly seen on the other." (R3, R2) Trouvelot was a well-known astronomer of his day and specialized in Saturn. He observed many of the strange shadow phenomena of ARL4. (WRC)

X3. Transient luminous phenomena. A Prof. Todd stated: "Near the extremities of the major axes of the bright outer ring of Saturn, with the aid of a powerful telescope, I have observed a certain sparkling flocculence which I interpreted to be a dissipation of the ring." (R5, R4)

X4. Theoretical study of ring materials. From an abstract in Science: "Although Saturn's rings are within the Roche zone, the accretion of centimeter-sized particles into large aggregates many meters in diameter occurs readily, on a time scale of weeks. These aggregates are disrupted when tidal stresses exceed their very low strengths; thus most of the mass of the ring system is continually processed through a population of large 'dynamic ephemeral bodies,' which are continually forming and disintegrating. These large aggregates are not at all like the idealized ice spheres often used in modeling Saturn's ring dynamics." (R6) Such processes might play roles in some of the other ring phenomena in this section. (WRC)

References

R1. "Rings of Saturn," Science, 2:660, 1883. (X1)
R2. "Changes in the Rings of Saturn," Scientific American, 53:4, 1885. (X2)
R3. Trouvelot, M. E. L.; "Changes Observed in the Rings of Saturn," Sidereal Messenger, 4:78, 1885. (X2)
R4. "Saturn's Ring," English Mechanic, 95: 85, 1912. (X3)
R5. "Possible Changes in Saturn's Rings," Nature, 88:388, 1912. (X3)
R6. Davis, Donald R., et al; "Saturn Ring Particles as Dynamic Ephemeral Bodies," Science, 224:744, 1984. (X4)

ARL10 The "Gaps" between the Rings

Description. The presence of annular gaps nearly devoid of material between Saturn's rings and ringlets. While matter is essentially absent in some gaps, in others, such as Cassini's Division, there exist many low density ringlets that become apparent only when backlit by the sun. Tens of thousands of rings and ringlets can be counted with (obviously) just as many gaps.

Data Evaluation. Pioneer and Voyager spacecraft photography. Rating: 1.

Anomaly Evaluation. At the very least, resonance theory cannot account for the thousands of gaps---there are not nearly enough resonances. Indeed, some astronomers ask whether resonances can really explain any gaps. Sweeper moon might plow out some gaps, but the Voyage photographs do not reveal these postulated satellites. More ominously for celestial mechanics, the complex, dynamic nature of the rings seems beyond the power of Newtonian mechanics to explain and may require a whole new theoretical structure. (See X4.) Although the seriousness of this situation may be relieved somewhat by the apparent success of the theory of density waves (induced by resonances) to explain the spiral structure of some portions of the rings. Rating: 1.

Possible Explanations. Resonances involving Saturn's moons and moonlets obviously play some poorly understood role in generating and preserving gaps. Sweeper moons, if they exist, may contribute, too. The theory of density waves, as induced by resonances, also shows promise in explaining spiral structures. Generally speaking, though, we seem to be at a theoretical impasse.

Similar and Related Phenomena. The "gaps" in the asteroid belt (AAO); the fine structure of Saturn's rings (ARL10).

Examples

X0. The "resonance" theory that is supposed to account for the gaps in Saturn's rings as well as those in the asteroid belt. "In 1867, Daniel Kirkwood provided a theory of subdivisions in the rings that has been the standard textbook explanation for a hundred years. The classical application of Kirkwood's theory to rings is the explanation of the gap between the A and B rings first observed by Jean Dominique Cassini in 1675. Kirkwood pointed out that ring particles in the Cassini division, each of which would follow its own circular orbit about Saturn, would repeatedly overtake the more distant, slower moving satellite Mimas as the same point in its orbit. That is, Mimas and these particles would be in resonance because their orbital periods have the whole-number ratio 2:1. Each time a ring particle overtook the immensely larger Mimas, it would feel the satellite's gravitational tug at the same point in its orbit. These periodic tugs, like the well-timed pushes that send a swing ever higher, perturb particles in resonance with Mimas into noncircular orbits, causing them to collide more often with particles outside the resonance. The result, according to simple resonance theory, is the clearing of a gap. It seemed to work for some qualitative aspects of the asteroid belt and the Cassini division." (R4)

X1. Although still widely employed to explain the gaps in Saturn's rings, many astronomers deny that resonance theory does any such thing.

T. C. Van Flandern: "It has always seemed strange to me that astronomers have spoken of the Cassini Division in Saturn's rings as a resonance phenomenon. The theory that the gap was cleared by Mimas, for which the 2:1 resonance position lies close to one edge of the Cassini gap. Yet it is well known that secular perturbations do not behave in this fashion. For example, the Kirkwood Gaps in the asteroid belt cannot be explained in this way, even the gap in 2:1 resonance with Jupiter, since the perturbations simply do not make it less probable for an orbit to occupy the gap. (The explanation of the Kirkwood Gaps is still considered to be unknown.) (R1)

J. Wisdom:" It has long been known that the Kirkwood gaps in the distribution of semimajor axes of the asteroids are associated with mean motion commensurabilities of Jupiter, yet no one has been able to demonstrate exactly why these commensurabilities should lead to gaps. The greatest obstacle is that there is no real understanding of the long-term dynamics near commensurabilities in the elliptic restricted three-body problem." (R6)

R. A. Kerr in a Science review of the Saturn ring problems: "Theorists are a bit troubled that they do not know exactly why one resonance produces a spiral density wave, another a sharp (ring) edge, and a third one of the more subtle narrow bands or gaps seen at resonance locations. 'That is a delicate question,' as (Scott) Tremaine puts it. The strength of the resonance, which is proportional to the satellite's mass and distance, seems to be an important but not the sole determinant. The 7:6 resonance of the larger of the two co-orbiting satellites (not 1980S28, as previously supposed) forms the sharp outer edge of the A ring, Tremaine notes, but its nearly equivalent 6:5 resonance produces only a minor feature." (R4)

In sum, there is an obvious connection between resonances and some ring gaps as well as other features, but no one knows just why this should be so. (WRC)

X2. To help explain the huge number of gaps in the rings, which greatly exceed the number of resonances of any consequence, the concept of "sweeper" moons was introduced. These small objects were supposed to gravitationally, perhaps electrostatically, clear ring material out to make the observed gaps. Unfortunately, the most thorough studies of Voyager photographs have failed to reveal sweeper satellites.

"The interior of the Cassini Division, once thought to be nearly devoid of matter, is seen in Voyager images to consist of a regular sequence of bright and dark features. Between the ringlets, each a few hundred kilometers across, lie empty gaps. Before Voyager 2, it seemed extremely likely that these gaps were cleared of particles by embedded satellites 10 to 30 km in diameter. Although no such objects were detected, some dynamical process involving perhaps smaller embedded 'moonlets' is still favored by many to explain this structure." (R4)

X3. Some "classical" ring gaps are not gaps at all. As mentioned in the final paragraph of X2 above, the Cassini gap is far from a void. "It was equally disturbing that the gap predicted by classical theory was found to be only slightly less dense. It was filled with a different size of particle that looked dark when lit from the front by the sun, but was quite bright (and equally complex) when seen from backlighting." (R3) Nevertheless, the gaps are definitely "different", and this requires an explanation. Apparently neither resonances nor sweeper moonlets can do the job, although they may be part of the answer. (WRC)

X4. Theoretical implications of the rings (and their gaps): "Several scientists, among them researchers at the Fusion Energy Foundation, have proposed that the complex ring structure is caused by a breakdown of Newton's equations that is similar to what happens in the three-body problem. It is well known that when more than two bodies interact via gravitational forces, a complex and unstable field is set up that can exhibit highly pathological behavior..... In a mathematical sense, the results of the Princeton work show that the gravitational field in which Saturn's rings exist is indeterminate. That is, it is mathematically unstable in the sense that a small change in the initial conditions will result in totally different final states. ..... The problem is not that the individual particles in the rings do not obey Newton's equations, or that they obey the equations in some different way than do the Sun and Saturn. It is rather that the causality of the physical system in question is determined on a different level from that of the individual particles. The so-called collective interactions in the ring have created a qualitatively new structure, which has its own laws of cause and effect---laws that are not mathematically reducible to the interaction of the millions of particles that make up the system. These new laws have yet to be discovered, but results from the moons of Saturn and Saturn's atmosphere point in a provocative direction." (R3)

References

R1. Van Flandern, T. C.; "New Saturnian Satellites?" Observatory, 99:8, 1979. (X1)

R2. Morrison, David; "The New Saturn System," Mercury, 10:162, 1981. (X2, X3)

R3. Bardwell, Steven; "Latest Voyager Results Overturn Newtonian Physics," Fusion, p. 56, December 1981. (X1-X4)

R4. Kerr, Richard A.; "Planetary Rings Explained and Unexplained," Science, 218:141, 1982. (X0-X2)

R5. Franklin, F.A., et al; "A Possible Link between the Rotation of Saturn and Its Ring Structure," Nature, 295:128, 1982.

R6. Wisdom, Jack; "Chaotic Behavior and the Origin of the 3/1 Kirkwood Gap," Icarus, 56:51, 1983. (X1)

ARL11 Ring Asymmetries and Eccentricities

Description. The presence in the Saturn ring system of eccentric ringlets, eccentric borders on the major rings, and intensity asymmetries. This category may include more than one phenomenon; and different causal mechanisms may be at work. Until more is known, though, they will be lumped together. Some, perhaps all, of these asymmetries and eccentricities are time-variable.

Data Evaluation. The ring intensity asymmetry is a telescopic anomaly of long standing. The ring eccentricities are imaged clearly in Voyager photographs. Rating: 1.

Anomaly Evaluation. Although explanations for these phenomena are noted below, we have no direct evidence that these hypotheses are valid; i.e., we have no measurements of meteor fluxes and no photos of perturbing moonlets. These anomalies will likely vanish when the Saturn system is explored more thoroughly. Rating: 3.

Possible Explanations. Ring intensity asymmetry has been attributed to sporadic meteor bombardment of the rings. Ring eccentricities are thought to be the consequences of resonances, although the necessary moonlets have not been found in most instances.

Similar and Related Phenomena. The rings of Jupiter, Neptune, and Uranus may display similar asymmetries and eccentricities, but close-up photos are not yet available.

Examples

X1. East-west asymmetry of the intensity of Saturn's rings. This phenomenon was recorded as early as 1714 by Maraldi. It is basically the occasional increased intensity of one ansa when compared to the other ansa. At some times, the difference in intensities is large, as in 1925 and 1969. Usually the eastern ansa is more intense, but sometimes the situation reverses. Piironen and Lukkari have correlated the degree of ring asymmetry with meteor activity. (R1)

X2. Eccentric ringlet in the outer gap of the C ring. "The outer gap is at a radius of about 87,100 km and the eccentric ringlet is 35 km wide and 100 km from the outer edge of the gap at its periapse. The ringlet is optically thick when viewed against Saturn's disk. At the apapse the ring is 90 km wide and 50 km from the outer edge of the gap. The ring has one periapse and one apapse, indicating that its eccentricity is not induced by resonance with an outer satellite. The line of apsides was not observed to move over a period of several days. This is consistent with a long precession period (~ 24 days) for the eccentric ring, probably due to the quadrupole moment of Saturn. It seems likely that this ringlet must continually have its eccentricity forced by some mechanism, probably a small satellite or satellites in the vicinity. Such forcing would seem to argue for an external resonance, but this feature is located near only a 2:1 resonance with the tiny satellite 1980S528." (R3, R2) Porco notes that there are actually two eccentric ringlets in the C ring; one at 1.29 Saturn radii, the other at 1.45 radii. (R6)

X3. Eccentric ringlet in inner Cassini division gap. "This ringlet precesses under the influence of Saturn's oblateness and does not appear to be directly tied to an external resonance. The radial position of this feature varies by at least 40 km, with the ringlet being slightly wider at its greatest radius." (R5, R4)

X4. The F ring. This ring is also eccentric. (R2) The F ring is actually ringlet size.

X5. Eccentric boundaries of the major rings. "The inner or B-ring edge of the Cassini division is a sharp discontinuity located at the 2-to-1 orbital resonance with Mimas, where a ring particle will orbit Saturn exactly twice for each orbit of Mimas. This boundary is eccentric, being in some places about 70 km closer to Saturn than in others. It is unique in that, although the shape of the boundary is elliptical, the center of the planet is not at one focus of the ellipse but rather at its center. This remarkable shape is taken as strong evidence that the boundary is indeed created and controlled by the Mimas resonance." (R4, R5, R7)

References

R1. Piironen, J.O., and Lukkari, J.; "The East-West Asymmetry of Saturn's Rings: A Possible Indicator of Meteoroids?" The Moon and the Planets, 23:373, 1980. (X1)

R2. "Saturn: More and More...," Science News, 118:307, 1980. (X2-X4)

R3. Smith, Bradford A., et al; "Encounter with Saturn: Voyager 1 Imaging Science Results," Science, 212:163, 1981. (X2)

R4. Morrison, David; "The New Saturn System," Mercury, 10:162, 1981. (X3, X5)

R5. Smith, Bradford A., et al; "Voyager 2 Encounter with the Saturnian System," Science, 215:499, 1982. (X3, X5)

R6. Porco, C., et al; "The Eccentric Saturn Ringlets at $1.29R_S$ and $1.45R_S$," Icarus, 60:1, 1984. (X2)

R7. Porco, C., et al; "Saturn's Nonaxisymmetric Ring Edges at $1.95R_S$ and $2.27R_S$," Icarus, 60:17, 1984. (X3)

ARL12 Hyperion's Chaotic Rotation

Description. The unpredictable (chaotic) rotation of Hyperion.

Data Evaluation. Analysis of Voyager spacecraft photographs. Note that we have here a deduction rather than direct observation. Rating: 2.

Anomaly Evaluation. Unpredictable motion, such as Hyperion's spin, implies either that we do not understand the physical laws in operation or that we do not know the forces applied. Assuming the latter to be the case: Rating: 3.

Possible Explanations. None.

Similar and Related Phenomena. None known at present.

Examples

X1. Analysis of Voyager-2 photos. After a discussion of Hyperion's nonspherical shape. "Even more bizarre, however, was Hyperion's motion. Other solar-system moons that are close enough to their planets have had the speeds with which they turn on their axes damped by tidal friction until they match the times required to go around their orbits. The result, as with earth's moon, is such that a satellite always keeps the same face toward its planet. But Hyperion does not appear to be following the rules. The only rule it does follow, in fact, says Jack Wisdom of the University of California at Santa Barbara (UCSB), may be the rule of chaos.

In the days and weeks following the Voyager encounter, it became apparent that determining the satellite's spin axis and the orientation in space of its axis was not going to be an easy matter. One early idea was that Hyperion was perhaps still 'wobbling' from a major impact, since its irregular shape suggested that it could be a fragment of a larger body, but such a wobble would be expected to settle down into some stable position. And in Hyperion's case, according to Wisdom, with UCSB colleague Stanton J. Peale and Francois Mignard of CERGA in Grasse, France, that may never happen.

Instead, says Peale, 'Hyperion appears to have been slowed by tidal friction to a state where it is forced to tumble chaotically, with large, essentially random fluctuations in its spin rate, in the direction of its spin axis relative to inertial space, and in the orientation of the spin axis relative to the body itself.' In fact, adds Wisdom, in fewer than two trips around Saturn, Hyperion's spin rate may very from a rotation every 10 days to no rotation at all." The conclusion that Hyperion rotates chaotically comes from analysis of Voyager data rather than direct observation. (R1)

References

R1. "Hyperion: A Moon in Chaos," Science News, 124:59, 1983. (X1)

ARL13 Irregular Density Trend of Saturn's Moons

Description. The unpredicted and unpredictable trend in the densities of Saturn's moons.

Data Evaluation. Analysis of the Voyager spacecraft tracking data. Rating: 1.

Anomaly Evaluation. Pre-Voyager expectations were that Saturn satellite densities would be either about the same or decrease with distance from Saturn. The "explanations" of the actual irregular distribution, as noted below, do not really reveal any basic understanding of the phenomenon ---they are too vague. Rating: 2.

Possible Explanations. The proto-Saturn nebula was inhomogeneous; Saturn's moons have had their order shuffled; the moons may include interloped from elsewhere.

Similar and Related Phenomena. The noble gas anomalies of the inner planets (AVW4, AMW5).

Examples

X1. Voyager-1 data. "Before Voyager 1's flyby, says Laurence Soderblom of the U.S. Geological Survey, there were two contrasting expectations of how the moon's densities might vary: Either they would get less dense with increasing distance from Saturn, based on the idea that increasing proportions of lightweight volatile materials would have been available farther from the hot center of the proto-Saturn nebula when the moons were formed, or they would all be the same, in the event that the proto-Saturn's heat was too little to differentiate the material mixture. Yet an early look, says Soderblom, suggests that neither may be true. Only three precise densities were initially available from Voyager data (though more will follow when gravitational perturbations in the spacecraft's trajectory can be factored in), but their implication, in Soderblom's view, is 'weird.' Close to Saturn is Mimas, at 1.2 ± 0.1 grams per cubic centimeter. Two moons out is Tethys, which is indeed less dense at 1.0 ± 0.1, but next in line is Dione, at 1.4 ± 0.1. Calculations for the other satellites are necessary to fill in the picture (some must wait for Voyager 2), but it seems at least possible that there will turn out to be no regular progression at all. Recent studies of isotopic anomalies in meteorites have hinted that the early nebula from which the solar system formed was not completely homogeneous or smoothly graduated; perhaps the message of Saturn's moons is that similar irregularities existed on a planetary scale." (R1, R2)

Voyager-2 data confirmed the irregular density trend: beyond Dione is Rhea at 1.33; then Titan at 1.88; then Iapetus at 1.16.(R3)

References

R1."Voyager 1's Saturn: Moonwatch," Science News, 118:325, 1980. (X1)

R2. Smith, Bradford A., et al; "Encounter with Saturn: Voyager 1 Imaging Science Results," Science, 212:163, 1981. (X1)

R3. Smith, Bradford A., et al; "Voyager 2 Encounter with the Saturnian System," Science, 215:499, 1982. (X1)

ARL14 Fine Structure of Saturn's Rings

Description. Spiral geometry, propagating waves, stacked layers, and similar detailed ring structures.

Background. By "fine structure" is meant geometrical details much smaller than the classical letter-bearing rings and formally named gaps, such as Cassini's division. An important aspect of fine structure was introduced in ARL10, where the many thousands of ringlets that comprise the major rings were described. Some of these ringlets are apparently not independent but rather parts of a continuous spiral. It is possible, as explained below, that spiral structures may be explained in terms of spiral density waves---a theoretical construct. Thus, the ARL10 anomaly, where there are not nearly enough resonances to account for all the ringlets, may be partially relieved by the spiral density wave theory.

Data Evaluation. Voyager spacecraft photographs and radio physics experiments. Some of the fine structure, especially that from radio physics experiments, is difficult to interpret. Rating: 2.

Anomaly Evaluation. An appealing theory of resonance-induced spiral density waves may explain both spiral and wavelet structures. The apparent layered structure of the rings, as determined by radio physics measurements, is more mysterious. But even here a wave interpretation may be possible. The fine structure seems to be more amenable to theoretical explanation than gross ring-gap structure. Rating: 3.

Possible Explanations. Gravitational resonances may induce spiral density waves, which create spiral structures that superficially resemble closely spaced ringlets. Electrostatic forces may be involved in generating fine structure, particularly the layered structures of X1.

Similar and Related Phenomena. The multiplicity of gaps between ring and ringlets and the paucity of resonances (ARL10); other solar system ring structures (Jupiter, Uranus, Neptune).

Examples

X1. Spiral groove geometry. "Another unexpected discovery came from a close analysis of the ringlets within the main inner ring, the B-ring. In some places, the spacing of the ringlets changes in a systematic manner, the ringlets further from Saturn becoming steadily more closely-spaced. Clearly they are not independent of one another, but are being influenced in a common way. A particularly striking case occurs just outside the point where a ring particle goes round Saturn in exactly half the period of the larger of Saturn's pair of co-orbital satellites. According to a theory originally developed to explain spiral galaxies, a tightly-wound spiral 'density wave' pattern can wind outwards from such a resonance---and the successive ringlets recorded by the photopolarimeter do fall exactly where one would expect to find the successive turns of such a spiral. The team is confident that the successive 'ringlets' here are in fact just one narrow spiral 'arm', like the grooves on a record." (R6, R3, R5, R7)

X2. Direct observation of wave-like structures. "Several wavelike features were observed by Voyager 2. The wavelike patterns in the A ring gap vicinity were observed with improved sensitivity and resolution from the lit face; some of these features appear to exhibit azimuthal asymmetry as well. Similar wavelike features were observed in the outer A ring near closest approach. None of these patterns has yet been satisfactorily explained. The wavelike pattern observed in the outer Cassini division by Voyager 1 was observed only very faintly by Voyager 2; a combination of unexpectedly extreme darkening of the region due to particle scattering properties and loss of several important unlit face observations prevented immediate detection of these features." (R4, R2)

X3. Radio detection of layers and waves while passing through the ring plane; Voyager PRA (Planetary Radio Astronomy) instrument. "A detailed look at the PRA data during the crossing, says Joseph Romig of Radiophysics, Inc., shows not only an intense peak centered right at the ring plane, but 'at least five separate regions,' with two lesser peaks above and two below the plane at similar spacing. Together, they represent a span extending from about 750 km above the plane to 750 km below it, almost as though Voyager 2 had encountered multiple rings arrayed like a stack of pancakes. This would be hard to swallow even for Voyager veterans of the anything-is-possible Saturn system, and the PRA team suggests that an alternative might be that the instrument was recording some sort of wave phenomenon, propagating outward through a cloud of particles to the north and south of the ring in the middle." (R1)

References

R1. "Saturn: Scanning the Unseen Scene," Science News, 120:182, 1981. (X3)

R2. Morrison, David; "The New Saturn System," Mercury, 10:162, 1981. (X2)

R3. Kerr, Richard A.; "Planetary Rings Explained and Unexplained," Science, 218: 141, 1982. (X1)

R4. Smith, Bradford A.; "Voyager 2 Encounter with the Saturnian System," Science, 215:499, 1982. (X2)

R5. "Saturn's Spirals," Science 82, 3:7, April 1982. (X1)

R6. Henbest, Nigel; "Saturn's Rings within Rings---Ad Infinitum," New Scientist, 93: 235, 1982. (X1)

R7. Kerr, Richard A.; "Rings, Rings, What Makes the Rings?" Science, 216:1210, 1982. (X1)

ARL15 Varying Crater Densities on Saturn's Moons

Description. The strikingly different crater densities measured on the small Saturnian satellites.

Data Evaluation. Voyager spacecraft photography. Rating: 1.

Anomaly Evaluation. One implication of the different crater densities is that after their accretion and bombardment, the satellites and their surfaces in particular evolved differently, with different fractions of their primordial crater populations being obliterated. If this is what actually happened, scientists cannot identify a reasonable energy source that would drive surface evolution. Rating: 2.

Possible Explanations. The satellites might have accreted in an inhomogeneous nebula, with some getting enough radioactive materials to cause varying degrees of surface melting. On the other hand, the satellites may not have originated in the Saturnian system at all, being captured at various times. Tidal and impact energies have been suggested as heat sources.

Similar and Related Phenomena. The enigmatic origin of earth's moon (ALE); the apparent origin-by-capture of the Martian satellites; the varying surfaces of Jupiter's Galilean satellites, which are much larger than the Saturnian satellites considered here.

Examples

X1. General observations. "Our data indicate that the surfaces of the Saturnian satellites are not uniformly cratered and hence not of uniform age. Significant variations in density occur at all diameters but are most noticeably in the distribution of large diameter (> 50 km) craters. The differing surface ages and tectonic features argues that these satellites have evolved significantly beyond their post-accretional state. The range in surface age suggests that the duration and degree of evolution varies between satellites. Such extensive post-accretional activity seems remarkable in view of the small size and mass of the Saturnian satellites." (Data analysis omitted.) "The problem then is the source of the energy responsible for evolutionary processes." (R1)

X2. Geological condition of the surface of Enceladus. The analysis reported in X1 omitted the moon Enceladus. Voyager photographs of the surface of this moon show it to be the most geologically evolved and youthful of all the Saturnian satellites. At least five different types of terrain can be identified; crater densities vary widely among them. Enceladus has a radius of only 255 kilometers---less than three of the four satellites studied in X1. The energy source that caused its surface evolution is unknown. (R2)

References

R1. Plescia, J. B., and Boyce, J. M.; "Crater Densities and Geological Histories of Rhea, Dione, Mimas and Tethys," Nature, 295:285, 1982. (X1)

R2. Smith, Bradford A., et al; "Voyager 2 Encounter with the Saturnian System," Science, 215:499, 1982. (X2)

ARL16 Youthful Features of Saturn's Rings

Description. The dynamic character, seemingly retarded dispersal, and apparent lack of erosive effects of Saturn's rings.

Data Evaluation. The "youth" of the features mentioned above is implied from various telescopic and spacecraft observations---"youth" is not observed directly! Rating: 3.

Anomaly Evaluation. Astronomers generally believe that Saturn's rings were formed at the same time Saturn accreted. Youthful rings would be highly anomalous. Rating: 2.

Possible Explanations. The conclusions about "youth" may be only misinterpretations of the data. Actually, Saturn's rings might have been formed recently after gravitational forces broke up a captured body, such as an asteroid.

Similar and Related Phenomena. The possible youth of lunar and Martian craters (ALE20 and AME18).

Examples

X1. Apparent retarded dispersal of the rings. "The most fundamental problem considered at the colloquium was one of the oldest in ring studies---why is there an A, a B, and a C ring? The mechanism that frustrates the natural spreading of ring particles at the outer edges of A and B seems to be strong resonances. But these do not work at the inner edges of A and B, or at the boundary of B and C. And the outer-edge resonance of A is only a temporary fix; nothing so puny as the larger co-orbital satellite can hold back the spreading of the rings without being pushed outward itself. The problem is that not enough spreading has occurred to fit theory, unless the near-ring satellites have been able to pass on angular momentum from the rings to a larger body. Goldreich conceded that he had spent a month looking for a resonance relation that would connect these satellites to Mimas and finally to Tethys, a suitably massive anchor for the rings. He found none, and no one else had even a tentative suggestion. The alternatives are that the rings are much younger than the 4.5-billion-year age of the solar system, an idea philosophically abhorent to most, or that theorists do not fully understand satellite resonances." (R1) The latter alternative is basically that suggested in ARL10. (WRC)

X2. Lack of evidence of erosive effects. "Erosive forces due to ultraviolet radiation and micrometeorite bombardment on icy particles must be at work also, and the latter in particular could destroy the ring system within a million years." (R2)

X3. Historical evidence of rapid changes in the rings. Referring to a century of micrometer measurements of Saturn's rings, V. Clube and B. Napier state: "If the measurements are taken at face value a very recent evolution or origin for the Saturnian ring system is implied." (R2)

X4. Ring thickness. V. Clube and B. Napier contend that inelastic collisions among the small icy particles in the rings should have shrunk the ring thickness down to 100-200 meters; and that the current thickness (1-3 kilometers) is consistent with an age of only 10-100 million years. (R2) Actually, ring thicknesses have been revised downward since R2 was published. (WRC)

References

R1. Kerr, Richard A.; "Planetary Rings Explained and Unexplained," Science, 218: 141, 1982. (X1)

R2. Clube, Victor, and Napier, Bill; "Short-Period Phenomena in the Solar System: An Interstellar Connection?" The Cosmic Serpent, New York, 1982, p. 69. (X2-X4)

AU INTRODUCTION TO URANUS

Key to Categories

AUB ORBITAL AND ORIENTATION PHENOMENA
AUF URANIAN INTRINSIC RADIATION
AUL RING AND SATELLITE PHENOMENA
AUO TELESCOPIC ANOMALIES

Uranus and Neptune are smaller and have higher densities than the solar system giants, Jupiter and Saturn. In a sense Uranus and Neptune are intermediates between the giants and the even smaller and denser terrestrial planets that ply the inner solar system. Such differences should result in fascinating new anomalies, but Uranus and Neptune are so far away that only a few challenges to astronomical theories have been found to date.

The Uranian anomalies number only four: (1) The uncomfortably high inclination of the planet's axis of rotation to the plane of the ecliptic; (2) The strange complement of rings; (3) The changing microwave emissions; and (4) The periodic changes of brightness. When spacecraft visit Uranus, we can be sure from past experience that this list will be multiplied in length several times.

AUB ORBITAL AND ORIENTATION PHENOMENA

Key to Phenomena

AUB0 Introduction

Uranus' axis of rotation and entire satellite system are inclined (almost grotesquely so) to the plane of the ecliptic. This is the only anomaly in this category at present, but it is a serious one. Not formally recognized here is a modest unaccounted-for residual in Uranus' orbital motion. This has been attributed to the presence of an undiscovered planet beyond Pluto. Neptune's residual is much larger, and this subject will be treated more thoroughly in ANB1.

AUB1 Anomalous Inclination of Uranus's Axis

Description. The 98° inclination of Uranus' axis of rotation to the pole of the ecliptic plane. The planet's satellites orbit in similarly inclined planes.

Data Evaluation. Long-established astronomical measurements. Rating: 1.

Anomaly Evaluation. Since no good explanations are at hand, this anomaly signals that there is something important that we not know about planetary and/or satellite formation. Rating: 2.

Possible Explanations. The oft profferred collision hypothesis does not explain the inclination of the satellite orbits to the plane of the ecliptic.

Similar and Related Phenomena. The retrograde spin of Venus (AVB1).

Examples

X1. Axial inclination. "The most extraordinary thing about Uranus is the fact that the inclination of its axis of rotation with respect to the pole of the ecliptic plane is 98°. Uranus and Venus are the only two planets with retrograde rotation. The rotation period of Uranus (10 hours) is apparently normal, being very similar to that of Jupiter, Saturn and Neptune, all of which have their axes of rotation nearly perpendicular to the ecliptic, and it is very difficult to understand why the direction and not the rate of rotation is anomalous. The only explanation that has been suggested is that, at the very end of the accretionary phase of the formation of the planet, a chance collision with another proto-planet (with perhaps 10% of the mass of Uranus) knocked the axis of rotation awry. But there is a severe difficulty with this hypothesis. The five satellites of Uranus all move in regular, nearly circular orbits in the plane of the planet's rotation rather than in the plane of the ecliptic. As discussed by R. Greenberg (U. of Arizona)...., there seems to be no possibility that the satellites could have been twisted into their present orbits by tidal effects; rather, they must have been formed where they now are, sharing the anomalous tilt of the planet. It is difficult to see how such an inclined satellite system could have been formed if the tilt of Uranus resulted from a single collision late in its period of formation. At present, there seems no way out of this difficulty; either (1) our understanding of the orbital evolution of the satellites is wrong, or (2) our understanding of the origin of the satellite system is wrong, or (3) our ideas of how Uranus developed its high axial inclination are wrong." (R1)

References

R1. "The Trouble with Uranus," Mercury, 4:25, July/August 1975. (X1)

AUF URANIAN INTRINSIC RADIATION

Key to Phenomena

AUF0 Introduction

Astronomers generally believe that Uranus, unlike Jupiter and Saturn, does not radiate more energy than it receives from the sun. Uranus does, however, emit microwave energy, which is probably derived from sun-driven atmospheric or ionospheric processes. The anomaly introduced here concerns the rather substantial secular increase in microwave radiation followed by an equally enigmatic decrease.

AUF1 Secular Changes in Radio Emissions

Description. The slow rise in the intensity of microwave radiation from Uranus in the 1970s and subsequent decline. The rise was measured as 30% in 10 years---a very large change. The increase in radio emissions was frequently reported as a Uranian "temperature" rise, although measurements were made only in the microwave region of the spectrum. The increase was correlated with the change in the position of Uranus' axis relative to the earth.

Data Evaluation. Several series of radio astronomy measurements in the 2-6 centimeter region of the radio spectrum. Rating: 2.

Anomaly Evaluation. Since several explanations have been proposed (see below), and we cannot choose between them, we obviously have much more to learn about this phenomenon. Whatever the explanation turns out to be, it seems sure to strongly modify our present conception of Uranus. Rating: 2.

Possible Explanations. During the reported increase in radio emissions, the orbital motion of Uranus was slowing revealing the north polar regions of the planet for the first time in 84 years. One theory has it that the north polar atmosphere is more transparent to microwaves generated deep in the atmosphere. Another possibility is that Uranus' north polar region is actually "warmer" due either to an intrinsic heat source or some microwave-emitting structure confined to this area. Least likely seems to be a general increase and fall in the overall temperature of the planet.

Similar and Related Phenomena. The periodic brightness changes of Uranus (AUO1).

Examples

X1. Ten-year records of Uranian radio emissions. "Major changes occurring deep in the atmosphere of Uranus have been discovered by two radio astronomers at NASA's Jet Propulsion Laboratory in Pasadena, Calif. M. J. Klein, of JPL's Planetary Atmospheres Research Section, and J. A. Turegano, a visiting research associate at JPL from the University of Zaragosa, Spain, have found that radio emissions from deep in the Uranian atmosphere have grown 30% stronger in 10 years.

Klein and Turegano, who made their observations with NASA's 64 m radio antenna at Goldstone, Calif., explain that the radio waves, which penetrate the dense clouds on Uranus, originate deep in the atmosphere where pressures are thought to be $\sim$10 times greater than those at the Earth's surface. They believe their results can be explained if the Uranian atmosphere is either warming or becoming clearer to the passage of radio waves, and they point out that it is hard to accept that temperatures so deep in a planet's atmosphere could be come 30% higher in only 10 years. (A similar change on Earth would raise the average air temperature above 120° C.) It is more likely, they explain, that the change is related to the planet's unique orientation. Unlike any other planet, Uranus spins on its side as it orbits the sun. Every 84 yr the sun shines directly on the north pole; 42 yr later the northern hemisphere is dark and the south pole points sunward. The north pole is now turning toward the sun after 42 yr of darkness. It will point directly at the earth (and the sun) in 1987." (R2, R1)

X2. VLA (Very Large Array) studies. "Abstract. Observations of Uranus at wavelengths of 2 and 6 centimeters with the Very Large Array were made in 1980 and 1981. The resulting maps of brightness temperature show a subsolar symmetry at 2 centimeters but a near-polar symmetry at 6 centimeters. The 6-centimeter maps show an increase in temperature from equator to pole with some evidence for a warm 'ring' surrounding the north pole. The disk-average temperatures (147 ± 5 K and 230 ± 6 K at 2 and 6 centimeters, respectively) are distinctly lower than recently reported values; these results suggest that the secular increase in temperature reported during the last 15 years has been reversed. The variations in brightness temperature probably reflect variations in ammonia abundance in the planet's atmosphere, but the mechanism driving these variations is still unclear." (R3, R4)

References

R1. Hunten, D. M.; "New Surprises from Uranus," Nature, 176:16, 1978. (X1)

R2. "Changes in the Uranian Atmosphere," American Meteorological Society, Bulletin, 59:1463, 1978. (X1)

R3. Jaffe, Walter J., et al; "Uranus: Microwave Images," Science, 225:619, 1984. (X2)

R4. Thomsen, Dietrick E.; "Planetary Radiograms," Science News, 125:266, 1984. (X2)

AUL RING AND SATELLITE PHENOMENA

Key to Phenomena

AUL0 Introduction

The known moons of Uranus are five in number and, although they are difficult to see, they seem well behaved enough. Never seen through the telescope are the planet's rings---nine at last count. We know these rings of material are there because they dim the light of stars passing behind them. The analysis of photometric occultation records suggest that the rings are irregular, very dark, very narrow, and possibly inclined to Uranus' equatorial plane.

AUL1 The Mysterious Rings of Uranus

Description. The following implied properties of the Uranian rings are considered to be anomalous: their narrowness, their extremely low albedos, their inclinations to the equatorial plane of Uranus, and the irregularity of the broad, outer ring.

Data Evaluation. Although the rings have not been seen visually, many studies have been made of the rings' occultations of stars. All ring features discussed below have been implied from such occultation records. Rating: 2.

Anomaly Evaluation. Like the electron, the Uranian rings are derived from instrument readings; they have not been seen. The rings' inclination, narrowness, and odd shapes may, like the anomalies of Saturn's rings, eventually yield to the methods of celestial mechanics, particularly resonance theory. Their blackness, though, seems to require some novel type of material or state of matter. Rating: 2.

Possible Explanations. Two models of the Uranian rings have been proposed to help explain some of their unusual features: crescent-shaped rings shepherded by satellites; and toroids of gas shed by small, unseen moonlets. Both theories have been criticized in the literature.

Generally speaking, conventional resonance theory may explain many features once we know more about the rings.

Similar and Related Phenomena. The anomalies of Saturn's rings (ARL).

Examples

X0. Historical note. On March 4, 1787, William Herschel reported that he had observed two rings around Uranus at right angles to one another. On March 16, only a single ring was seen. Herschel occasionally thought he saw rings in later years, too, but he apparently decided that they were only telescopic ghosts. (R1, R7) Modern determinations of Uranus' rings are based on their occulting of stars; no one has ever seen these rings by reflected light. (WRC)

X1. Ring elucidation via stellar occultations. The first hint of the existence of the rings of Uranus came on March 10, 1977, while a stellar occultation by Uranus was being observed from several different spots on earth. "An amazing series of observations, lasting less than 10 minutes, has shown that Uranus, seventh planet in the solar system, has a ring of satellites similar to Saturn's famous ring system. On 10 March several observatories monitored an occultation of the star SAO 158687 by Uranus. Such events, in which a planet passes through our line of sight to a star, are used to investigate planetary atmospheres. On this occasion astronomers unexpectedly observed a string of secondary occultations while the star was well clear of the planet. This indicates that the planet has a retinue of numerous dark objects in addition to the five known satellites. (R2, R11)

Analysis of the photometric records of the several observing groups led to a tentative conclusion that there were five, perhaps seven rings of material around Uranus. Furthermore, the outer ring seemed to be irregular in shape because its outbound occultation record was markedly different from the inbound record. (R3)

As more stellar occultations were monitored, other anomalous ring properties were implied: (1) Most of the rings are very narrow; a difficult situation to understand because perturbations of the ring particles should spread them out radially in just a few decades; (2) The invisibility of the rings in the best telescopes, which is equally mystifying because it implies that the rings are composed of particles blacker than the blackest coal dust. (R6, R8)

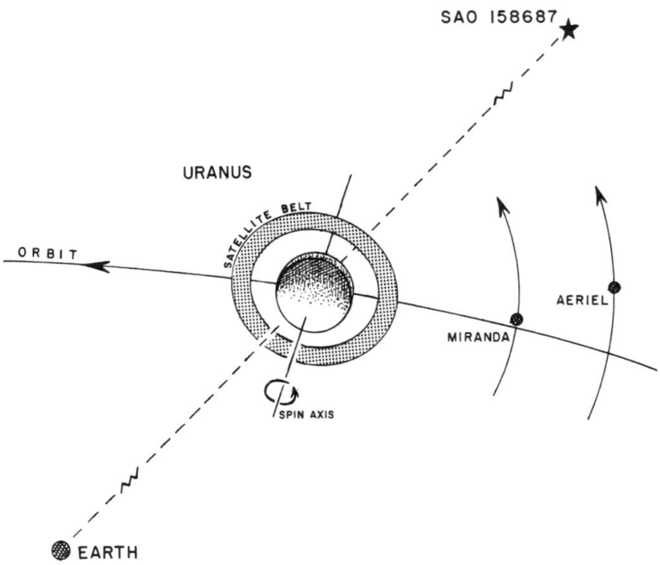

Geometry for the March 10, 1977, occultation of the star SAO 158687, during which a Uranian ring structure was detected. (X1)

After even more occultations, it was concluded that at least seven of the nine rings were slightly inclined to the equatorial plane of Uranus, where "well-behaved" ring particles are assumed to reside. (R10)

X2. Theories of the Uranian rings. The rings may be incomplete; that is, crescent- or horseshoe-shaped. (R4, R5) Some or all of the rings may be gaseous toroids, like the sodium torus trailing Jupiter's moon Io. The Uranian satellites presumed to be shedding material for the toroids would be in addition to the known complement of five, and have never been identified optically. (R6, R9)

References

R1. Baum, Richard; "William Herschel and the Ring of Uranus," The Planets, New York, 1973, p. 106. (X0)

R2. Hughes, David W.; "New Uranian Satellite Belt," Nature, 266:587, 1977. (X1)

R3. "The Strange Rings of Uranus," Science News, 111:245, 1977. (X1)

R4. "Uranus Rings May Be Crescents," New Scientist, 74:777, 1977. (X2)

R5. "Uranus's Luck Does Not Hold Out All Round," New Scientist, 80:607, 1978. (X2)

R6. Van Flandern, Thomas C.; "Rings of Uranus: Invisible and Impossible?" Science, 204:1076, 1979. (X1, X2)

R7. Capen, Charles F.; "Herschel and the Rings of Uranus," Astronomy, 7:42, January 1979. (X0)

R8. Villard, Ray; "More Rings for Uranus?" Star & Sky, 2:8, December 1980. (X1)

R9. Hunten, Donald M., et al; "Rings of Uranus: Proposed Model Is Unworkable," Science, 208:625, 1980. (X2)

R10. Kerr, Richard A.; "The Uranian Rings Get Stranger and Stranger," Science, 218:276, 1982. (X1)

R11. Mitton, Simon; "A Ring around Uranus," New Scientist, 73:685, 1977. (X1)

AUO TELESCOPIC ANOMALIES

Key to Phenomena

AUO0 Introduction

Uranus is too distant for terrestrial astronomers to identify anomalies involving atmospheric or surface details of the kind so common with the moon and inner planets. The only anomaly found so far in the literature is the periodic variation of the visible brightness of the planet-as-a-whole; the 8-year period of this phenomenon doesn't seem to correlate with anything.

AUO1 Periodic Brightening of Uranus

Description. The waxing and waning of Uranus' brightness after all effects of varying distances between earth and sun have been substracted out. An 8-year cycle has been established based on over a century of data. A study of recent vintage states that the planet's brightness peaks during the maximum of the solar cycle. (Of course, the sunspot cycle is very different from 8 years!)

Data Evaluation. Uranus is so distant that brightness measurements are relatively difficult, particularly in the sense that no detailed knowledge of the atmosphere and surface are available to hint at what is causing the changes in brightness. Rating: 2.

Anomaly Evaluation. The measured 8-year period is so different from the sunspot cycle (11 years), the period of rotation (10 hours), and the Uranian year (84 years), that the phenomenon is probably intrinsic---possibly atmospheric. In the absence of further insight, the anomaly rating is high. Rating: 2.

Possible Explanations. Probably an atmospheric phenomenon!

Similar and Related Phenomena. The secular increase in radio emissions (AUF1).

Examples

X1. Analysis of surface brightness. "An important analysis of all observations of the integrated brightness of Uranus from 1864 to 1932 has been carried out by W. Becker. He showed that after the effects of varying distance from earth and sun had been allowed for, and after a wave of amplitude of 0.29 magnitude and period 42 years arising from the oblateness of Uranus had been removed, there remained a well-defined periodic variation in the brightness of Uranus. This residual variation was characterized by an 8.4-year period, visual amplitude 0.31 magnitude, and a sinusoidal light-curve. The existence of the 8-year periodicity is confirmed by a series of visual observations of the brightness of Uranus made between 1936 and 1947 by the writer..... As Becker has pointed out, solar variability can be eliminated as the immediate cause of the 8-year cycle in the brightness of Uranus, because the light-variations of Mars, Jupiter, Saturn, and Neptune are quite different. The explanation of the 8-year cycle must therefore be sought in the atmosphere of Uranus itself." (R1)

A similar, but more recent, brightness analysis. "V.K. Balasubrahmanyan and D. Venkatesan have analysed luminosity data for the giant planets and found an increase in brightness corresponding to the most active period in the solar cycle, and a decrease toward solar minimum. Jupiter, Saturn, Uranus, and Neptune change in brightness as much as 20 per cent, according to a study by W. Becker 20 years ago. More recent observations of Uranus and Neptune over a short period (six years) indicate a variation of 2 per cent in the blue region of the spectrum..... The blue magnitude variations of Uranus and Neptune may be ascribed in part to changes in light-reflecting properties as the planet rotates. But the bulk of the variation remains something of a mystery. If one argues that the solar constant would be expected to decrease when the solar disc is most densely covered with sunspots, which are cooler than the normal photosphere, then it follows that there should be a slight drop toward solar maximum---in contradiction to the observed magnitude brightening." (R2)

References

R1. Ashbrook, Joseph; "Variations in the Brightness of Uranus," Astronomical Society of the Pacific, Publications, 60: 116, 1948. (X1)

R2. "Why the Solar Cycle Alters the Planets' Brightness?" New Scientist, 52:146, 1971. (X1)

AV INTRODUCTION TO VENUS

Key to Categories

AVB VENUSIAN ORBITAL AND SPIN ANOMALIES
AVE GEOLOGICAL PHENOMENA
AVF INTRINSIC RADIATION SOURCES
AVL POSSIBLE TEMPORARY SATELLITES OF VENUS
AVO ANOMALOUS TELESCOPIC OBSERVATIONS
AVW VENUSIAN ATMOSPHERIC ANOMALIES
AVX TRANSIT PHENOMENA
AVZ MAGNETIC FIELD OF VENUS

Venus, its crescent hanging bright in the evening sky, is like no other solar system planet. The planet's thick, hot atmosphere cloaks a parched surface seen only by a few spacecraft that have survived the descent through miles of searing gases. Despite its atmospheric shield, Venus presents many enigmas to the terrestrial telescope astronomer. Many of these mysteries seem to involve optical and subjective factors: the eerie ashen light, the blunted cusps or horns, the fleeting radial spoke system, the famous phase anomaly, and several other unexpected apparitions. Then there was Neith, the lost satellite of Venus, seen by many astronomers during the nineteenth century. Just what was this transitory companion of our sister planet? The accepted value for the planet's period of rotation has also had a curious history. It has been set at about 24 hours, 225 days, and more recently, 243 days. This uncertainty reflects the near-invisibility of the planet's surface features. The current 243-day (retrograde) period, derived from radar measurements, puts the axial rotation of Venus in synchronism with the earth's orbital motion around the sun. When the two planets are closest, at inferior conjunction, Venus always points the same face towards the earth. Why?

As if Venus-seen-through-the-telescope were not puzzling enough, several U. S. and Russian space probes have orbited Venus, sampled its atmosphere, and landed on its surface. Among the anomalies they have discovered are Venus's almost total lack of water as compared to nearby ocean-covered earth and a strange abundance of primordial rare gases (argon and neon, especially), again when compared to nearby earth. But then, argon-40, the radioactive product of potassium-40, a major producer of terrestrial heat, is in short supply on Venus. There is something amiss here. As for Venus's intrinsic magnetic field; it is almost non-existent, even though neighboring earth and Mercury boast magnetic fields. All in all, Venus seen close-up appears to be quite different from neighboring earth and Mercury, raising suspicions that it may have had a different evolutionary history.

It is also interesting to note that the space probes, when they examined Venus close-up, did not verify the reality of the radial spoke system, cusp phenomena, and some other telescopic anomalies.

AVB VENUSIAN ORBITAL AND SPIN ANOMALIES

Key to Phenomena

AVB0 Introduction

Early astronomers were convinced that Venus rotated on its axis either once per 225-day year or in just under 24 hours. This great discrepancy is not really surprising, because no one could telescopically see the least bit of solid planetary surface. It took modern radar measurements to demonstrate how wrong one could be if one tried to estimate Venus's period of axial rotation by observing the vague markings on thick clouds. Venus turned out to have a 243-day retrograde spin period. The 'retrograde' part was a shocking development. Something had obviously happened to Venus's primordial angular momentum. Perhaps less anomalous is Venus's tendency to rotate almost precisely four times on its axis between each inferior conjunction with earth. And then there is the residual advance of Venus's perihelion. Does it prop up the General Theory of Relativity or will solar oblateness turn it into negative evidence?

AVB1 The Length of the Venusian Day

Description. Venus spins on its axis very slowly in retrograde fashion. Its spin period of just over 243 days puts it almost exactly in resonance with the earth's orbit. Between the times when Venus is closest to earth (inferior conjunction) the planet spins almost precisely four times, presenting almost the same face of the planet at closest approach. The anomalous features of this situation are: (1) The slow retrograde spin, which is very unusual in the solar system (Mercury has a slow prograde spin, Uranus's spin is retrograde); and (2) the near-resonance with earth. Since both features may be linked in the same explanatory scenario, they are treated together here.

Background. Venus is an easy telescopic target but its thick atmosphere has always prevented astronomers from seeing its surface and measuring its true spin period. See X0 below.

Data Evaluation. Modern radar measurements of Venus's spin period are very precise. Rating: 1.

Anomaly Evaluation. If Venus was formed like the rest of the planets, it should have originally possessed appreciable prograde spin. Its current slow retrograde spin implies: (1) A different origin; or (2) A dramatic solar system event that despun the planet. Several theories have been proposed along these lines, but none has been generally accepted. All of these theories are rather radical in the context of solar system stability and astronomical uniformitarianism. In contrast, the earth-Venus resonance, which seems to be so remarkable, seems to be explicable in terms of tidal theory, although radical theories have been proposed here, too. Composite rating: 2.

Possible Explanations. Proposed despinning scenarios include: surface impacts of Venusian moons, the escape of a Venusian satellite (Mercury), and a prior association with Jupiter. Velikovskian scenarios envision Venus in close encounters with earth to explain the earth-Venus resonance. See X3-X7 below for more details.

Similar and Related Phenomena. Mercury's axial spin is very slow and in resonance with its orbit around the sun (AHB4); Uranus's retrograde spin.

Examples

X0. Historical background. Early estimates of the length of the Venusian day relied upon careful tracking of planetary features as they moved across the planet. This procedure assumed that the spots and other markings moved at the same velocity as the surface. Figures of just under 24 hours were most common. (R1-R4) One estimate, based on 11,000 micrometrical measurements of spot motion, was as astoundingly precise 23 hours, 21 minutes, 21.9345 seconds! (R5) Another astronomical faction believed Venus was in 1:1 resonance with its revolution around the sun. This yielded an axial spin period of 225 days. Even as recently as 1955, conflicting values existed. The famed planetary astronomer Gerard P. Kuiper could only say that the Venusian day was "not more than a few weeks." (R6) Then there were a few outcast measurements---one set as early as 1903---based on slight Doppler shifts in the Venusian spectrum. These indicated that Venus had a retrograde spin, which was clearly absurd. (R9)

X1. Permanent feature. The first radar-derived measurements had large probable errors: 239-293 days (JPL) and 248-258 days (Cornell). Both sets of data agreed that Venus had a retrograde spin. (R7)

Soon, however, more precise values became available: "Abstract. Combination of two types of radar data shows the orbital plane and equator of Venus to be included by less than 2 degrees, and the sidereal rotation period to be 243.09 $\pm$ 0.18 days (retrograde)---remarkably close to the 243.16-day period for which the spin would be in resonance with the relative orbital motions and Earth and Venus. In this resonance, Venus would make, on average, four axial rotations as seen by an Earth observer between successive close approaches of the two planets. Estimates of the instantaneous spin period, accurate within about 0.01 day, would provide important information on the difference of Venus's equatorial moments of inertia, on their orientation, and on the magnitude of the tidal torque exerted on Venus by the sun." (R8)

By now (1967), it was apparent that the planet's visible atmosphere spun far faster than the invisible surface below that reflected the radar waves. "Abstract. Optical observations of Venus have yielded various values of the rotation period extending from less than one to several hundred days. Radar observations give a retrograde rotation of the solid globe in 244 ± 2 days. Recent ultraviolet photographs, however, show relatively rapid displacements of clouds in the high atmosphere of Venus which suggest a retrograde rotation of only 5 days. The two rates seem to be physically incompatible." (R9) See AVW1 for a discussion of the "superrotation" of the Venusian atmosphere.

X2. General observations. "In earlier papers we examined the stability of a resonant rotation rate for Venus. In the resonant rotation state the planet presents the same face to the earth at each inferior conjunction. The present investigation is concerned with the capture probability at this resonance. It is shown that trapping at the resonance can be understood if Venus possesses a fluid core similar to the earth's. Maximum capture probability occurs if the core responds to changes in angular velocity of the mantle with a time lag of about 3 x 10^4 yr. If Venus is in the resonant rotation state, then mapping of its gravitational field will determine the direction of its primordial rotation and may also permit an estimate of its magnitude." (R10) If Venus's core is fluid, one might anticipate some dynamo action that would generate an intrinsic magnetic field; but Venus's magnetic field is very small (AVZ1).

X3. Theory. "The observed spin orbit resonance of Venus, whereby the same side of Venus faces the Earth at each inferior conjunction, cannot be explained adequately by gravitational interaction with the Earth alone. The expected solar tidal drag on the solid body of Venus would easily overwhelm the Earth's couple upon any reasonable permanent deformation of Venus. If there exists, however, a solar atmosphere tide, partly thermally induced and similar to that known on the Earth, its torque may counteract that due to the solar solid body tide at a particular rotation period. The small interaction with the Earth is then sufficient to lock the period to one of the resonances in the vicinity of that angular velocity." (R11) The same authors later noted that, "...the dynamical effects of convection will give the Venus atmosphere asymmetries in its mass distribution with respect to the Sun that are fully expected to provide an adequate couple for the Earth-resonance theory."

X4. Theory. Venus lost its primordial angular momentum when its ancient moon or moons crashed onto its surface. (R12)

X5. Theory. Velikovsky believed that the anomalies of Venus's spin could be explained by a scenario in which Venus escaped from Jupiter and, after brushes with earth, during which the Venus-earth resonance was established, finally settled down into its present orbit. (R13)

X6. Theory. Mercury was once a satellite of Venus. Mercury gradually escaped to its present orbit, reducing Venus's spin angular momentum in the process. (R14, R15)

X7. Theory. "It is possible that the Earth has captured Venus into a resonance in which the spin period is 243.165 d. However, for this resonance to be a stable equilibrium, either Venus must have a gravitational field that is ~ 10 times 'rougher' than the Earth's, a possibility that is largely ruled out by direct observation, or a third torque must balance the body torque due to the Sun. Tides in the atmosphere driven by periodic solar heating could supply the necessary third torque, but no quantitative theory has previously been published. Such a theory is presented here, in which we argue that the current rotation is a stable balance between atmospheric and solar body tides." (R16)

X8. Possible dynamic situation. I.I. Shapiro and colleagues note that radar observations accumulated from 1964 to 1978 have allowed them to pin down Venus's spin period with high precision. "They find it to be 243.01 ± 0.03 days. The 3 1/2 hour difference between this value and the resonance period of exactly 243.16 days; while very small, is statistically significant. On the other hand, the researchers point out that the probability of Venus' rotation period falling <u>by chance alone</u> within one-fifth of a day of a resonance period is under 1%. Therefore they suggest that Venus could either now be evolving toward such a resonance, or was once in resonance in the recent past. Venus may not, after all, be married to Earth, but whether the two are divorced or engaged---and in either case, why---has yet to be determined." (R17) The possibilities of a recent or imminent resonance is redolent of recent solar system instability. It would be interesting if "recent" means "within the time of man," to that there would after all be astronomical explanations of the many legends of celestial turmoil. (WRC)

References

R1. "Mercury and Venus," <u>English Mechanic</u>, 10:14, 1869. (X0)

R2. Flammarion, Camille; "The Rotation of Venus," <u>Observatory</u>, 17:354, 1894. (X0)

R3. MacEwen, Henry; "The Rotation of Venus," <u>British Astronomical Association, Journal</u>, 8:89, 1897. (X0)

R4. Roberts, C.; "Venus," <u>English Mechanic</u>, 62:14, 1895. (X0)

R5. Antoniadi, E.M.; "Notes on the Rotation Period of Venus," <u>Royal Astronom-</u>

ical Society, Monthly Notices, 58:313, 1898. (X0)
R6. "Solve Venus Mystery," Science News Letter, 67:151, 1955. (X0)
R7. "A Day on Venus Lasts 248 to 258 Earth Days," Science News Letter, 86:165, 1964. (X1)
R8. Shapiro, Irwin I.; "Resonance Rotation of Venus," Science, 157:423, 1967. (X1)
R9. Smith, Bradford A.; "Rotation of Venus: Continuing Contradictions," Science, 158:114, 1967. (X0, X1)
R10. Goldreich, Peter, and Peale, Stanton; "Spin-Orbit Coupling in the Solar System. II. The Resonant Rotation of Venus," Astronomical Journal, 72:662, 1967. (X2)
R11. Gold, Thomas, and Soter, Steven; "Atmospheric Tides and the Resonant Rotation of Venus," Icarus, 11:356, 1969. (X3)
R12. French, Bevan M., and Singer, S.F.; "How Did Venus Lose Its Angular Momentum?" Science, 173:169, 1971. (X4)
R13. Velikovsky, Immanuel, and Juergens, Ralph E.; "The Birth of Venus from Jupiter," Kronos, 2:3, no. 1, 1976. (X5)
R14. Van Flandern, Thomas C., and Harrington, Robert S.; "A Dynamical Investigation of the Conjecture That Mercury Is an Escaped Satellite of Venus," Icarus, 28:435, 1976. (X6)
R15. Hughes, David W.; "Spinning Comets and Planets," Nature, 273:100, 1978. (X6)
R16. Ingersoll, Andrew P., and Dobrovolskis, Anthony R.; "Venus' Rotation and Atmospheric Tides," Nature, 275:37, 1978. (X4)
R17. "Venus and Earth: Engaged or Divorced?" Astronomy, 7:58, October 1979. (X8)

AVB2 The Residual Advance of Venus's Perihelion

Description. The residual advance of Venus's perihelion, after the effects of all the other planets have been taken into account, using Newtonian celestial mechanics, is about 8.4 arc-seconds per century. In this calculation the sun is considered a perfect sphere.

Background. Little publicity has been given to the fact that, as in the case of Mercury, the perihelions of Venus and earth advance more than Newtonian celestial mechanics predicts, assuming a spherical sun. For further background, see (AHB1).

Data Evaluation. The residual advance of Venus's perihelion is considerably smaller than that of Mercury and apparently subject to more uncertainty. The two values found in the literature so far differ substantially. Rating: 3.

Anomaly Evaluation. If solar oblateness turns out to be significant in computing the advance of Venus's perihelion using Newtonian gravitation, the residual may change so much that General Relativity may no longer predict it accurately. General Relativity might not even be necessary. See AHB1. Rating: 1.

Possible Explanations. The General Theory of Relativity fully explains the latest value calculated for the advance of Venus's perihelion, based on a perfectly spherical sun. Some astronomers think the sun oblate enough to account for the advance without General Relativity.

Similar and Related Phenomena. The advance of Mercury's perihelion (AHB1); the advance of earth's perihelion; solar oblateness (AS).

Examples

X1. Permanent feature. The residual advance in Venus's perihelion is about 10 arc-seconds per century. A distribution of mass that would explain the Venusian anomaly would also explain Mercury's advance, making Relativity unnecessary. (R1)

X2. Permanent feature. "At the March 1956 meeting of the American Astronomical Society, R. L. Duncombe, of the U.S. Naval Observatory, announced results for the planet Venus derived from his recent analysis of many thousands of observations made from 1750 to 1949. For the first time the observational results have been accurate enough to detect the relativistic effect for Venus. Duncombe quoted the following values and probable errors for the advance of the perihelions of Mercury, Venus and Earth which are in excess of those predicted by all known Newtonian forces. G. M. Clemence derived the figures for Mercury, and Clemence, H. R. Morgan and Duncombe derived those for Earth. The observed discrepancies are followed by the predicted relativistic motions calculated by Clemence. All values are in seconds of arc per century. For each planet the two values agree within the probable errors. This close agreement between observed and predicted values gives considerable support to the general theory of relativity." (R2)

References

R1. Jeffreys, Harold; "The Motion of the Perihelion of Mercury," Nature, 101: 103, 1918. (X1)
R2. Morton, Donald C.; "Relativistic Advances of Perihelions," Royal Astronomical Society of Canada, Journal, 50:223, 1956. (X2)

	Mercury	Venus	Earth
Observed Discrepancy	43''.11 ± 0''.45	8''.4 ± 4''.8	5''.0 ± 1''.2
Relativistic Prediction	43''.03	8''.6	3''.8

AVB3 Supposed Changes in Venus's Orbit in Historical Times

Description. Some ancient historical accounts suggest that Venus changed its orbit within historical times.

Data Evaluation. Ancient historical accounts rarely appear in the scientific literature that forms the bulk of the data base used in compiling the Catalog of Anomalies. The few items that do find their ways into the journals are rarely taken seriously, even though they indicate that the ancient skies were full of "prodigies". A perusal of the works of Ignatius Donnelly, Immanuel Velikovsky, and Alfred de Grazia will underscore this point. Even so, the analysis of such data must take lower priority when judged against the mountain of scientific literature still remaining unexamined in the world's libraries. Consequently, the Catalog of Anomalies will incorporate ancient historical accounts only when they are swept up in the search of the "accepted" scientific literature. Finally, even though the pertinence of ancient historical data is recognized, they are frequently hard to interpret and have probably been distorted by centuries of retelling. Rating: 4.

Anomaly Evaluation. Any large-scale change in the location or shape of Venus's orbit in historical times would be an anomaly of the first order. Rating: 1.

Possible Explanations. The energies required for significant changes in the location or shape of Venus's orbit are so large that a massive interloper, perhaps from outside the solar system would have to be involved.

Similar and Related Phenomena. Most ancient heavenly "prodigies" are probably comets and large meteors. Still, in addition to Venus, some tales of celestial dislocations seem to involve the moon and Mars.

Examples

X1. Historical account. "In a resumé of the recent progress of astronomy contributed to an American work by Prof. Holden, of Washington, we note a reference to a communication made by M. Boutigny of the Academy of Sciences at Paris in December, 1877, calling attention to a passage in Varro, which describes the planet Venus, as having about the year B. C. 1831 (not B. C. 31, as misprinted in Prof. Holden's Report) 'changed its diameter, colour, figure, and course.' M. Boutigny had probably overlooked the circumstance that this story of Varro's had been brought into notice long before, in a French work of astronomical authority, the 'Cometographie' of Pingré, who believing that the fable was originated by some celestial phenomenon, considered it was most probably due to the appearance of a bright comet. Pingré thus gives the fragment, preserved by St. Augustin:--- 'There was seen, says Varro, a surprising prodigy in the heavens, with regard to the brilliant star Venus, which Plautus and Homer call, each in his own language---the evening star. Castor affirms that this fine star changed its colour, size, figure, and track, which had never occurred before, and which has not occurred since. Adrastus, of Cyzicus, and Dion the Neopolitan, refer this great prodigy to the reign of Ogyges.' Pingré's explanation will be found in 'Cometographie,' t. i. p. 247; the epoch he assigns for the phenomenon is 'vers 1770.'" (R1)

References

R1. "Varro's Story of the Anomalous Track and Figure of Venus," Nature, 20:351 1879. (X1)

AVE GEOLOGICAL PHENOMENA

Key to Phenomena

AVE0 Introduction

Even though Venus's solid surface is invisible to terrestrial telescope users, some of the planet's geological features are accessible through: (1) Surface photography; and (2)Radar mapping from spacecraft and from the earth. Surface photography has so far been carried out only by the Russian Venus landers in the Venera series. Venus is definitely not earth-like geologically speaking. Although volcanos, lava flows, chaotic terrain, rifts, plains, and other formations resembling terrestrial features have been discerned by radar mapping. The major terrestrial feature that is missing on Venus is water. There is no consensus on what happened to all of Venus's primordial water. Topographically, Venus does not seem to be peppered with craters of all sizes, like the moon, Mars, and Mercury. But there are huge ring structures of enigmatic origin. On the surface itself, the Venera cameras found a profusion of rocks, both smoothed and sharp-angled. Layered, sedimentary rocks are obvious in some of the pictures. The origin of these, too, is mysterious. Further mapping of the Venusian surface will doubtless add to this short list of anomalies---a list that could not have been made at all a few years ago.

AVE1 Sharp-Edged Angular Rocks

Description. The presence of sharp-edged, angular, apparently uneroded rocks on the surface of Venus. Some smooth rocks are also in evidence, as are layered rocks (AVE2).

Data Evaluation. High-quality photos from the Russian Venera spacecraft. Rating: 1.

Anomaly Evaluation. Many scientists expected highly erosive conditions on Venus's surface due to temperature fluctuations and winds in the dense atmosphere. The discovery of sharp-edged, fresh-looking rocks with few signs of erosion was initially a surprise. Other scientists were not caught unawares (see below). In terms of scientific expectations, then, this is only a minor anomaly. Rating: 3.

Possible Explanations. Even though many had anticipated a desert-like surface on Venus, with sand dunes and considerable erosion, C. Sagan has pointed out that temperature changes on Venus should be rather small. Therefore, winds should be slight and possess little erosive capability. (R2) Nevertheless, some of the Venera photos show layered rocks amidst sand and other debris. Are the layered rocks truly sedimentary in origin and, if so, were the sediments carried by wind? No one can tell for certain. Possibly the sharp-edged rocks are very young and not yet smoothed by the flow of sand and other erosive agents.

Similar and Related Phenomena. None.

Examples

X1. Permanent feature. Description of the Venera 9 photos. "The initial photo, apparently taken with the camera looking almost straight down (suggesting that mission officials wanted to ensure at least one picture before moving anything), contains remarkably clear view of some sharp-edged, angular rocks. According to Boris Nepoklonov, one of the mission scientists quoted by the Soviet news agency Tass, 'This seems to knock the bottom out of the existing hypothesis by which the surface was expected to look like a desert, covered with sand dunes because of constant wind and temperature erosion.' In fact, he says, 'even the moon does not have such rocks. We thought there couldn't be rocks on Venus---they would all be annihilated by erosion---but here they are, with edges absolutely not blunted. This picture makes us reconsider all our concepts of Venus." (R1)

But Venus has no seasons, and wind levels in the lower atmosphere should be an order of magnitude less than those measured on earth. "It may be useful to point out instead that it is the presence, not the absence of erosional mechanisms on Venus which is surprising." (R2)

References

R1. "Grand Unveiling of the Rocks of Venus," Science News, 108:276, 1975. (X1)

R2. Sagan, Carl; "Erosion and the Rocks of Venus," Nature, 261:31, 1976. (X1)

AVE2 Layered Sedimentary Rocks

Description. The presence of obviously layered sedimentary rocks of unknown origin on the surface of Venus. Ripples and other marks of flow are present.

Data Evaluation. High quality photographs from the Russian Venera series of spacecraft. Rating: 1.

Anomaly Evaluation. If the photographed sediments were laid down by water (now vanished), we would have an important anomaly because Venus's temperature is now so high that liquid water cannot exist. Much more probable are repeated aeolian and/or ballistic showers of fine material caused by winds, volcanos, and meteorite impacts. Based on these latter possibilities, the anomaly is not so remarkable. Still, the transporting power of Venusian winds is not known for certain. Splashes of fine material from meteor impacts are difficult to estimate because Venus's deep, thick atmosphere should decelerate meteors to low impact velocities. Rating: 2.

Possible Explanations. Wind deposits (assuming that the apparently low wind velocities have high carrying power by virtue of Venus's high atmospheric density); splashes of fine debris from meteorite impacts; volcanic dust.

Similar and Related Phenomena. The loess and other terrestrial aeolian deposits.

Examples

X1. Permanent feature. "Abstract. Venera 13 and Venera 14 transmitted almost complete panoramic views of their landing sites. Analyses of the photographs show the presence of rock formations undergoing geomorphic degradation. The formations display ripple marks, thin layering, differential erosion, and curvilinear fracturings. Some of them are interpreted as lithified clastic sediments. The lithification could have taken place at depth or at the surface, resulting in a type of duricrust. The origin of the sediments is unknown but could be aeolian, volcanic, or related to impacts or to turbidity currents." (R1)

References

R1. Florensky, C. P., et al; "Venera 13 and Venera 14: Sedimentary Rocks on Venus?" Science, 221:57, 1983. (X1)

Two views of the Venusian surface from the Venera spacecraft, showing layered rocks. (X1)

AVE3 Lack of Water on Venus

Description. The great scarcity of water on Venus in comparison to the earth. Venus's water inventory is 10^4 to 10^5 less than that of the earth. Venus and earth are similar in size and supposedly formed from much the same primordial materials. Venus, therefore, "should" have much more water than it does.

Data Evaluation. The Venera photographs and temperature measurements confirm the absence of water on the Venusian surface. Various types of instruments on the Russian Veneras and American Pioneer spacecraft report little water in the atmosphere. (See also AVW2.) It should be pointed out, however, that water vapor measurements are rather difficult to make, even in the earth's atmosphere. Rating: 2.

Anomaly Evaluation. This anomaly is one of unfulfilled expectations. Venus, earth's "twin" planet "should" have more water than it does. The accepted theory of solar system formation gives Venus about the same quantities of primordial constituents as the earth. What special mechanism or event deprived Venus of its initial water inventory? Or is the situation even more serious, with Venus having been formed elsewhere? Rating: 2.

Possible Explanations. Sunlight may have dissociated Venusian water, allowing the hydrogen to escape; the water may have reacted with reduced gases vented from the planet's interior; the water may have reacted with iron and other compounds in magmas; the water may have been recycled back into Venus's interior. These forgoing processes also could have occurred on earth to some degree. A major difference between Venus and the earth is the presence of life. The rapid development of life on earth may have siphoned off carbon dioxide before the Greenhouse Effect took hold. This fact could have prevented earth's temperature from rising to Venusian levels. Earth's water, then, was not thermally dissociated and its hydrogen driven off into space in this view.

Similar and Related Phenomena. The terrestrial Greenhouse Effect; the anomalous abundance of argon in the Venusian atmosphere (AVW4); the lack of water in Venus's atmosphere (AVW2).

Examples

X1. Permanent feature. "Where has all of Venus' water gone? Theorists have asked this question for years. It doesn't make sense to them that a planet so like the earth in size and distance from the Sun should have 10, 000 to 100, 000 times less water. After all, the pair have comparable amounts of carbon dioxide and nitrogen, so the water was probably there at the outset but has somehow disappeared over geologic time." (R1)

References

R1. Beatty, J. Kelly; "Venus: The Mystery Continues," Sky and Telescope, 63:134, 1982. (X1)

AVE4 Large Ring Structures

Description. The presence of large, concentric ring structures on the Venusian surface. These structures range from several hundred to a thousand kilometers in diameter.

Data Evaluation. Radar data from the Russian Venera spacecraft, supplemented by terrestrial radars. The resolution of the Venera radars is now about 1.3 kilometers. Rating: 1.

Anomaly Evaluation. The origin of these immense circular structures is unknown, although they may be analogous to the lunar ringed mares and some terrestrial domed structures. If they are actually impact structures, one wonders why smaller craters are not detected, like those seen on Mars, Mercury, and the moon. Rating: 3.

Possible Explanations. Meteorite impact structures; magma "plumes" beneath the Venusian surface.

Similar and Related Phenomena. The lunar ringed mares; highly eroded terrestrial ring structures (ET).

Examples

X1. Permanent features. "One family of features that Venera imaging has sharpened without identifying includes concentric circular rings in the rolling uplands that range up to several hundred kilometers and even 1000 kilometers in size. (Harold) Masursky reports that Soviet researchers are split as to whether they are scars of ancient impact craters or some kind of center of crustal deformation and movement. He notes their superficial resemblance at least to terrestrial mantle gneiss domes, thought by some to be traces of volcanic hot spots of Earth's first several billion years." (R1)

References

R1. Kerr, Richard A.; "Venusian Geology

Coming into Focus," Science, 224:702, 1984. (X1)

AVF INTRINSIC RADIATION SOURCES

Key to Phenomena

AVF0 Introduction

As our most prominent evening and morning star, Venus reflects a high proportion of the sunlight hitting its outer layers of clouds. Still, some solar radiation is absorbed and eventually reradiated in the infrared portion of the spectrum---just as in the case of the other inner or terrestrial planets. Venus also radiates internally generated energy. In fact, it radiates so much more energy than it intercepts from the sun that planetologists are wondering if Venus is chemically similar to Mercury, earth, and Mars. Venus may even have had a different place of origin!

Scientists also have problems---less serious ones, fortunately---in explaining two other sources of intrinsic radiation: (1) A variety of infrared-emitting structures in the Venusian atmosphere; and (2) The light that casts sharp shadows on the surface of the planet.

AVF1 Sharp Shadows of Rocks on Venusian Surface

Description. The casting of sharp shadows of surface rocks on Venus. Sharp shadows are typical of strong, single light sources. Given Venus's dense, cloudy atmosphere, a diffuse natural lighting without shadows was expected.

Data Evaluation. According to the literature sampled so far, sharp shadows were reported only for the Venera 9 mission. Because the spacecraft did not survive long on the Venusian surface, it is not known whether the phenomenon is permanent or transient, or perhaps even an artifact of the spacecraft itself. Rating: 3.

Anomaly Evaluation. If the sharp shadows are merely due to a rift in the cloud cover, the anomaly is a minor one; but Venus's atmosphere is so dense and cloudy that this explanation seems unlikely. On the other hand, other directional light sources are very speculative. (See below.) We really don't know what sort of phenomenon we have here. Rating: 2.

Possible Explanations. A cloud rift that admitted sunlight while Venera 9 was taking photos; continuous lightning or electrical discharge in a single direction; rock fluorescence, perhaps stimulated by heat or electricity.

Similar and Related Phenomena. "Summer" lightning on earth (GLL); earthquake lights (GLD8).

Examples

X1. Apparent permanent feature. Venera 9 results. "An important question is why the surprisingly sharp rocks also seem to have surprisingly sharp shadows. If the Venusian atmosphere diffuses incoming sunlight as broadly as has been expected, why are not the shadows either faint or multidirectional, if not completely absent? Soviet officials gave no immediate indication of whether the landing craft carried their own lighting, or even of the wavelengths at which the images were made." (R1)

"Another mystery in the Venera 9 pictures is the apparent shadows cast by the rocks. Avduevsky points out that as the lander descended it took continual measurements of the illumination from all sides. It recorded the sort of diffuse light expected under a cloud cover. 'Then it landed, and all of a sudden these shadows.' If they are shadows, they would indicate a directed light source in the Venus atmosphere, possibly a rift in the clouds or something more exotic." (R2) The Russian testimony seems to indicate that the sharp shadows had nothing to do with spacecraft lights. (WRC)

References

R1. "Grand Unveiling of the Rocks of Venus," Science News, 108:276, 1975. (X1)
R2. "The Unveiling of Venus: Hot and Stifling," Science News, 109:388, 1976. (X1)

AVF2 Bright Infrared Structures

Description. Large, infrared-bright, cloud-like features detected moving over Venus. These features extend for thousands of kilometers and move at different speeds than the ultraviolet cloud-like features. One well-defined, fairly permanent feature is called the polar dipole. Other infrared features superficially resemble the ultraviolet clouds. (This category could have just as legitimately been filed with the atmospheric anomalies (AVW).)

Data Evaluation. The data come from an infrared radiometer on the Pioneer Venus Orbiter and the telescopic examination of Venus's dark side with an infrared photometer-spectrometer. Rating: 1.

Anomaly Evaluation. Evidently these infrared structures are anomalously warm regions in the Venusian atmosphere. Their origins are matters of speculation at present; and it is uncertain why they do not move in concert with the ultraviolet clouds. Rating: 3.

Possible Explanations. None.

Similar and Related Phenomena. Venus's bright bands (AVO5); Venusian lineaments (AVO6).

Examples

X1. Permanent feature. "The Venusian polar dipoles are long-lived, elongated, warm features seen in images of thermal emission from the polar cloud tops of the planet. They are almost 4,000 km across, are centred close to the pole, and appear to rotate with a period of ~3 days retrograde. The northern hemisphere dipole was first identified as such by the Orbiter Infrared Radiometer (OIR) on the Pioneer Venus Orbiter (PVO), and in this study we use OIR images at 11.5 μm to investigate its detailed rotation. Its rotation rate is observed to change steadily over the 72-day data set, and there is some evidence for oscillatory variations superimposed on this trend." (R1)

X2. Changing patterns. Abstract. "Observations of the dark side of the planet Venus at infrared wavelengths between 1.5 and 2.5 μm have shown it to be anomalously bright in portions of this waveband, and to exhibit structured cloud patterns whose rotation period is longer than that of any other clouds." (R2)

References

R1. Schofield, J.T., and Diner, D.J.; "Rotation of Venus's Polar Dipole," Nature, 305:126, 1983. (X1)

R2. Allen, David A., and Crawford, John W.; "Cloud Structure on the Dark Side of Venus," Nature, 307:222, 1984. (X2)

AVF3 Intrinsic Heat Radiation from Venus

Description. The radiation by Venus of more heat that it receives from the sun. One estimate puts the excess at 15%. Most of this radiation, of course, is in the infrared.

Background. Planets can radiate more heat than they receive from the sun due to (1) Intrinsic heat generation by natural radioactivity, by chemical phase changes, by gravitational contraction, and by natural fission and fusion reactions; and (2) Residual heat of formation; i.e., the planet has not cooled down completely.

Data Evaluation. Measuring net energy flow on a distant planet is rather tricky. Radiation over the entire electromagnetic spectrum must be summed up, including incident and reflected sunlight. In the case of Venus, radiometers on orbiting spacecraft and terrestrial telescopes have consistently indicated a net outflow of heat from Venus. Rating: 2.

Anomaly Evaluation. A small amount of net heat radiation from Venus would be understandable in terms of natural radioactivity, just as potassium-40 generates heat within the earth. A large net outflow, such as the 15% estimate from X1 below, would imply (1) An anomalously large inventory of natural radioactivity in comparison with the other inner planets; or (2) Venus has not cooled down as much as the other inner planets and may have had a different origin. Either situation would be very disturbing to current theories of solar system origin and evolution. Rating: 2.

Possible Explanations. Estimates of net heat flow may be off the mark, thus saving the situation; Venus originated elsewhere and possesses a different chemical makeup; Venus in much younger than the other inner planets. The last possibility has Velikovskian overtones, but it must be considered for the sake of completeness.

Similar and Related Phenomena. Radioactive heat production within the earth; natural fission reactors (the Oklo phenomenon, EC); the large net heat outflows from Jupiter (AJF1) and Saturn (ARF2).

Examples

X1. Permanent feature. "Perhaps the most perplexing of the atmospheric problems lingering after Pioneer is the 460°C temperature at the bottom of the atmosphere. The much ballyhooed greenhouse effect of Venus's carbon dioxide atmosphere can account for only part of the heating, and evidence for other heating mechanisms is now in a turmoil. The question concerns how the sun's energy behaves once it penetrates the highest clouds. When Pioneer Venus's probes looked at the amount of radiant energy passing through the atmosphere, each one found more energy being radiated up from the lower atmosphere than enters it as sunlight." (R1)

Working with data from the Pioneer Venus Orbiter, F.W. Taylor "...found that Venus radiates 15 per cent more energy than it receives. To keep the surface temperature constant, Venus must be producing this extra heat from within. All the inner planets, including the Earth, produce internal heat from radioactive elements in their rocks. But Taylor's observations of Venus would mean that the planet is producing almost 10 000 times more heat than the Earth---and it is inconceivable, according to present theories of planetary formation, that Venus should have thousands of times more of the radioactive elements than the Earth does. At last week's meeting, Taylor's suggestion met with scepticism---not to say sheer disbelief---from other planetary scientists." (R2)

References

R1. Kerr, Richard A.; "Venus: Not Simple or Familiar, But Interesting," Science, 207:289, 1980. (X1)

R2. "The Mystery of Venus's Internal Heat," New Scientist, 88:437, 1980. (X1)

AVL POSSIBLE TEMPORARY SATELLITES OF VENUS

Key to Phenomena

AVL0 Introduction

Venus certainly has no satellite of appreciable size today, but during the last 350 years there have been numerous sightings of objects close to Venus that may well have been temporary moons of that planet. Although the observations of supposed satellites of Venus are actually of rather good quality, they have been eroded by the centuries and accumulated astronomical disdain for "temporary" phenomena.

AVL1 Bright Objects Resembling Venusian Satellites

Description. The appearance close to Venus of bright objects that could have been temporary satellites of the planet. These objects frequently show the same phase as Venus.

Background. The supposed satellite of Venus was an important astronomical problem of the 1800s, along with the equally elusive intramercurial planet. So real was the Venusian satellite that an orbital period was computed for it (11-12 hours), and it was listed in some astronomical tables. The name Neith was given to the moon of Venus after the goddess whose veil no mortal has ever lifted. The last "good" series of observations of Neith occurred in 1761 and 1764, although there were a few isolated sightings in the 1800s. Now, with almost a full century devoid of observations, and with much better telescopes, it must be concluded that nothing substantial orbits Venus. This does not mean that Venus never had a temporary satellite or that other objects (asteroids or comets) never passed close enough to Venus to be temporarily mistaken for a Venusian satellite. In this context, see the discussion of the intramercurial planet (AEO) in another volume of this Catalog.

An emotional element is present in both the histories of Neith and Vulcan (the intramercurial planet). To dispose of the Neith problem once and for all, the assertion of M. Stroobant that all sightings of Neith were actually those of stars or, in one case, Uranus, has been adopted by almost all astronomers. This century-old "solution" of the problem does not even come to grips with the emphatic claims of the early observers that Neith had the same phase as Venus. Stars and Uranus do not go through phases.

Data Evaluation. Over thirty observations of bright objects close to Venus were made between 1645 and 1892. Many of the observers were very competent and employed excellent instruments. The 1761 and 1764 observations involved several sightings each by several different astronomers. The earlier observers, in particular, remarked that the object had the same phase as Venus. Although there are no comparable sightings in this century, we have many records of unidentified objects close to the sun (AEO). It is also pertinent that no satellites of Venus have been seen during its transits of the sun. Rating: 2.

Anomaly Evaluation. A temporary bright object near Venus, even with the same phase, is not especially anomalous, because it is generally recognized that considerable debris occupies the inner solar system. Rating: 3.

Possible Explanations. (1) A temporary satellite of Venus (R4); (2) Asteroids passing close to Venus and exhibiting the same phase (R7); (3) Stars not identified as such by the observers, as concluded by M. Stroobant (R12, R14, R15, R18); (4) Telescopic ghosts (R13); (5) Comets.

Similar and Related Phenomena. Intramercurial planets (AEO); temporary satellites of earth (AGL); the slow, retrograde spin of Venus and its possible association with the escape of a satellite (AVB1).

Examples

X1. November 15, 1645. F. Fontana at Naples. "Fontana, much given to delicate observations, perceived distinctly above Venus a small star, of feeble brightness, which presented exactly the same phase as the planet; it looked like a miniature of Venus. He believed he had discovered a satellite, similar to those Galileo had found about Jupiter. But he was surprised at no one having seen it before, and also at being unable to detect it afterwards, however careful his search." (R4, R1, R11, R18) Some writers conclude that Fontana's instrument was faulty without saying why. (R7, R9)

X2. January 25, 1672. Giovanni Cassini. "From fifty-two minutes after six in the morning to two minutes after seven, when the brightness of twilight made it disappear. Venus was then horned; and this phenomenon, the diameter whereof was nearly a fourth part of the diameter of Venus, was of the same shape. It was distant from the southern horn of Venus a diameter of the planet on the western side." (R17, R1-R4, R7, R8, R11)

X3. August 28, 1686. Giovanni Cassini. "At fifteen minutes after four in the morning, looking at Venus with a telescope of thirty-four feet, I saw at the distance of one third of her diameter eastward a luminous appearance of a shape not well defined, that seemed to have the same phase with Venus, which was then gibbous on the western side. The diameter of this phenomenon was nearly equal to a fourth part of the diameter of Venus. I observed it attentively for a quarter of an hour, and having left off looking at it for four or five minutes I saw it no more; but daylight was then advanced." (R17, R1-R4, R7, R8, R11, R18)

X4. October 23, 1740. James Short. "Directing a reflecting telescope of 16.5 inches focus (with an apparatus to follow the diurnal motion) towards Venus, I perceived a small star pretty nigh her; upon which I took another telescope of the same focal distance, which magnified about fifty or sixty times, and which was fitted with a micrometer in order to measure its distance from Venus, and found its distance to be about 10° 2' 0" (sic). Finding Venus very distinct, and consequently the air very clear, I put on a magnifying power of 240 times, and to my great surprise found this star put on the same phasis with Venus. I tried another magnifying power of 140 times, and even then found the star under the same phasis. Its diameter seemed about a third, or somewhat less, of the diameter of Venus; its light was not so bright or vivid, but exceeding sharp and well defined. A line, passing through the centre of Venus and it, made an angle with the equator of about eighteen or twenty degrees. I saw it for the space of an hour several times that morning; but the light of the sun increasing, I lost it altogether about a quarter of an hour after eight. I have looked for it every clear morning since, but never had the good fortune to see it again." (R3, R1, R2, R7, R8, R11, R17)

X5. May 20, 1759. Meier. "...about 8h. 45m. 50s., I saw above Venus a little globe of far inferior brightness, about 1 1/2 diam. of Venus from herself. Future observations will show whether this little globe was an optical appearance or the satellite of Venus. The observation was made with a Gregorian telescope of thirty inches focus. It continued for half an hour, and the position of the little globe with regard to Venus remained the same, although the direction of the telescope had been changed." (R3)

X6. February 10 through August 13, 1761. A series of sightings by several European astronomers, beginning with LaGrange on February 10. On four nights in May, beginning on May 3, Montaigne, at Limoges, saw a small crescent 20' from Venus. On following nights (May 4, 7, 11), it was seen again in different positions but still with the same phase as Venus. In early August, two Danish astronomers, Roedkioer and Boserup reported a "satellite" near Venus. (R3, R1, R2, R7, R8, R11, R12, R14, R15, R17, R18)

X7. March 1764. Roedkioer, Horrebow, in Denmark, record a Venusian satellite on several days, beginning March 3. On March 15, 28, and 29, Montbarron, at Auxerre, also sees the object. (R7, R2, R3, R8, R11, R15, R17)

X8. January 3, 1768. Roedkioer again reports an observation of a satellite of Venus. (R7, R12, R15)

X9. May 22, 1823. T.W. Webb. "I turned to Venus, then high in the S.W., and saw a star, exactly resembling Mercury, or a miniature Venus, p or sp the planet, at a short distance, perhaps 20' or 30', and 1/3 or 1/4 of its diameter...." (R3)

X10. January 3, 1878. Observation by Horrebow. (R14)

X11. February 3, 1884. "...at 6 o'clock in the evening, M. Stuyvaert, of the Brussels Observatory, observed on the disk of Venus, near the illumined border, an extremely brilliant point, that recalled the aspect of the satellites of Jupiter as they transit the planet. The interest of this observation is increased by another made a few days later, on the 12th of the same month, at 8 o'clock

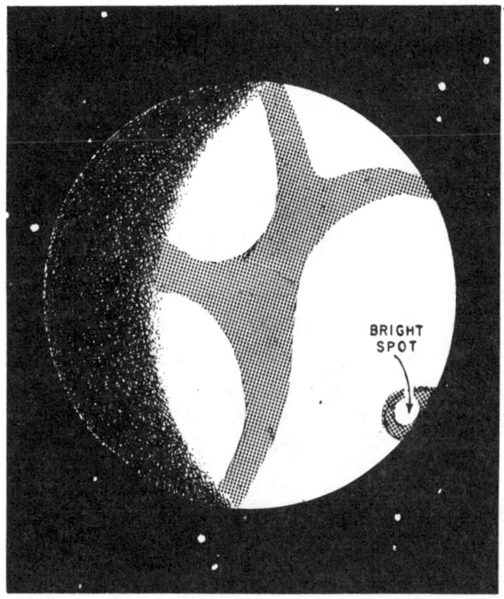

Drawing of Venus made on February 3, 1884, showing a bright point of light. (X11)

in the evening. M. Niesten then saw, a little south of Venus, a small star that seemed to be composed of a nucleus and a very faint nebulosity. He looked in vain for the star on the succeeding evenings. Has Neith, the problematic planet, deigned to reappear after an absence of more than a century?" (R5, R6, R8, R10)

X12. August 13, 1892. E.E. Barnard. "While examining Venus on that date with the 36 inch of the Lick Observatory---as I had done at other times previously---I saw a star in the field with the planet. This star was estimated to be of at least the 7th magnitude. The position was so low that it was necessary to stand upon the high railing of a tall observing chair. It was not possible to make any measures, as I had to hold on to the telescope with both hands to keep from falling. This star was estimated to be 1' south of Venus and 14^s-preceding, at 4^h50^m a.m. Standard Pacific Time.... There seems to be no considerable star near this place and the object does not agree with any BD. star. The observation was made in broad daylight, a half hour before sunrise. Unless this was one of the brighter asteroids (not Ceres, Pallas, Juno, or Vesta, however, which were elsewhere) I am unable to account for the observation. The elongation of Venus from the Sun was about 38° which would exclude the possibility that the object was an Intra-Mercurial planet,, but it does not preclude the possibility of its being a planet interior to Venus, though such is not probable." (R16)

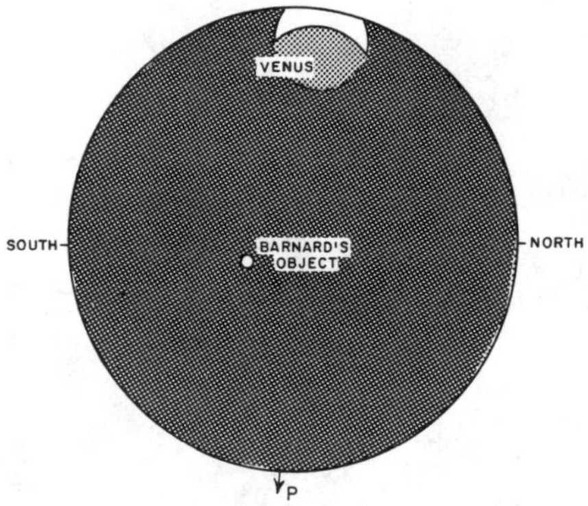

Barnard's Object seen near Venus through the Lick 36-inch telescope. (X12)

References

R1. Blacklock, Arthur W.; "The Satellite of Venus," Astronomical Register, 6: 196, 1868. (X1-X4, X6)

R2. Webb, T.W.; "Supposed Satellite of Venus," Astronomical Register, 6:224, 1868. (X2-X4, X6, X7)

R3. Webb, T.W.; "The Satellite of Venus," Nature, 14:193, 1876. (X2-X7, X9)

R4. Houzeau, M.; "On Certain Enigmas of Astronomy," English Mechanic, 29:32, 1879. (X1-X3)

R5. "The Problematic Planet Neith," Scientific American, 51:145, 1884. (X11)

R6. "The Problematic Satellite of Venus," Knowledge, 5:452, 1884. (X11)

R7. "The Problematical Satellite of Venus," Observatory, 7:222, 1884. (X1-X4, X6-X8)

R8. "Does a Planet Exist Immediately Exterior to Venus?" English Mechanic, 39: 294, 1884. (X2-X4, X6, X7, X11)

R9. Sadler, H.; "Satellite of Venus," English Mechanic, 39:345, 1884. (X1)

R10. Gore, J.E.; "The Supposed Planet 'Neith'," English Mechanic, 39:369, 1884. (X11)

R11. "The Planet Neith," Science Gossip, 22:178, 1886. (X1-X4, X6, X7)

R12. "The 'Satellite' of Venus," Nature, 36:543, 1887. (X6, X8)

R13. Lynn, W.T.; "Montaigne's Alleged Observation of a Satellite of Venus in 1761," Observatory, 10:73, 1887. (X0)

R14. "The Supposed Satellite of Venus," Observatory, 10:363, 1887. (X6, X10)

R15. "The Supposed Satellite of Venus," Sidereal Messenger, 6:357, 1887. (X6-X8)

R16. Barnard, E.E.; "An Unexplained Observation," Astronomische Nachrichten, 172:25, 1906. (X12)

R17. Dennett, Frank C.; "Venus: The Planet of Mystery," Knowledge, 9:304, 1912. (X2-X4, X6, X7)

R18. "The Satellite of Venus," Sky and Telescope, 13:333, 1954. (X1, X3, X6)

AVO ANOMALOUS TELESCOPIC OBSERVATIONS

Key to Phenomena

AVO0 Introduction

Venus is one of the brightest objects in the night sky. It is so bright, in fact, that irradiation and other psychological effects play roles in explaining anomalous telescopic observations. Such effects may be the key to understanding why Venus somtimes displays phenomena that are almost identical to those recorded for Mercury; viz., cusp blunting, terminator irregularities, darks lineaments, and faint rings of light around the dark limb. Perhaps the famous phase anomaly of Venus should also be included in this list of parallel phenomena occurring on both of the solar system's inner planets, but so far no one has offered convincing evidence that Mercury's actual phase also leads and lags the theoretical phase.

The frequent appearances of Venus's "ashen light"---that faint glow of the dark side of the planet---reminds us that, in contrast to Mercury, Venus is swathed in a hot, thick atmosphere that is chemically and electrically active. No one knows whether these processes play any role in generating the ashen light, but they are leading candidates. Venus's complex atmosphere gives us many more opportunities for physical explanations of cusp blunting, terminator irregularities, etc., than we have for nearly identical phenomena on Mercury. Venusian clouds and electrical storms certainly seem made-to-order for explaining transient telescopic appearances. Unfortunately, these phenomena cannot be used to account for the parallel anomalies seen on Mercury.

Like Mercury, Venus has occasional darkish lineaments. These lines often arrange themselves around a hub to produce a spoke-like system. The central problem here, however, does not seem to be geometry, but rather the fact that many observers cannot see the spokes under any circumstances, nor do the spokes appear on spacecraft close-up pictures. Yet, some very competent astronomers swear the spokes are there, just as some still see Martian canals. This sort of problem arises again and again in solar system astronomy. The customary solution is to call the lineaments illusory and forget about them.

AVO1 Venusian Cusp Phenomena

Description. Temporary dark and bright features observed around the southern and northern cusps or horns of Venus. Five varieties of phenomena are recognized here: (1) Bright spots, hoods, or "polar caps"; (2) Blunted or truncated cusps; (3) Extended cusps that reach well beyond 180°; (4) Detached bright spots near the cusps; and (5) Dark or dusky notches, indentations, collars, and bands located just below the cusps. These phenomena definitely favor the southern cusp. Generally, the features persist for several days---even a month or more.

Background. These anomalous dark and bright features were employed historically to determine the planet's period of rotation. Such estimates were always far-removed from the present-day radar-derived figure. Many early astronomers took these phenomena as indicators of Venusian mountains or craters. Little effort is devoted today to elucidating these fascinating phenomena.

Data Evaluation. Just about all serious students of Venus, from the earliest astronomers to modern amateurs, have noticed deformations of Venus's cusps. The older literature contains many pertinent observations; but one finds little in the modern journals. Rating: 1.

Anomaly Evaluation. If the cusp phenomena are merely the consequence of shifting cloud formations or psychological contrast effects, they would hardly be anomalous. But they seem to be more permanent than Venus's fast moving atmosphere would allow. Then, too, Mercury also displays some similar cusp phenomena; and its atmosphere is negligible. The real cause(s) of the cusp phenomena are unknown. Certainly, the space probes have provided few clues relevant to these telescopic features. Rating: 2.

Possible Explanations. Shifting clouds in the Venusian atmosphere, perhaps affected to some degree by the terrain below; auroral phenomena stimulated by solar activity; atmospheric light sources of a chemical or electrical nature; volcanic activity (especially the dusky markings; psychological contrast effects.

Similar and Related Phenomena. Mercury cusp phenomena, in particular the blunting of the southern cusp (AHO1); the annular phase of Venus (AVO2); the ashen light of Venus (AHO3); the Venusian spoke marks (AVO6); Venus's polar dipole (AVF2).

Examples

X1. Bright spots, white hoods, polar caps. The cusps or horns of Venus frequently seem much brighter than the rest of the sunlit portion of the planet. These bright spots are often lenticular in shape, looking much like the polar caps of Mars. The phenomenon is not constant; and some well-known astronomers have never remarked on these bright markings.

December 29, 1813. Franz von Paula Gruithuisen, at Munich, seems to have been the first to record the polar hoods. He compared them to the Martian polar caps. (R20)

1871. Voegel and Lohse saw the phenomenon clearly from Bothcamp Observatory. (R20)

May and June, 1876. White spots seen near the cusps. (R2)

December 15, 1877. E.S. Holden, at Lick Observatory, observes the spots. (R20)

December and January, 1877 and 1878,

respectively, G. V. Schiaparelli made an extensive study of the southern white spot and concluded that Venus rotated one in 225 days. (R20) (Note that the southern white spot is almost always much brighter than the northern. WRC)

November 13, 1877 to February 7, 1878. Description by E. L. Trouvelot: "From Nov. 13, 1877, till Feb. 7, 1878, two remarkable white spots, strongly reminding me of those seen on Mars, have been observed on the opposite limbs, near the extremity of the cusps. The southern spot, which always appeared the brightest, became very prominent from Jan. 6, 1878, till Feb. 5, and appeared then to be composed of a multitude of bright peaks forming on its northern border a row of brilliant star-like dots of light. After inferior conjunction, which occurred a few days later, the white spots were no more visible." (R1, R7, R20)

August 17, 1884. I. P. Guldenschuh, of Rochester, New York, observed a brilliant white lenticular spot on Venus---like a snow cap. (R3)

February 18 and 20, 1891. More observations of the white spots by Trouvelot. (R5)

The Venusian "polar caps" drawn February 20, 1891. (X1)

Summer 1892. More observations reported by Schiaparelli. (R20)

Early 1900. Drawings of Venus by S. Bolton reveal considerable cusp blunting and bright areas at each pole. (R10)

February 10, 1913. Bright sector seen near the southern cusp. (R13)

July 16, 1919. C. Flammarion reports polar caps on Venus. (R15)

May and June, 1961. The Venusian cusp caps seemed brighter soon after solar flares and may represent some kind of aurora on Venus. (R19)

General observations. "Cusp Caps. Observers of Venus often note that one or both cusps are abnormally bright compared to the remainder of the disk.... For the present time we will call these anomalous brightenings 'cusp caps', though this term suggests a physical interpretation that is unproved. (James) Bartlett showed that one or both cusp caps were visible 54% of the observations (477 out of 830 observations), that the south cap alone was visible 11% of the time, the north 7%, and both caps simultaneously 35%. These figures correspond closely to those obtained when Bartlett's 221 observations or Ranck's 158 observations are considered alone. I believe that we can attach considerable significance to these values, particularly because both Bartlett and Ranck had such a long and continuous series of observations. This is a prerequisite to a statistical study of features like the cusp caps. Bartlett also found that the cusp caps appeared at virtually all phases of the planet, from which he concluded that seasonal effects on the planet are not important in governing the appearance of the caps, since all positions in the orbit were represented by cusp cap observations. This last statement does not follow, however, since phase is not uniquely related to Venus' heliocentric longitude alone, but also to the difference in heliocentric longitudes of Venus and the Earth. Bartlett's work indicates that the south cap was most often the larger of the two. Some reports of details seen in the cusps were mentioned, but are probably not reliable enough for useful analysis. Dusky cusps were also discussed, but these are reported far less often than the bright ones."(R18)

General observations. Several reports give brief descriptions of the "polar caps" of Venus, but no dates of observations. (R11, R12, R17)

X2. The occasional blunting or rounding-off of the southern cusp of Venus, while the northern cusp remains sharp. The phenomenon is sometimes associated with terminator irregularities (AVO10). Mercury's southern cusp exhibits the same effect (AHO1).

December 28, 1789; January 31, 1790; and December 25, 1791. J. H. Schroeter notices blunting of the southern cusp. Detached points of light beyond the blunted cusps also seen. (R20)

Between 1833 and 1836, J. H. Maedler confirms the cusp truncation or blunting phenomenon. (R20)

Winter 1853-1854, James Breen, at Cambridge, frequently noticed blunting of the southern cusp. (R20)

Early 1900. Drawings of Venus by S. Bolton reveal considerable blunting of the southern cusp. (R10)

May 22, 1927. Pronounced blunting of the southern cusp in conjunction with terminator irregularities. (R16)

General observations. A drawing of Venus by S. Bolton shows a bright spot on the southern cusp. The cusp also seems to be blunted at first glance, but Bolton did see the complete point of the cusp, but it was very faint. (R11)

X3. Abnormal extension of the cusps well beyond the 180° dictated by geometry. Minor extensions are rather common. On rare occasions, a thin ring of light encircles almost the entire planet. Complete rings of light, the so-called "annular phases" of Venus are covered in AVO2.

May 13, 1881. E. L. Trouvelot draws Venus with the crescent covering about 250°. (R5)

January 2, 1886. I. G. Lohse, using a 15-inch refractor at Scarborough, England, notices a narrow streak of light, in length about one-twentieth of Venus's diameter, extending the northern cusp but still separated from it. (R20)

January 28, 1894. The cusps of horns of Venus extend well beyond the planet's axis, showing that a refractive gas surrounds the planet. (R6)

July 4, 1895. "... regarding the extension of light in the southern part of the planet... I made a similar observation on the 4th inst. Referring to the inclosed diagram, ab represents a bright bluish extension, gradually getting darker towards the dotted line cd, which defined the northern limit of this shading. The full line is the outline of Venus, and the dotted line shows the dim partial outline of the dark side." (R8) One could also consider this situation as an incomplete manifestation of the ashen light of Venus (AVO3). (WRC)

July 4, 1895. Bluish-purple extension of Venus's southern cusp. (X3)

March 15, 1913. Drawings of Venus by A. Southgate shows prolonged cusps. (R14)

April 1948. Prolongations at both cusps reported by T. L. Cragg on several successive days. (R20)

X4. Detached luminous points and fragments hovering off the cusp extremities. May be an incomplete manifestation of X3; i.e., cusp extension.

December 28, 1789; January 31, 1790; and December 25, 1791. J. H. Schroeter detects detached points of light off the southern cusp of Venus, which was blunted all three times. He believed the bright spots to be sunlight reflected from lofty mountain peaks, in analogy with a similar lunar phenomenon. (R20)

Detached point of light off the southern blunted cusp seen by Schroeter on December 28, 1789. (X1, X4)

April 2, 1870. The only observation of a detached bright speck hovering off the northern cusp was made by A. P. Holden. (R20)

March 30 and 31, 1881. The Rev. T. E. Espin reported 'tiny brilliants' near the southern cusp resembling detached mountain peaks. (R20)

September 30, 1959. R. Baum draws Venus's thin cusp extremity as a row of detached bright points. He wonders if the broken line of light is due to unsteady air. (R20)

X5. A notch or indentation just below either cusp, giving them a hooked appearance. Included here are observations of a collar or dark band that sometimes appears in the same regions.

March 9, 1790. J. H. Schroeter reports a hook-shaped southern cusp. This condition continued through March until it disappeared on the 30th. (R20)

March 29, 1873. W. F. Denning observes an indentation near the northern cusp. (R20)

1876. Arcimis, the Spanish astronomer, observes a prominent notch in the southern cusp. (R20)

December 8, 1877 through January 6, 1878. Schiaparelli draws Venus with a dark band around the bright spot on the southern cusp. (R9)

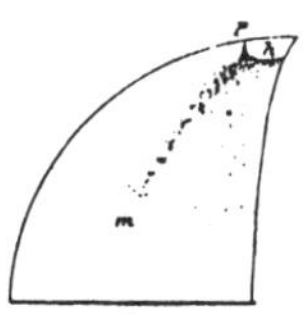

In 1877, Schiaparelli saw a dark band around Venus's southern cusp, plus a dark radial streak. (X5)

March 30 through April 5, 1881. W. F. Denning describes a dusky indentation not far from the northern cusp as "...extremely small, and looks like a crater." (R20)

1881-1895. Dusky band noted around the southern cusp frequently at the Royal Observatory, Brussels. (R20)

March 9, 1889. Prominent notch just below the southern cusp. (R4)

April 18, 1900. Dark indentations near both cusps. (R10)

Another notch and cusp band, plus a dark radial streak, 1889. (X5)

February 15, 1913. F. Sargent, Bristol, England, discovers a pronounced notch just below the northern cusp. The appearance of the notch changed considerably during the next few days. (R20)

February 23, 1921. Dusky band seen. (R20)

Early March 1953. "A well marked hemispherical depression was reported in the terminator below the southern cusp by several American and English observers during the first week of March 1953. Clearly seen

May 13, 1900. Blunted southern cusp and notch. (X1, X5)

on the evening of March 10 by the writer, it seemed bounded on its northern and southern flanks by dim projections and was preceded on the disk by an oval bright spot. "(R20)

December 17, 1970. A feature like that of 1953 seen by W. J. Wilson and R. Baum, when Venus was in the opposite phase. (R20)

General observations. "In addition to bright cusp caps, observers often report dusky bands or collars near the cusps which define or border the caps. P. Moore suggests that the bands are simple contrast effects of the bright cusps, saying that his own observations 'indicate that the collars can only be seen when the caps are unusually prominent'. I suspect that many observers would disagree that they are seen only in the presence of bright cusps. I do." (R18, R11)

References

R1. Trouvelot, L.; "White Spots on Venus," Observatory, 3:416, 1880. (X1)
R2. Russell, H. C.; "White Spots on Venus," Observatory, 3:574, 1880. (X1)
R3. "Is There a Snow Cap on Venus?" English Mechanic, 40:129, 1884. (X1)
R4. Fard, Al; "Bluntness of the Southern Cusp of Venus," English Mechanic, 49: 91, 1889. (X5)
R5. "The Planet Venus," Scientific American Supplement, 34:14066, 1892. (X1, X3)
R6. "Venus," English Mechanic, 59:11, 1894. (X3)
R7. "The Planet Venus," Scientific American Supplement, 37:15236, 1894. (X1)
R8. MacEwen, Henry; "Venus," English Mechanic, 61:567, 1895. (X3)
R9. Antoniadi, E. M.; "Notes on the Rotation Period of Venus," Royal Astronomical Society, Monthly Notices, 58:313, 1898. (X5)
R10. Bolton, Scrivan; "Venus," English Mechanic, 75:249, 1902. (X1, X2, X5)
R11. Bolton, Scrivan; "Venus," English Mechanic, 81:129, 1905.
R12. Pickering, William H.; "The Planet Venus and Its Problems," Harper's Magazine, 125:96, 1912. (X1)
R13. Allison, F. B.; "Venus---Southern Horn," English Mechanic, 97:83, 1913. (X1)
R14. Southgate, A.; "Venus," English Mechanic, 97:199, 1913. (X3)
R15. Flammarion, Camille; "Les Caps Polaires de Venus," English Mechanic, 110:7, 1919. (X1)
R16. Mead, W. J.; "Venus---An Observation," English Mechanics, 2:120, 1927. (X2)
R17. "Polar Cap of Venus," Royal Astronomical Society of Canada, Journal, 48: 199, 1954. (X1)
R18. Cruikshank, Dale P.; "A Review of Some ALPO Venus Studies, "Strolling Astronomer, 17:202, 1963. (X1, X5)
R19. Pither, C. M.; "The Origin of the Cytherean Cusp Caps," British Astronomical Association, Journal, 73:197, 1963. (X1)
R20. Baum, Richard; "The Himalayas of Venus," The Planets, New York, 1973, (X1-X5)

AVO2 Ring of Light around Venus

Description. The appearance of a thin ring of light around the dark side of Venus. This phenomenon occurs primarily when the planet is near inferior conjunction; that is, it is backlighted by the sun. An identical effect sometimes happens when Venus begins and/or ends a transit of the sun (AVX1). The more frequently observed extension of Venus's cusps well beyond 180° is doubtless an incomplete manifestation of this phenomenon (AVO1).

Data Evaluation. From the older astronomical literature, several excellent observations by respected astronomers; from the last 80 years, nothing. Rating: 1.

Anomaly Evaluation. The rating given below assumes that the annular phase of Venus is only an optical phenomenon; i.e., atmospheric refraction. It should be noted, though, that a similar effect is sometimes seen when Mercury makes contact with the sun during a transit--- and Mercury's atmosphere would seem to be too thin to account such refraction. Rating: 3.

Possible Explanations. Refraction of sun by the Venusian atmosphere, possibly enhanced by psychological contrast.

Similar and Related Phenomena. Luminous ring around the new moon (ALO); luminous ring around Mercury during transit (AHX3); Venus's ashen light (AVO3); the extension of the cusps of Venus (AVO1); the partial ring seen during transits (AVX1).

Examples

X1. December 1842. During inferior conjunction, Guthrie, in Great Britain, saw a narrow fringe of light around the whole disk of Venus. (R2)

X2. December 10 and 12, 1866. C.S. Lyman, New Haven, Connecticut. "Some days before the conjunction, it was apparent that the crescent formed more than a semi circle---on the 7th, full 40° more by measurement. On the 10th, it formed a complete circle---bright, thin and delicate (the crescent proper), on the side toward the sun, but on the opposite side, a mere faint line of light, very difficult to be seen, on account of the strong light in the field, and the atmospheric disturbance. Yet, by glimpses, it was distinctly perceived as a ring, by several observers, and constantly as more than three-fourths of a circle.

The appearances were similar, though perhaps a little better seen, on the 12th, the day after the conjunction. Yet, the planet was then only half a degree farther from the sun, and the full ring could be made out only in the more favorable moments with respect to light and atmosphere ---particularly, when the light, both of the sun and of the planet, was partially cut off from the object glass, by the shutter of the observatory. Such a compromise between sun-light and planet-light gave generally the best views, except twice, about noon, when, fortunately, a passing cloud left the planet in sight for a few seconds, while yet the sun was obscured. The background was then comparatively dark, and the thread of light around the limb opposite to the sun perfectly distinct and complete. The northern portion of the crescent proper, however, did not diminish uniformly in brightness, or apparent thickness, toward the cusp, but a considerable space, between 25 and 50 degrees, from the vertex to the left, by estimate, was very perceptibly fainter than a like portion of the circumference next beyond toward the right, whence it gradually narrowed to the mere faint line of light before mentioned. These observations were made between half past 11 and half past 1 o'clock." (R1, R2, R4, R7)

X3. December 8, 1874. C.S. Lyman, at New Haven, Connecticut, after summarizing his 1866 observations (X2). "No opportunity has since occurred of repeating these observations, until the day of the recent transit. On Tuesday, December 8th, Venus was again in close proximity to the sun; and the writer had the satisfaction of watching the delicate silvery ring enclosing her disk, even when the planet was only the sun's semidiameter from his limb. This was at 4 P.M., or less than five hours before the beginning of the transit. The ring was brightest on the side towards the sun---the crescent proper. On the opposite side the thread of light was duller and of a slightly yellowish tinge. On the northern limb of the planet, some 60 or 80 degrees from the point opposite the sun, the ring for a small space was fainter and apparently narrower than elsewhere. A similar appearance, but more marked, was observed on the same limb in 1866." (R3)

During the actual transit, J. Tebbutt, in New South Wales, Australia, saw the dark limb edged in light, especially at egress. (R5) See AVX1 for further discussion.

X4. December 5, 1882. At this transit, a ring of light formed around the dark portion of Venus shortly after first contact. See AVX1. (R6)

X5. March 4, 1894. P.H. Kempthorne, England. "I was looking at Venus on March 4, at 11.35 a.m. (Calver mirror, 8 1/2 in., power 60), and endeavouring to discern the unilluminated portion of the orb, when suddenly, for a second or two, I saw it encircled by a faint ring of light. It was possible to note that the ring was clearly defined towards the planet's limb, but ill-defined on the side remote from it.... The phenomenon was as beautiful as it was to me unexpected. The dark part of the orb was undistinguishable except at these moments. The cusps of the crescent extended beyond the half-circle...." (R8)

X6. November 29, 1906. H.N. Russell and Z. Daniel, Princeton University Observatory. "On 1906 November 29, 5h 7m G.M.T., Venus about $1^{\circ}49'$ from the Sun's center, was observed with the 5-inch finder of the 23-inch telescope. In moments when the air was steady the complete outline of the planet was distinctly seen. On the side nearest the sun it was bright and easily visible, but on the opposite it was very faint and could be seen only for a few seconds at a time. When the complete circle was seen, the space within it always seemed a shade darker than without. I suspect, however, that this was a subjective effect, as it was not noticed when the fainter part of the ring disappeared through bad seeing. No other marked peculiarities were noticed though a bright spot was several times suspected in the bright part of the ring." (R9-R11)

References

R1. "Observations of Venus near Its Inferior Conjunction," American Journal of Science, 2:43:129, 1867. (X2)
R2. Schafarik, A.; "On the Visibility of the Dark Side of Venus," Report of the British Association, 1873, p. 404. (X1-X2)
R3. Lyman, C.S.; "On Venus as a Luminous Ring," Philosophical Magazine, 4:49:159, 1875. (X3)
R4. Neison, E.; "The Ring of Light around Venus during the Late Transit," Astronomical Register, 13:196, 1875. (X2)
R5. Tebbutt, John; "Ring of Light around Venus," Astronomical Register, 13:222, 1875. (X3)
R6. Keeler, J.E.; "The Ring of Light Surrounding Venus," Sidereal Messenger, 1:292, 1883. (X4)
R7. Gibbs, Lewis R.; "Annular Phase of Venus," Science, 16:303, 1890. (X2)
R8. Kempthorne, P.H.; "Venus," English Mechanic, 59:81, 1894. (X5)
R9. Russell, Henry N., and Daniel, Zaccheus; "Venus as a Luminous Ring," Popular Astronomy, 15:516, 1907. (X6)
R10. "Venus as a Luminous Ring," Knowledge, 4:205, 1907. (X6)
R11. "Venus as a Luminous Ring," Scientific American, 97:461, 1907. (X6)

AVO3 The Ashen Light Phenomenon

Description. A soft glow observed over the dark side of Venus. The radiated light is usually described as grayish, but the adjectives greenish and coppery have also been applied. A few of the apparitions of the ashen light have coincided with the appearance of a brighter ring of light around the dark side (AVO2); i.e., the so-called annular phase of Venus. Although there are weak correlations claimed with the inferior conjunctions and solar activity, the ashen light seems to occur almost anytime.

Data Evaluation. Once again, most specific observations of the phenomenon are rather old. The more recent literature acknowledges that the ashen light is a real and frequent phenomenon, but provides few eye-witness acounts. Spacecraft instruments, however, have detected emission lines originating from Venus's dark limb. These areas are irregular. All in all, we have here a real, objective, far-from-rare phenomenon. Rating: 1.

Anomaly Evaluation. The popular current explanation of the ashen light relies upon the dissociation of molecular oxygen by sunlight, its transportation to the dark side by atmospheric motion, and subsequent recombination. This process should always be underway, but the ashen light is far from being a permanent thing. Furthermore, the ashen light is almost always described as a fairly uniform glow over the entire dark limb, not the patchy, highly variable source expected from atmospheric convection. An auroral explanation seems negated by the uniform disposition of the ashen light. In sum, no convincing explanation seems at hand, although there are good candidates. Rating: 3.

Possible Explanations. Recombination of sunlight-dissociated molecules; auroral phenomena; unrecognized electrical and chemical phenomena in the Venusian atmosphere.

Similar and Related Phenomena. Venus seen as a ring (AVO2); the phase anomaly of Venus (AVO4); flickering light on Venus's dark limb (AVO9); earthlight reflected from the moon; terrestrial airglow.

Examples

X1. 1520. Observed by M. Maestlin. (R5)

X2. January 9, 1643. Riccioli reports a reddish hue on the dark portion of Venus. (R5)

X3. May 2, 1715. W. Derham, canon of Windsor, sees the "secondary light" of Venus during the total solar eclipse. (R14)

X4. June 7, 1721. C. Kirch, at Berlin, reports that the bright crescent of Venus seemed larger than the faintly shining dark side. (R2, R5)

X5. March 8, 1726. Ashen light again seen by Kirch. (R2, R5)

X6. October 20, 1759. A. Mayer observes Venus 10° from the sun and sees its whole disk "like the crescent moon which reflects the light of the earth." Phenomenon again reported by Kirch. (R2, R5)

X7. 1790. W. Herschel sees part of the dark limb shine with a faint light. (R2)

X8. Spring and summer 1793. F. Hahn of Mecklenburg sees the phenomenon repeatedly, in twilight as well as daylight. (R2)

X9. January and February 1806. Schroeter observes the ashen light on February 14. C. L. Harding, at Gottingen, sees the phenomenon with "utmost sharpness and distinctness". On February 28, he reports the light is reddish gray. Apparently he also viewed the ashen light on January 24 and 28, February 20, and March 1, 1806; but there seems to be some confusion in the literature on these dates. (R2, R5)

X10. February 10, 1822. J.W. Pastorff, Buchholz, Prussia, distinguished bright and dark patches on Venus in a faint gray light. (R2)

X11. June 8, 1825. Gruithuisen, at Munich, witnesses the phenomenon at 4 A.M., in almost full daylight. (R2)

X12. September 27 and 28, 1855. At 11 A.M., in broad daylight, G.A. Jahn, at Leipzic (sic), observes the dark side of Venus. (R2)

X13. January 14, 1862. Liverpool, England. A Mr. Berry reports the ashen light. (R2)

X14. 1863. Price, at Uckfield, England, sees the dark part of Venus. (R6)

X15. October 22, 1863. J.F. Barber, Nottingham, England. "Turning my telescope (6 in. aperture by Cooke) upon Venus on the 22nd October last, I was much surprised to see almost the whole of the unilluminated disc of the planet; it was so striking in appearance, that I thought it must be the reresemblance to the new Moon, which made me <u>fancy</u> that I could see the unilluminated portion. My sister-in-law, whom I called to witness the Planet, but without telling her what to look for, said she instantly saw the whole disk. The atmosphere was beautifully clear, but still the planet was so far past conjunction, that I should scarcely have imagined the phenomenon would be visible." (R1)

X16. April 20, 1865. W. Englemann, of the Leipzic Observatory, notes that the dark side of Venus is greenish gray---a little brighter than the sky. (R2, R5)

X17. April 22, 1873. H. Sadler, Sherborne, England, sees the dark limb distinctly. It was lit with a grayish hue. On the 24th, at intervals, there seemed to be a brighter glow darting quickly across the dark limb from southwest to northeast. (R3)

X18. 1877. Banks, Green, and Noble report the phenomenon. No further details found. (R5)

X19. March 30, 1878. S. Mills reports that the dark limb could be seen with remarkable distinctness. It possessed a warm neutral tint. (R4) Nobel and Webb also report the phenomenon in 1878. (R5)

X20. January 8, 1883. The whole disk of Venus was "splendidly visible even to the unaccustomed eye of entirely uninstructed persons....But the most important feature of the observation was the ring, that I could detect all round the disk (dark part and cresent), of brownish-red colour, more pronounced on the illuminated side than on the dark part of the limb, but of a peculiar coppery hue, the close resemblance of which to the coppery hue the Moon's disk assumes when totally eclipsed was very striking." (R5)

X21. 1894. Observations of the phenomenon by T.P. Battersby. (R7)

X22. July 12, 1895. R.K. Sale and others see the entire disk of Venus, like "the old moon in the young moon's arms". It was two months away from inferior conjunction. (R8)

X23. March 31, 1905. J. Willis, of Folkestone detects the unilluminated portion of Venus's disk. (R9)

X24. October 18, 1906. Observations by T.P. Battersby, at Gibraltar. "To my great surprise, I perceived the unilluminated part of the planet's disc clearly for the first time in all my observations of Venus. I called my wife to confirm my observation, and she at once perceived it and described it exactly as it appeared---namely, as a faint segment, very much smaller in appearance than the completion of the circle of the bright part of the planet. She observed a thin bright ring round the dark segment; but this I did not personally see." (R11) Evidently, the so-called annual phase of Venus and the ashen light sometimes go together. The larger appearance of the bright limb is probably due to a type of contrast phenomenon termed "irradiation", wherein bright object seem larger than identical dark objects. (WRC)

X25. January 12, 1910. R. Killip describes Venus as similar in appearance to the old moon in the young moon's arms. (R12)

X26. December 29, 1957 and the following two weeks. The ashen light was unusually prominent during this period. (R13)

X27. Instrumented studies of the ashen light. Telescopic spectrograms taken in 1959, during a four-month interval that included inferior conjunction with Venus, revealed no emission spectra. (R13)

The Russian Venera 9 and Venera 10 spacecraft detect a glow from Venus's dark side in 1976. The glow was strongest 56 mi miles above the planet's surface. The glow consisted of emission bands due to the recombination of oxygen molecules in the upper atmosphere. The glow was strongest on the dark side near the terminator, but was also patchy and variable. The dissociated oxygen may have been formed on the bright limb through the action of sunlight and then carried to the dark side by general atmospheric circulation. (R17) See below for additional speculations about the origin of the ashen light. (WRC)

X28. Theoretical speculations. The idea that the ashen light of Venus is reflected earthlight, as in the case of the old moon in the young moon's arms, is uniformly rejected by all who compute the light intensities. (R2, R10)The ashen light has often been thought to be an auroral phenomenon. (R2) One modern study shows "that there was a pronounced maximum in geomagnetic activity at inferior conjunction during times of ashen light occurrence during the 1959-1962 period. This result in turn suggests an auroral origin of this glow." (R16) But, as mentioned in X27, space probe results tend to favor a mechanism similar to the earth's airglow. (R17)

X29. January 24, February 28, March 1, 1806. D. L. Harding, at Gottingen, sees the dark side of Venus. Once it was a reddish-gray; another time greenish-blue. (R18)

X30. January 14, 1862. According to Berry, at Liverpool, the unilluminated hemisphere shone with a faint light, like the moon's lumiere cendree. (R18)

X31. September 1863. Near inferior conjunction, C. Leeson Prince saw the whole disk slightly illuminated, with "a phosphorescence flitting of light around the edge of the entire disk." (R18)

X32. February 5, 1870. R. Langdon, of Silverton Station, reported a splendid manifestation of the ashen light. (R18)

X33. February 22, 1870. The phenomenon was recorded by W. Noble. (R18)

X34. February and March 1870. J. Browning sees the planet's dark side on at least twenty evenings, always in bright twilight and always darker than the sky. (R18)

X35. September 25, 1871. At noon, A. Winnecke, of Karlsruhe, observed the phenomenon with a heliometer. (R18)

X36. October 15 to November 12, 1871. Voegel and Lohse saw the dark side lit by a secondary light extending about 30° from the terminator on seven mornings in bright twilight. (R18)

X37. August 9, 1871. Whole disk of the planet perceived by A. Schafarik. (R18)

X38. March 22, 1873. Luminescence of the dark side reported by T. G. Elger. (R18)

X39. April 19, 1873. Whole body of the planet seen at inferior conjunction by R. Langdon. (R18)

X40. July 1876. Dark side of Venus suffused with an ashen glow on several occasions. (R18)

X41. 1876. Banks, Grover, and Arcimis see the phenomenon during this year. (R18)

X42. September 30 to October 14, 1876. C. V. Zenger, of Prague, detected a faint grayish light, sometimes tinted reddish, covering part of the disk near the terminator. (R18)

X43. January 31, 1878. T. W. Webb sees the ashen light for the first time. (R18)

X44. July 11, 1884. W. Noble saw the phenomenon at inferior conjunction. (R18)

X45. January 2 and 3, 1886. Dark part of

Venus distinctly seen by Lohse and Wigglesworth. (R18)

X46. October 21 and 26, 1887. Grayish gleam from the planet's nocturnal side. (R18)

X47. March 16, 1889. Phosphorescence very plain according to S. M. B. Gemmill. (R18)

X48. October 22, 1895. Dark side visible. H. McEwen. (R18)

X49. July 25 and August 5, 1895. Dark hemisphere observed by N. S. Aldis in New Zealand. (R18)

X50. 1895. Throughout the whole summer, Antoniadi, at Constantinople, found the dark side clearly visible as inferior conjunction approached. (R18) Others also observed the phenomenon during this period. (R18)

References

R1. Barber, J. F.; "Visibility of the Dark Side of Venus," Astronomical Register, 1:190, 1863. (X15)
R2. Schafarik, A.; "On the Visibility of the Dark Side of Venus," Report of the British Association, 1873, p. 404. (X4-X13, X16, X28)
R3. Sadler, H.; "Astronomical," English Mechanic, 17:327, 1873. (X17)
R4. Mills, Samuel; "Venus," English Mechanic, 27:117, 1878. (X19)
R5. Zenger, C. V.; "On the Visibility of the Dark Side of Venus," Royal Astronomical Society, Monthly Notices, 43:331, 1883. (X1, X2, X4-X6, X9, X16, X18-X20)
R6. Lynn, W. T.; "The Visibility of the Unilluminated Part of Venus," Observatory, 11:155, 1888. (X14)
R7. Battersby, T. Preston; "Visibility of Dark Side of Venus," English Mechanic, 58:464, 1894. (X21)
R8. Sale, Robert Killip; "Visibility of the Dark Side of Venus," British Astronomical Association, Journal, 6:33, 1895. (X12)
R9. Willis, John; "Venus," English Mechanic, 81:221, 1905. (X23)
R10. Holmes, Edwin; "Earthshine on Venus," English Mechanic, 81:288, 1905. (X28)
R11. Battersby, T. Preston; "Visibility of Dark Side of Venus," English Mechanic, 84:304, 1906. (X24)
R12. Sale, Robert Killip; "Dark Side of Venus," English Mechanic, 91:12, 1910. (X25)
R13. Weinberg, J. L., and Newkirk, G.; "Airglow of Venus: A Reexamination," Planetary and Space Science, 5:163, 1961. (X26, X27)
R14. "Ashen Light of Venus," Sky and Telescope, 17:123, 1958. (X3)
R15. Cruikshank, Dale P.; "A Review of Some ALPO Venus Studies," Strolling Astronomer, 17:202, 1963. (X28)
R16. Levine, Joel S.; "On the Possibility of Auroral Activity on Venus," Strolling Astronomer, 19:149, 1965. (X28)
R17. "Venus Light Confirmed," Astronomy, 5:65, April 1977.
R18. Baum, R. M.; "The Principal Observations of the Nocturnal Hemisphere of Venus 1643-1900," British Astronomical Association, Journal, 67:242, 1957. (X29-X50) Baum also lists many examples between X1 and X28.

AVO4 The Phase Anomaly of Venus

Description. The retardation of the phase of Venus in its western elongations and the acceleration of the phase in its eastern elongations. The proportion of Venus that should be lit by the sun can be calculated quite easily. Actual observation, however, shows a complex schedule of difference between observed and theoretical phase. At theoretical dichotomy or half-phase, the observed phase is either 4-6 days ahead or behind schedule, depending upon which side of the sun the planet is on. This discrepancy between theory and observation is present to some degree at all phases.

Background. J. H. Schroeter seems to be the first astronomer to have noticed this phenomenon. He mentioned it in 1793. For this reason, it is often called the Schroeter Effect.

Data Evaluation. Venus's phase anomaly is obvious to all students of this planet. There is general agreement on the major aspects of the phenomenon. Venus's precise phase, however, is difficult to measure because of irradiation and the cusp and terminator irregularities. Some controversy, therefore, exists about the detailed history of the phenomenon throughout its cycle. The phenomenon, in fact, may vary somewhat from one apparition to the next. Rating: 2.

Anomaly Evaluation. Hardly a vestige of an explanation of Venus's phase anomaly exists--- no theory seems to work, especially when the asymmetry of the phenomenon is considered. While this sounds serious, the stakes are not really very high. Probably some atmospheric and/or subjective effect will account for the observations. Rating: 2.

Possible Explanations. Irradiation and atmospheric refraction have been proposed in the way of explanations, but they seem helpless to account for both leading and lagging phases. A psychological explanation relying on contrast seems most likely.

Similar and Related Phenomena. Prolongation of the cusps of Venus (AVO1); the ashen light (AVO3); Venus's irregular terminator (AVO10).

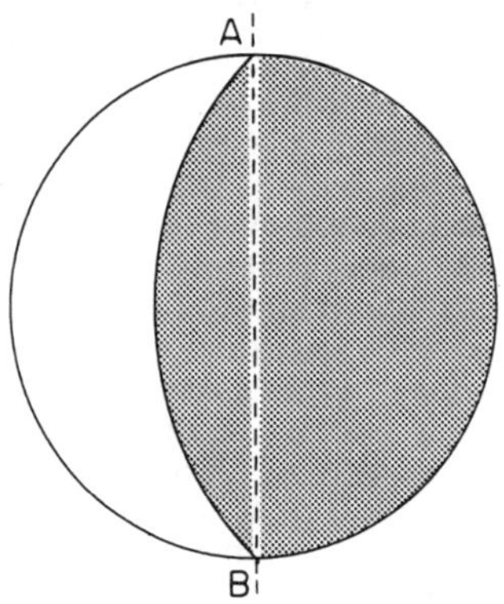

At theoretical phase dichotomy, the sun should illumine exactly half the disk of Venus, but actual observations lag or lead theory considerably.

Examples

X1. Micrometer studies of Venus's phase. "Henry McEwen, a past Section Director of the British Astronomical Association, made over 900 measurements of the phase variation using a slade micrometer mounted on a 5-inch altazimuth telescope. The eyepieces employed were a Ramsden of 156 diameters and a Zeiss orthoscopic of 180 diameters. All measurements were made in daylight in order to reduce the effects of irradiation. A discussion of the observations made during the period 1919-1927 appeared in the Journal of the British Astronomical Association." The figure "shows the results of McEwen's work. Many observers claim that the retardation of the phase is greater in western elongations (or apparitions) than its acceleration in eastern ones. The graph for

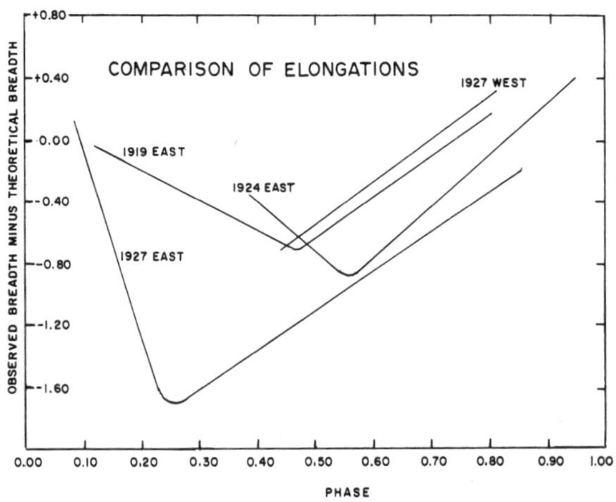

Graphical comparison of the observed phase of Venus with its theoretical phase, according to the micrometric measurements of H. McEwen. (X1)

the western elongation of 1927 contradicts this belief, for it in no way differs radically from the graphs of the other elongations. However, it must be stressed that the graph represents only twenty-five observations during one elongation. These curves... indicate a relatively constant rate of change during the gibbous phase of the planet with the maximum deviation occurring near the time of dichotomy, although the evidence is insufficient to give more than a strong indication of this condition..... The variation between the times of observed and theoretical dichotomy, although only a small part of the more general problem, is the best observed part of the deviation from theory. Apparent dichotomy is invariably late in western elongations (apparitions) and early in eastern ones, but the amount of the deviations from theory is highly variable. However, the average time lag is about 3 1/2-4 days." (R2)

X2. Measurements of drawings of Venus. Following a discussion of the general character of the phase anomaly: "In general there is a time difference of about four to six days between times of theoretical and observed dichotomy. This phase anomaly has been termed 'Schroeter's Effect' by Moore. The observational situation is, however, much more complex than Schroeter's Effect alone. It is obvious that a difference in observed and calculated phase at the time of dichotomy must be accompanied by incorrect apparent phases at other times, near dichotomy, but it was not until a book by Bronshten recently appeared that we knew just how large the (O-C)s are for phases other than 0.5. The results quoted in Bronshten's book are from measurements made by N.N. Michelson and V.N. Petrov. These results are depicted schematically in Fig. 2. Unfortunately only observational drawings were used by Michelson and Petrov, it would of course be preferable to have a long series of direct micrometric measurements of the phase of Venus. I have measured a large number of drawings myself and arrive at identical conclusions to those shown in Figure 2, but with some additional points to be discussed later in this paper. It can be seen from Fig. 2 that the (O-C) phase can be as large as 0.10 for phases near 0.1 and 0.9." (R4)

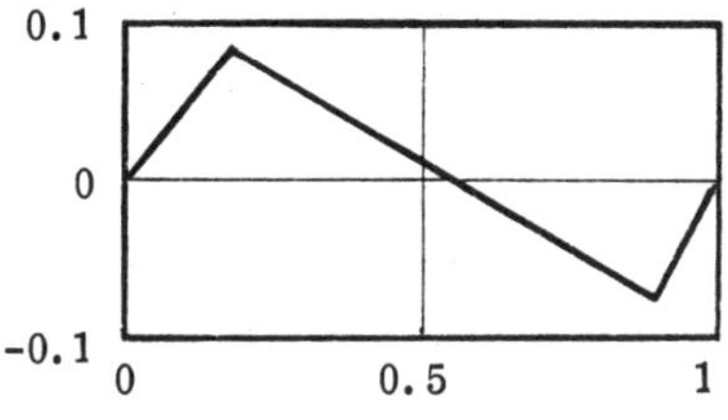

The phase anomaly of Venus. (X2)

X3. Measurements of drawings of Venus. "Theoretically it should be possible accurately to predict (to within a few hours) the date when exactly half of Venus is lit by the sun. This should occur at greatest elongation, when the angle between the earth and sun is exactly 90° with respect to Venus. However, this is found almost never to be the case, and there is a discrepancy of several days both at eastern and western elongations. No satisfactory explanation for this has been given as yet.

In the past two evening apparitions, 1960-61 and 1962, a special study of the planet's phase was undertaken by the Montreal Centre. In the absence of micrometer measurements, the phase was measured directly from drawings. Although this method may seem crude, it was found that the portrayal of phase by the more experienced observers was sufficiently consistent to be used for this purpose. The average difference in phase estimated by three observers, G. Wedge, G. Gaherty and K. Brasch, was from 1 to 4 per cent.

The observed values for both apparitions were plotted on graphs together with the curves of the predicted phases. In 1960-61, from a total of 32 observations made before dichotomy it was found that the observed phase was about 5 per cent less than the theoretical value, i.e. the observed disk had about 5 per cent less illumination than expected. This corresponds to a dichotomy 8 days earlier than the predicted date of January 30, 1961. The observed half-phase was indeed 7 to 10 days early, occurring between January 21 and 23, 1961. There is considerable scatter among the values obtained early in this apparition; this can probably be attributed to the lack of experience of the observers.

In 1962 there is a higher degree of mutual agreement among the values obtained. From a total of 42 observations before dichotomy, the average deviation was 6.7 per cent, indicating that the half phase would be approximately 14 days early. It was observed to occur about 11 to 13 days early, between August 21 and 23, compared with the predicted date of September 2, 1962."(R3)

X4. General observations. "Venus was interesting in 1962 when, during July and August, it became clear that the phase was decidedly less than theoretical....Of the present writers, Brinton used his 12-inch reflector, together with one observation with the 18-inch refractor at the Pino Torinese Observatory in Italy, while Moore used his 8 1/2-inch reflector and, on occasions, Brinton's 12-inch. During observations made after dichotomy, it was noticed that there was an increased difficulty in determining phase.

Upon examination, it appeared that this was due to a discrepancy between the visible and theoretical shapes of the telescopic shapes of the telescopic image of the planet. Being close to dichotomy, the image should have been almost exactly a semi-circle. In fact, it appeared as a segment bounded by a chord smaller than a diameter. Such an appearance could only result from viewing a sphere which is less than half illuminated, which seemed, on the face of it to be absurd.

In the absence of an immediately available micrometer we made a search of photographs of Venus. Many of these showed the effect to a marked degree. On the other hand photographs of the Moon and Mercury sometimes showed a similar, though lesser, effect. This suggested that photographic evidence is unreliable." (R5)

X5. Possible explanations. Irradiation (R1); distortion of the visible portion of Venus by its tidal bulge. (R6)

References

R1. Antoniadi, E. M.; "Notes on the Rotation Period of Venus," Royal Astronomical Society, Monthly Notices, 58:313, 1898. (X5)

R2. Rushton, Minick; "On the Variation of the Phase of Venus from Theory," Strolling Astronomer, 15:49, 1961. (X1)

R3. Brasch, Klaus R.; "A Study of the Phase of Venus," Royal Astronomical Society of Canada, Journal, 57:264, 1963. (X3)

R4. Warner, Brian; "The Phase Anomaly of Venus," British Astronomical Association, Journal, 73:65, 1963. (X2)

R5. Brinton, Henry, and Moore, Patrick; "On the Phase Anomaly of Venus," British Astronomical Association, Journal, 73:119, 1963. (X4)

R6. Ellis, E. L.; "The Dichotomy of Venus," British Astronomical Association, Journal, 84:351, 1974. (X5)

AVO5 Venus Seen Darker than the Sky

Description. The dark limb of Venus appears, on rare occasions, to be darker than the surrounding sky. In a sense, this seems the antithesis of the ashen light phenomenon (AVO3).

Data Evaluation. Several rather subjective observations. No photometric confirmations of the phenomenon. Rating: 3.

Anomaly Evaluation. Although one immediately suspects a simple contrast phenomenon, Flammarion's arguments in X1 seem to overturn this explanation. If contrast and other psychological effects can be eliminated, temporary sky brightenings behind Venus offer another possibility. A sunlit dust cloud would suffice. Rating: 3.

Possible Explanations. Sunlit dust behind Venus, resembling the zodiacal light, would make the dark limb of Venus appear darker than the sky.

Similar and Related Phenomena. The ashen light of Venus (AVO3); the zodiacal light (AZO).

Examples

X1. August and September 1895. Juvisy, France. "It is certain that total darkness does not exist in the starry sky. (Speaking for myself, I can always see the time by my watch by the sole light of the stars, and yet the hands are very thin.) A planet is never entirely dark, even if we admit that it has no phosphorescence, no light of its own. But the results of observations made this year at Juvisy during the months of August and September puts the question in another light.

It appeared several times to M. Antoniadi, M. Georges Mathieu and myself that the interior of the crescent of Venus, that is to say, the part which is not lighted up, was darker than the sky.

It was not the effect of the contrast produced by the luminous crescent, for on the outside of the crescent this aspect was not noticed, but inside the disk became darker and darker toward the edge. The observations were made in full sunlight, Venus being on the meridian, about noon at the time of the conjunction and at all hours from nine A. M. to three P. M.

Now is it possible that a black body can be seen through the lighted atmosphere?

The tone of the darkened disk was slightly violet, but appeared rather darker than the sky than lighter.

The case is not, therefore, the same as that of the gray light of the moon.

It might almost be believed that it was merely an optical illusion, a sort of negative visibility, a tendency of the sight to prolong the fine points of the slender crescent, and theoretically complete the form of the disk.

But this explanation will not suffice either, for on hiding the luminous crescent the same somber disk is still seen, and then the interior of the disk is certainly darker and of another tone than the sky outside.

May not the entire globe of Venus project itself on a slightly lighted background, on the zodiacal light, on a very widespread solar atmosphere?

To sum up, it seems certain to us that the visibility of the dark hemisphere of Venus is not an optical illusion, and it seems to us that the explanation is to be attributed in the first place to the solar atmosphere, to the zodiacal light, before which the planet forms a screen." (R1)

X2. General observations. "I should like to add that I have also seen the whole of Venus when crescent-shaped, the unilluminated part being darker than the sky around..." (R2)

References

R1. Flammarion (Camille); "Visibility of the Dark Hemisphere of Venus," Scientific American Supplement, 40:16597, 1895. (X1)

R2. Sale, Robert Killip; "Dark Side of Venus," English Mechanic, 91:12, 1910. (X2)

AVO6 The Venusian Spoke System

Description. Dark, spoke-like lineaments that often converge toward the subsolar point. The fully developed phenomenon includes a rim, producing a definite wheel-like appearance. More often only disorganized straight lines are seen. Some observers see the spokes tinged with yellow. The lines may be sharp and clear; but many see them as fuzzy. Most astronomers do not see them at all! The spokes may persist for days, retaining their focus on the subsolar point as the planet moves around the sun.

Background. Although a few early astronomers reported isolated lineaments on Venus, it was Percival Lowell who sketched the first spoke systems in 1896. The phenomenon seems to have been forgotten in the years following Lowell. Only in 1951 did R. M. Baum "recover" the spoke system and once again bring it to the attention of astronomers.

Data Evaluation. As with the lineaments on Mercury and Mars, the Venusian spokes have been seen by too many competent astronomers using a wide variety of telescopes for anyone to deny that a small fraction of Venus observers do perceive something of a spoke-like nature when observing Venus. The data are abundant and of high quality; it is the interpretation of the data that is controversial. Rating: 1.

Anomaly Evaluation. Systems of lineaments have been detected on several planets and satellites by a small fraction of astronomers. These observations are not verified by most astronomers and, more importantly, by spacecraft. This is an important perceptual anomaly, because science is based on all observations being accessible to all normal people. It is not scientific to dismiss well-founded data as "illusory". The anomaly may be psychological rather than astronomical, but it remains an anomaly. Rating: 1

Possible Explanations. Venus does display linear cloud patterns, particularly in the ultraviolet. The cloud patterns, however, do not form spoke systems and move much more rapidly across the planet's disk than the spoke system, which seems to remain centered on the subsolar point. It is possible that the eyes of a minority of observers are more than normally sensitive, allowing them to perceive unappreciated atmospheric structures on Venus. The Venusian spokes, like the Martian canals, may be mentally generated. If so, why do different people see basically the same spoke system and Martian canals?

Similar and Related Phenomena. Mercury's lineaments (AHO4); the canals of Mars (AMO1).

Examples

X1. Pre-Lowell observations. Schroeter (in 1801), Gruithuisen, Fournier, Perrotin, Mascari, and other astronomers reported linear markings on Venus. (R17)

X2. Observations during the Lowell period (1896-1916).

Lowell's first systematic study of Venus. "Such was the state of our knowledge when in August I began systematic observations upon the planet with the 24-inch refractor. I had no sooner begun observing than it was apparent that there were markings upon the disk and before long it was furthermore apparent that the markings bore no resemblance to the indefinite patches commonly depicted. So distinct and definite did the markings prove that they have resulted in the detection of the planet's rotation and of the fundamental physical characteristics of the planet's surface. I shall give in this paper a first summary of what they have disclosed, together with reproductions of some of my drawings and those of my assistant Mr. Drew.

To begin with, then, the markings are both distinct and well defined; their contours standing out sharply against the lighter parts of the disk. In shape they are of two kinds; long, relatively narrow, and, generally speaking, straight markings; and spots. They are all of them permanent and permanently visible, our own air alone ever obscuring them. Indeed the seeing must be distinctly bad to have the more prominent among them not discernible. Such phenomena show that they belong to the actual surface of the planet itself." Lowell concluded that the planet's period of rotation was identical with its orbital period. (R2) An incorrect conclusion. (WRC)

Many of Lowell's drawings of Venus appeared various journals for two decades. (R4, R7, R8, R11, R13, R15)

Lowell's observations of canal-like features on Venus confirmed in 1897 by Foulkes on Malta. (R3) Leo Brenner also sees linear markings on Venus, but attributes them to the earth's unsteady atmosphere. (R5) E. M. Antoniadi, one of the best-known astronomers of the day, was certain that the Lowell markings are totally illusory. (R6) However, other observers, such as Griffiths draw spoke systems very similar to those of Lowell. (R6)

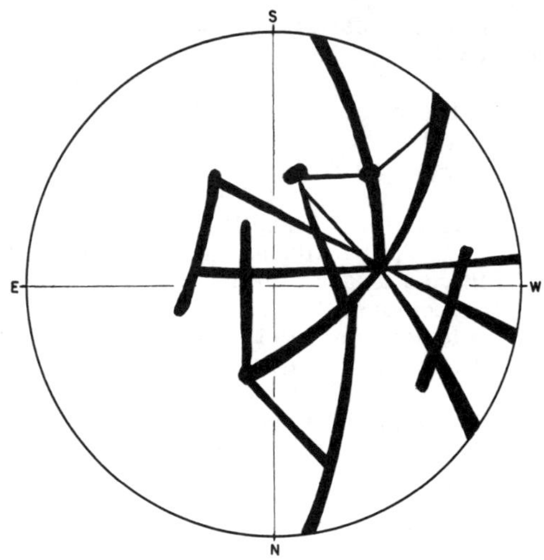

Venusian spokes drawn by Percival Lowell in 1896. (X2)

Meanwhile astronomers at the Lowell Observatory follow the course set by Lowell. Here are observations by A. E. Douglass. "In the last six years many thousands of hours have been spent by us at telescopes of 13, 18, and 24 inches aperture and their smaller finders, when seeing was sufficiently good for profitable work on the finest known planetary detail. Expressed in standard terms, the seeing was practically always such that in a 6-inch aperture the spurious disc of the interference pattern was well defined, and a very large part of the time the rings of the same pattern were unbroken. I

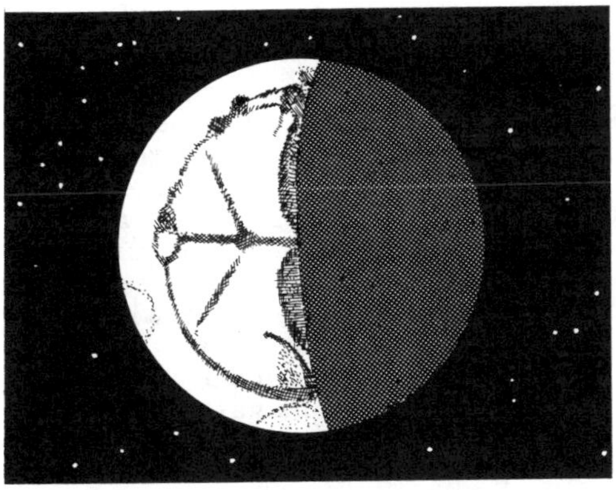

Griffith's sketch of the spoke system, made on February 17, 1897. (X2)

consider that any astronomer who cannot say the same for the seeing during his hours of work, and whose hours of work do not reach a commendable number, has no right to criticise our results; for he lacks the experience by which alone he becomes capable of judging.

Under proper conditions of air and aperture the markings on Venus are absolutely certain. Under proper conditions they are to me about as easy or difficult to see as the irregularities on the terminator of the Moon when it is near the first quarter, viewed by the naked eye. I have on a few occasions seen a large projection perfectly distinct. So it is with Venus. At the best seeing the markings are visible at the first glance.

To say that no markings save M. Antoniadi's symmetrical shadings of atmospheric contrast exist, or that the detail seen here is due to pressure on our objective, or to defective densities in the eye piece, or to our own eyes, or to the imaginings of our brains; or, most ridiculous of all, to our looking all day at some map and then seeing it on the planet, is to offer suggestions too absurd to be taken seriously.

We use the telescope in both positions, normal and reversed: that shows that the markings are not in the lens. We use different eye-pieces and twist them in varying position angles: that shows that the markings are not there. We sit in different positions, so that the markings cannot be in our eyes; and different persons in perfect independence find the same detail, so it is not a mental phenomenon."

Several drawings of spoke-like structures follow with appropriate discussions. See the accompanying sketches.

Douglass then continues his explanation of why some see the linear markings and others do not. "The first reason why other observers have not seen these markings is bad atmosphere. When I began observing the third satellite of Jupiter, for days, even weeks, I drew nothing but hazy indefinite markings or belts, such things as M. Antoniadi describes as appearing to him on Venus. But one night after making several drawings of that character the seeing suddenly became superb, the curtain rose as it were, and I saw sharp distinct black lines about which uncertainty was impossible. The very same thing happened on the fourth satellite four days later. I had been drawing the same indefinite shadings, when one night the seeing improved, the curtain again rose, and I perceived sharp definite lines. After once thoroughly understanding the character of the object sought, I could see them and profitably study them under conditions of seeing formerly prohibitory.

The experience on Venus has been similar. On the day succeeding my first good view, I spent nearly the whole afternoon without catching a single certain glimpse. Suddenly the seeing improved for an instant, and I saw the same markings unmistakably. If it had not been for that glimpse I would

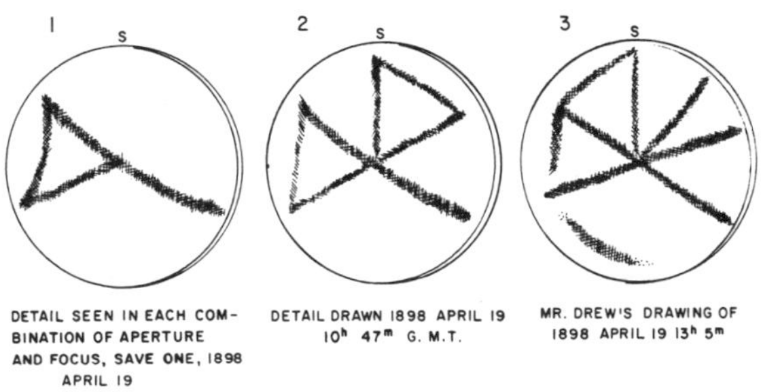

Various spoke-like markings observed at the Lowell Observatory in 1898. (X2)

have gone away perfectly ready to believe that no markings existed. I am not surprised that other astronomers doubt them." (R7, R8)

Of all Lowell's critics in the matter of Venusian spokes, Antoniadi was the most persistent. In 1898, he wrote: "To believe in the reality of Mr. Lowell's cytherean canals means to believe that the globe of Venus, with its atmosphere, is imprisoned in a cage of black hoops meeting about a common diameter directed along the visual ray. Under such circumstances we cannot help considering the whole of this anomalous canal system as entirely illusory, and the central black spot as merely the 'pilula' of Fontana." (R9, R10)

In 1902, Lowell apparently had doubts about the Venusian spoke system and wrote in the Astronomische Nachrichten (no. 3823) that he had reason to believe that it was an optical illusion. (R12). Yet, in 1904, in Lowell Observatory Bulletin No. 6, he stated that "these markings came out at times with a definiteness to convince the beholder of an objectiveness beyond the possibility of illusion." (R17)

Lowell died in 1916, but "Lowell-type" observations of spokes did not vanish completely. "On the afternoon of April 23 at 3h 5m summer time I happened to turn the 8 1/4 inch to Venus, which was less than half an hour east of the meridian and in high northern declination (23°), and I was greatly surprised to find the 'seeing' excellent. During the past forty or forty-five years I have never seen Venus so sharp and well defined ---the most difficult of all the planets to get a satisfactory view of. For fully ten minutes the planet was perfectly steady and sharply defined and I am very sure that I could see faint traces of the spoke-like markings near the center of the disk. The markings were very much like those seen and described many years ago by the late Dr. Percival Lowell." (R14)

C. H. Housden believed in the reality of the spoke marks and considered them to be artificial zones of vegetation constructed by intelligent Venusians. Housden based his theory on the supposed fact that Venus always kept the same face pointed toward the sun. This face was hot but the opposite face was cold and ice-covered. Where the two zones met, the sun melted the ice, and artificial canals conducted the water into the sunlit side. (R16)

X3. R. M. Baum's observations, commencing in 1951. Baum essentially saw the spoke marks on Venus as Lowell saw them. He began his study of Venus on February 5, 1951, but perceived no fine detail until February 19, when a dark spot appeared near the subsolar point. "Then followed a hiatus to March 1, thanks to inclement weather, when another spot was observed in about the same position---possibly the same spot ---which was the radiant of 'an extremely curious system of dark longitudinal streaks'. It may be noticed that on that date the diameter of the disc was still only 11".36 and the Ephemeris value of k was approximately 0.905. Thus Baum got his first view of the streak system well within the period which Lowell found best for studying it.

It is important also to notice at this point that Mr. Baum had not then read either Lowell's or Schiaparelli's papers on the planet, which adds significance to his surprising discovery. At any rate both March and April proved so favorable that he was able to keep the system in view for 61 days and on all occasions when the streaks were well seen he found them to possess a faint color, a sort of yellowish gray. Baum continued to observe the central spot and its associated streaks well into the crescent phase, i.e., past greatest elongation east. In a figure dated June 13, at 19h 15m, U. T., the central spot is still to be seen, being then almost bisected equally by the terminator; and in a figure dated July 7, at 19h 15m, U. T., the streaks are still evident though the central spot was then on the night side of the terminator and so invisible. After inferior conjunction indications of the streaks were seen on the very narrow crescent as early as October 10 at 6h, while they had become unmistakable by October 23.

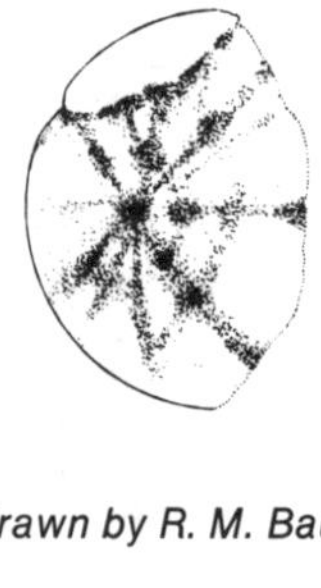

Radial spokes drawn by R. M. Baum, April 26, 1951. (X3)

From a close study of these surprising features Baum drew certain conclusions. The observed spots, especially the straight streaks, are not superficial but permanent features as are certain polar features, notably the dark band around the southern cusp cap (Baum, like Lowell, regards the south cusp cap as marking the actual pole); and from his study of the streak system he considered that the rotation is very slow, lying between 195 days and 225 days, and that the axis is sensibly perpendicular to the plane of the orbit. He further considered that the central spot from which the streaks radiated represents 'an enormous column of hot air' rising from the sub-solar point and drawing into it currents of colder air from all quarters of the disc, thus agreeing with the conclusions reached by Lowell. Whether such an explanation of the streak system is valid will depend, of course, upon the validity of the long rotation period and the assumed inclination. This writer is of the opinion that despite the careful work of Mr. Baum, and the admitted weight of Percival Lowell, neither can be regarded as established---there are too many observations looking another way. But we are not so much concerned with a physical theory to account for the streak system, as we are to account for its plain visibility to some and its complete invisibility to others.

R. M. Baum is an indefatigable observer and a fine draughtsman. Reading his report of the streak system one cannot doubt that here is a simple record of what was actually seen; yet it has to be admitted that other equally careful and experienced observers, e.g. Patrick Moore, have never been able to see the slightest indication of such a system. The present writer occupies a median position, having seen enough of the system to be convinced that pronounced linear markings do exist on Venus; but having seen it so imperfectly as to be quite unable to define the whole.

A little study of the work of various observers indicates that with respect to Venus they tend to fall naturally into two groups. In the one group is found preeminently R. M. Baum who sees the markings principally as a system of streaks, while in the other are those who see them always as large, diffuse, dusky areas. Between the two are those who partake of the characteristics of both, seeing sometimes linear markings, sometimes diffuse spots, and sometimes both in association. To the latter category belong O. C. Ranck, D. P. Avigliano, and the writer." The author believes that the solution to this mystery lies in the different color sensitivities of the different observers. (R17)

Patrick Moore, another English astronomer and a friend of Baum, responded to the preceding article, maintaining that the spoke must be rejected as illusory because only a few observers can see it. Quoting Moore: "Venus is the most difficult problem observationally of all the planets, and we must be doubly wary. It is my contention that the 'spoke system' of Lowell is merely an optical effect, seen by modern observers simply because of the inadequacy of their equipment. I do not believe that any observer, however skillful, can see with a 3-inch refractor details which the great Barnard and Antoniadi missed with the giant telescopes of Lick and Meudon respectively." (R18) Actually, Lowell's equipment and location in Arizona were excellent. (WRC)

Baum replied to the above critique by Moore, defending the Venusian spoke system as an objective entity. He also added two pertinent bits of data: (1) Ultraviolet photographs of Venus taken in 1927 with the 60-inch Mt. Wilson reflector revealed bands slightly convergent on the subsolar point. These streaks had much in common with the streaks drawn by Schiaparelli, Lowell, and many other. (2) In 1953 Moore and Baum had their eyes tested for color sensitivity. Baum was able to see farther into the blue end of the spectrum than Moore. (R20)

X4. More recent hints of a Venusian spoke system. In R21, Cruikshank reviews the Baum-Moore controversy, and continues as follows: "While it would not be proper for me to enter the argument actively at this late date, it is fitting, I believe, to re-examine the so-called spoke system in the light of recent observations and on theoretical grounds. A. Dollfus has made important planetary observations under optimum conditions with large apertures in recent years. He reports that the dusky markings on Venus generally have the pattern of a radial system with the center at the subsolar point. Dollfus finds that the general appearance of the markings varies greatly from day to day but that the radial pattern or portions of it recur and that this represents the undisturbed condition of the markings observed. He offers a general map of the planet showing these permanent features. In many respects, this pattern reported by Dollfus is consistent with a model of the circulation of the Venus atmosphere proposed by Yale Mintz. The Mintz model is based on convection in the planet's atmosphere (assuming a very slow rotation) with the principal circulation from the subsolar to the antisolar point..... In contrast with visual observations of the radial pattern of dusky features (and hard, linear features) we have the well-known ultraviolet photographs of Ross, Kuiper, etc. The markings shown in the ultraviolet (and more recently in the visual blue---W. K. Hartmann, unpublished) do not correspond to a radial pattern." (R21) Further details on the work of Dollfus and his map may be found in R19.

Still more recently, certain "characteristic" cloud patterns on Venus have been generally accepted as real. Such markings are also confirmed by spacecraft close-up photographs. "Characteristic forms of dark ultraviolet markings on the clouds have been studied from Earth. These are probably the same as the optical markings noted by early observers. Horizontal psi-shaped features have an extension of the equatorial bar through arms which are sometimes angular and at other times circular. Features that look like a reversed letter C appear more often near the evening terminator than the morning terminator. Horizontal Y-shaped features sometimes have a tail stretching round the planet. Sometimes there are two parallel equatorial bands. Patterns are almost always symmetrical about the equator. Arms of the various features open in the direction of their retrograde motion, which varies between 50 and 130 m/sec (164 and 427 ft/sec). A big question about the cloud motions was whether they resulted from actual movement of masses of atmosphere or were merely a wave motion." (R22) The velocities of these markings far exceed those of the classical spoke systems, which apparently rotated only once in 225 or more days. See AVW1 for a discussion of the possible superrotation of the Venusian atmosphere. (WRC)

References

R1. Brenner, Leo; "Venus, 1895, Observed at Manora Observatory," English Mechanic, 62:357, 1895. (X1)

R2. Lowell, Percival; "Detection of Venus' Rotation Period and of the Fundamental Physical Features of the Planet's Surface," Popular Astronomy, 4:281, 1896. (X2)

R3. MacEwen, Henry; "Canals on Venus," British Astronomical Association, Journal, 7:461, 1897. (X2)

R4. Lowell, Percival; "Determination of the Rotation Period and Surface Character of the Planet Venus," Royal Astronomical Society, Monthly Notices, 57:148, 1897. (X2)

R5. Brenner, Leo; "Canals on Venus," Observatory, 20:208, 1897. (X2)

R6. Antoniadi, E. M.; "Illusions," British Astronomical Association, Journal, 8:94, 1897. (X2)

R7. Douglass, A. E.; "The Markings on Venus," Royal Astronomical Society, Monthly Notices, 58:382, 1898. (X2)

R8. Douglass, A. E.; "The Markings on Venus," English Mechanic, 67:426, 1898. (X2)

R9. Antoniadi, E. M.; "Notes on the Rotation Period of Venus," Royal Astronomical Society, Monthly Notices, 58:313, 1898. (X2)

R10. Antoniadi, E. M.; "The 'Canals' of Venus," English Mechanic, 67:474, 1898. (X2)

R11. "The Planet Venus," Scientific American, 78:85, 1898. (X2)

R12. Wesley, W. H.; "Markings on Mercury and Venus," Knowledge, 4:228, 1907. (X2)

R13. Watson, H.; "Markings on Mercury---Venus---Mars," English Mechanic, 90:116, 1909. (X2)

R14. Seagrave, Frank E.; "Venus," Popular Astronomy, 27:406, 1919. (X2)

R15. "Venus," English Mechanics, 1:443, 1927. (X2)

R16. Hale, Wm. G.; "Markings on Venus," English Mechanics, 2:102, 1927. (X2)

R17. Bartlett, James C., Jr.; "The Radial Markings of Venus and Their Modern Resurrection," Strolling Astronomer, 9:2, 1955. (X2, X3)

R18. Moore, Patrick; "The Radial Markings of Venus: Another Point of View," Strolling Astronomer, 9:50, 1955. (X3)

R19. "The Dusky Markings of Venus," Sky & Telescope, 15:397, 1956. (X4)

R20. Baum, Richard M.; "The Radial Markings of Venus: A Rejoinder," Strolling Astronomer, 9:82, 1955. (X3)

R21. Cruikshank, Dale P.; "A Review of Some ALPO Venus Studies," Strolling Astronomer, 17:202, 1963. (X4)

R22. Fimmel, Richard O., et al; "Pioneer Venus," NASA SP-461, 1983. (X4)

AVO7 Bright Spots

Description. Transient bright points or patches seen on either the dark or sunlit limbs. These spots shine out well away from the cusps, where similar luminous phenomena are rather common. Spots on the sunlit limb must be very bright indeed to stand out.

Data Evaluation. Observations of this phenomenon are very scarce. Rating: 3.

Anomaly Evaluation. Since several possible explanations are at hand, this phenomenon is probably not very anomalous. Rating; 3.

Possible Explanations. Highly reflective clouds, like those seen on Mars; active atmospheric electrical processes, such as glow discharge and intense thunderstorm systems; atmospheric ducting of sunlight.

Similar and Related Phenomena. Venusian cusp phenomena (AVO1); bright Martian clouds.

Examples

X1. March 15, 1868. Observing with a 10 1/4 inch reflector. "About half-past four p.m., the sun becoming partially obscured by some trees, I set Venus off upon the circles of the equatorial, and looking along the tube I saw that she was plainly visible to the naked eye, although the sun was shining brightly. Viewing the planet with a power of 185, I found definition above the average. The feature that first attracted my attention was a curious oblong white nebulous spot, of considerable dimensions, shining with far greater brilliancy than any other part of the disc. This spot was on the edge of the disc, and fully 80° from the southern horn. From the resemblance it bore to the cloud-like patches I have described on Mars ("Intellectual Observer," September, 1867), I have little doubt that this spot was a highly reflective cloud in the atmosphere of Venus." (R1)

X2. 1868. Mr. De La Rue noticed a cluster of bright spots at the southern limb about 40° from the terminator. (R1)

X3. April 17, 1873. Observation by R. Langdon. "...at 8 o'clock, P.M., I was viewing the planet with one of Mr. Browning's excellent achromatic eyepieces, when I saw two exceedingly bright spots on the crescent---one close to the terminator, towards the eastern horn, and the other in the centre of the crescent. These spots appeared like two drops of dew; they were glistening in such a manner as to cause the surrounding parts of the bright crescent to appear dull by contrast." (R2)

X4. August 17, 1884. Using a 7-inch reflector, I.P. Guldenschuh observed a bright spot on the bright limb of Venus in the position indicated in the sketch. (R3)

August 17, 1884. Bright spot seen on Venus. (X4)

X5. January 14, 1910. B. Wymer reports two diffuse bright spots on Venus's dark limb. A portion of the dark limb's edge was also easily visible. (R4)

X6. March 17, 1953. Observation by R.M. Baum using a 6.5-inch reflector and a 3-inch refractor. At the time of the observation the ashen light phenomenon was occurring. "On this occasion the dark side was seen to be brighter that the sky, as is the case when the planet is projected against a dark field, and to be veritably glowing with a strong phosphorescent glow of a deep ruddy hue. Upon closer inspection the whole of the dark hemisphere was found to be not uniform in hue but actually mottled over with minute nebulous granules. These were seen better by averted vision. It was whilst studying this strange structure, that my attention was deflected by what appeared to be a speck of light of about the 8th magnitude situated in the southwest quadrant of the disk, i.e., in the southern part of the dark regions. The position angle of this feature measured round from the north point through west (in this case the right hand limb, taking west as right and the bottom as north, as with the Earth) was 171°. Its estimated distance from the terminator taken in angular measure from the drawing made was about 49°, where the disk is imagined flat. Nothing certain was seen by looking directly at the spot but by averted vision it was seen with certainty---additionally I could not be sure but that it was scintillating, which could have been pure fancy though. Thinking that perhaps the object was due to an optical fault in the reflector, which was the instrument in use at the time, I switched over to the aforementioned refractor. A much sharper image was obtained and the dark side shone out with greater clarity. The fine mottling was also more distinctly perceived. Above all however out shone the stellar speck." (R5)

References

R1. Browning, John; "Markings of the Planet Venus," English Mechanic, 7:186, 1868. (X1, X2)

R2. Langdon, R.; "Observations of the Planet Venus with a 6-inch Silvered Glass Reflector," Royal Astronomical Society, Monthly Notices, 33:500, 1873. (X3)

R3. Shearman, T.S.H.; "A Bright Spot on Venus," English Mechanic, 50:482, 1890. (X4)

R4. Wymer, Bert; "Venus," English Mechanic, 91:36, 1910. (X5)

R5. Baum, Richard M.; "On the Observed Appearance of a Remarkable Light Spot on the Night Side of Venus," Strolling Astronomer, 10:30, 1956. (X6)

AVO8 The Maedler Phenomenon: Brushes of Light

Description. Brushes or fans of light emanating from the two cusps of Venus and diverging in the direction of the sun. An extremely rare phenomenon.

Data Evaluation. Besides the classic observation of Maedler himself, we have two vague, highly questionable descriptions from the 17th. century. Rating: 3.

Anomaly Evaluation. Since none of the proposed explanations (see below) is very convincing, the Maedler Phenomenon, if real, is extremely puzzling. Rating: 2.

Possible Explanations. Some unappreciated interaction of the solar wind with Venus that results in aurora-like luminosity. Here, however, one would not expect the brushes of light to diverge in the direction of the sun. Optical phenomena in the terrestrial atmosphere akin, perhaps, to sun pillars. Once more, the geometry and orientation of the Maedler Phenomenon does not seem reasonable.

Similar and Related Phenomena. Sun pillars; skewed halo displays (GEH9).

Examples

X1. January 29, 1686. John Gadbury, a careful chronicler of natural phenomena, notes in his diary that Venus appeared like a comet. (R1)

X2. February 21, 1686. Gadbury again records that Venus is comet-like. R.M. Baum wonders if Gadbury might have seen some type of halo phenomenon depending upon ice crystals in the atmosphere. (R1)

X3. April 7, 1833. "One of the strangest observations ever recorded of Venus was made by the reknowned German astronomer Johann Heinrich Maedler with a four-inch refracting telescope on April 7, 1833. At the time Venus stood east of the Sun and was well placed for observation. In his Beitraege (1841) Maedler tells us how on that evening numerous brushes of light were seen to emanate from the illuminated limb of the planet, then a crescent, and to diverge in a sunward direction. These brushes of rays were roughly fan-shaped, and invested Venus with the look of a broad multi-tailed comet. According to the sketch which accompanies Maedler's account, the brushes originated at the extremities of the crescent and extended along the limb to affect an arc of between 50° and 60° from the tip of each cusp. No brushes are depicted in the segment between, i.e., the center of the limb. Maedler watched for some considerable time and tested the appearance for illusion by rotation of the ocular and by change of magnification; but although he attempted no explanation, he concluded that the appearance was at least real." (R1)

X4. General observations. A comet-like interaction of Venus with the solar wind has been described. (R2) This phenomenon may not be related to the Maedler phennomenon at all because the brushes of light described by Maedler diverge toward the sun. (WRC)

Diverging brushes of light from each cusp of Venus. (X3)

References

R1. Baum, Richard; "The Maedler Phenomenon," Strolling Astronomer, 27: 118, 1978. (X1-X3)

R2. Wallis, Max K.; "Comet-Like Interaction of Venus with the Solar Wind III: The Atomic Oxygen Corona," Geophysical Research Letters, 9:427, 1982. (X4)

AVO9 Flickering Light on the Dark Limb

Description. A flickering light playing over the unillumined side of Venus. The light resembles terrestrial heat lightning but is steadier.

Data Evaluation. A single observation over a century old. Rating: 3.

Anomaly Evaluation. Since Venus is now thought to play host to frequent electrical storms, there is an excellent source of flickering light available. The anomaly rating is therefore low. Rating: 3.

Possible Explanations. Planet-wide electrical discharges in the Venusian atmosphere or, alternatively, concentrated electrical storms that produce light that is transmitted over wide areas by the atmosphere.

Similar and Related Phenomena. The ashen light (AVO3).

Examples

X1. January 18, 1878. "...from 5 p.m. to 5.30 p.m., I saw the unilluminated side of Venus, referred to by Mr. Webb in letter 13938, p. 527, and there appeared upon it a kind of flickering light, resembling flashes of lightning, only slower in its movements. I covered half the field lens of the eyepiece with black paper, so as to keep the bright crescent out of sight, but the flickering light was still seen." (R1)

References

R1. Langdon, R.; "Astronomical Notes," English Mechanic, 26:575, 1878. (X1)

AVO10 Terminator Irregularities

Description. Temporary departures of the terminator (the line dividing dark and sunlit limbs) from the smooth arc dictated by optics. Four types of terminator irregularities are recognized here: (1) Smooth, gentle undulations; (2) Serrations; (3) A reversed-S shape; and (4) Thin lines of light separated from but parallel to the inside edge of the crescent.

Background. Mercury's terminator also displays irregularities, some of which seem identical to those of Venus. Since Mercury and Venus are physically very different planets, it may be that terminator irregularities have nothing to do with the planets themselves.

Data Evaluation. Many terminator irregularities are probably seen frequently---the reversed-S effect, for example---but are seldom reported. In general, the literature contains few well-described examples of the phenomenon at hand. Rating: 2.

Anomaly Evaluation. Although terminator irregularities are perplexing, they can probably be accounted for in terms of subjective effects (irradiation), bad seeing, and variations in the atmosphere of both earth and Venus. The fact that Mercury's terminator irregularities resemble those of Venus so closely underscore the reasonableness of subjective and seeing factors. Rating: 3.

Possible Explanations. As intimated above under Background, the explanation of terminator irregularities may be subjective (irradiation) or geophysical (bad seeing). As for Venus itself, changes in the reflectivity of the upper atmosphere could also contribute.

Similar and Related Phenomena. Mercury's terminator irregularities (AHO2).

Examples

X1. November 23, 1877. A drawing of Venus made by E.L. Trouvelot depicts the terminator as a gentle reverse-S curve. (R1, R2)

X2. 1881-1895. Unusual curvature of the terminator during this period reported by Breen, Maedler, Niesten, and Stuyvaert. (R3) Just what "unusual" means is not specified in R3. (WRC)

X3. May 14, 1892. F. Porro, Turin Observatory, using an 11-inch refractor, saw the terminator humped in the center, with depressions on either side. (R3)

X4. 1892. Long, low swellings and depressions along Venus's terminator seen by Alexander, Mascari, Poro, and Zona. (R3)

X5. January 28, 1910. "I enclose a sketch of Venus taken on Jan. 28 at 5 p.m., with the 3 1/2 in. Short's Gregorian, power 140. The horns were sharp. but what struck me was a broken line of light near the terminator, a little nearer to it than shown in the drawing, about 2" or less away, commencing from the south horn. Has this been observed before?" (R4) The broken line of light runs parallel to the thin crescent for almost 90°. (WRC)

X6. September 30, 1959. R.M. Baum, using a 4.5-inch equatorial, draws the terminator with distinct serrations. He believes this effect may be due to unsteady air. (R3)

References

R1. "The Planet Venus," Scientific American Supplement, 34:14066, 1892. (X1)
R2. "The Planet Venus," Scientific American Supplement, 37:15236, 1894. (X1)
R3. Baum, Richard; "The Himalayas of Venus," The Planets, New York, 1973. (X2, X3, X4, X6)
R4. Matthews, Edgar; "Venus," English Mechanic, 91:34, 1910. (X5)

AVW VENUSIAN ATMOSPHERIC ANOMALIES

Key to Phenomena

AVW0 Introduction

The heavy mantle of the Venusian atmosphere cuts off our direct view of the planet's surface. As if to compensate for this barrier to the scientific study of Venus's surface, the atmosphere itself presents us with several interesting and potentially important anomalies.

First, the outer layers of the atmosphere "superrotate"; that is, they rotate once around the planet in just four days, while the solid surface takes 243 days to make a complete turn. One scientist also wonders whether the whole atmosphere may not expand and contract on a four-day schedule due to the periodic waxing and waning of the intensity of CO_2 absorption lines.

The scarcity of water vapor and the abundance of primordial rare gases in the Venusian atmosphere in comparison to the earth's atmosphere pose serious questions about the validity of current models of solar system formation. The atmospheric veil that has made Venus "the planet of mystery" is itself a mystery.

AVW1 Superrotation of the Venusian Atmosphere

Description. The rotation of the Venusian cloud tops around the planet every four days in comparison to the 243-day period of axial rotation of the solid planet.

Background. The 4-day cloud-top period is dervived from modern ultraviolet photographs.

Historically, the telescopic observation of visible features from the earth led to estimates of 24 hours and 225 days for the planet's spin period.

Data Evaluation. Thousands of ultraviolet photographs and close-up experiments carried by space probes confirm the reality of the 4-day period of rotation for the outer clouds. Rating: 1.

Anomaly Evaluation. No generally accepted physical mechanism has been found that can drive the Venusian atmosphere at such high angular velocities relative to the planet's solid surface. Rating: 2.

Possible Explanations. Most astronomers assume that the driving force for Venus's superrotation is thermal, although no acceptable models are at hand. Since Venus apparently has an internal energy source (AVF3), the energy source propelling the atmosphere need not be entirely solar. Another possibility is that magnetohydrodynamic forces, generated either internally or by the solar wind, may keep the atmosphere spinning.

Similar and Related Phenomena. The superrotation of the earth's atmosphere (AGL1).

Examples

X1. Terrestrial observations of retrograde superrotation. V.M. Slipher (in 1903) and C.E. St. John and S.B. Nicholson (in 1923) reported spectroscopic studies of the Venusian atmosphere suggesting retrograde circulation. These data were ignored because Venus was expected to rotate in the same sense as the rest of the inner planets. (R1)

The first modern indications of retrograde superrotation were seen in ultraviolet photographs taken in 1961. (R3) A French photographic study of Venus in the ultraviolet confirmed the 1961 results in 1965. (R1)

In 1967 a study by a group at New Mexico State University provided additional support, as follows: "Throughout May and June of this year the evening terminator was turned toward the earth and a number of series of ultraviolet images of Venus were obtained by T. Pope and A. Murrell, using the new 61-cm reflector at the New Mexico State University Observatory. The unusual quality of these photographic images allowed several well-defined features to be measured on many of the series, with intervals extending from 1.9 to 4.7 hours. From the most casual examination it was immediately evident that the motion of the clouds was toward the evening terminator, away from the subsolar point, and that retrograde motion was being observed in the high-level ultraviolet clouds. Moreover, the same general motion was observed in several clouds distributed over the visible disk. The displacement rates of the clouds varied only slightly from day to day, and if it were to be assumed that their motion is generally unchanging around the circumference of the planet, these observed displacements would give rise to a mean rotation rate of somewhat less than 5 days with an uncertainty of about 1/4 day..... The greatly different rotation periods of Venus as determined by both optical and radar observations are difficult to reconcile with one another, for such would require a persistent and widespread planetary wind system having speeds in excess of 300 km/hr with respect to the solid surface; and although the terrestrial jet stream will attain these speeds, it is basically a narrowly confined zonal wind and, therefore, quite different from that observed on Venus. We are, unhappily, confronted with two rather widely divergent rotation periods for the atmosphere of Venus and its solid surface, 5 days and 244 days, respectively." (R1)

Other terrestrial studies of the motions of the ultraviolet markings on Venus invariably support retrograde superrotation. (R3)

X2. Summary of contradictory data. From the abstract of R2: "Spectroscopic, photometric, and radiometric evidence against a 4-day atmospheric rotation is also reviewed. The bulk of the somewhat contradictory evidence seems to favor slow motions, on the order of 5 m/sec, in the atmosphere of Venus; the 4-day 'rotation' may be due to a travelling wavelike disturbance, not bulk motions, driven by the uv albedo differences." (R2) The contradictory evidence is not from ultraviolet studies, such as those summarized in X1. (WRC)

X3. Radar data questioned. "The contradiction between the short atmospheric and the long radar axial period remains unresolved. The *Pioneer* data add to its difficulty inasmuch as all the recorded air movements are substantially from east to west. Since such movements arise from Coriolis forces, this would imply that all atmospheric flow is from the equator to the poles, which is absurd and is not borne out by the simultaneous records of meridional air flow. It is suggested that the origin of the scale is wrong, and the true axial period is close to four days, the radar period originating in a counter-rotating belt of plasma." (R4) Here the radar data are questioned, while in X2 the reality of the high speed circulation is doubted. (WRC)

X4. Close-up space probe results. Both Russian and American space probes to Venus have confirmed atmospheric superrotation; viz.: "Pioneer Venus Orbiter significantly extended observations of the ultraviolet patterns in the clouds of Venus. While Mariner 10 obtained 8 days of pictures, Pioneer Venus obtained hundreds of days of pictures to provide a greatly improved record of the bulk motions of the cloud tops. A question arising from the Mariner 10 observations was whether the features that move around in a 4-day period were bulk movement of masses of atmosphere or wave motions in the atmosphere. The Pioneer probe results indicate that the air is actually moving at the indicated speed of about 100 m/sec. Below the clouds, probe data show that the velocity starts to decrease to a very small value at the surface. The large features, especially the Y and C markings, can be regarded as waves of a special kind that move around the planet at the same speed as the air. All four probes, and some Soviet probes as well, showed the same westward motion with little or no north-south motion." (R6) Thus, the anomaly of superrotation is confirmed. One also wonders whether the common Y and C markings sometimes may blossom temporarily into spokes (AVO6). (WRC)

X5. General observations. "Unlike the study of Venus's surface, the study of some aspects of its atmosphere has been stymied by insufficient or irreconcilable data. The 360-kilometer-per-hour winds that carry the cloud tops around the entire planet in 4 days have fascinated researchers for years, but the dozen U.S. and Soviet probes that have penetrated the atmosphere and thousands of U.S. cloud photographs have not revealed what drives the winds. Pioneer scientists believe that the driving forces involve winds that blow from the equator toward the poles and winds in the form of eddies, but both of these are so slow that they have not been detected yet. In fact, no winds in any direction have been reliably detected below an altitude of 10 kilometers, where 50 percent of the atmosphere's mass lies. The best guess at the moment is that two conveyor-like patterns of circulation, called Hadley cells, slowly carry hot air from the equator toward the poles and back. The cells, one above the other, appear to be kept entirely separate by a stable layer just below the cloud layer." (R5)

References

R1. Smith, Bradford A.; "Rotation of Venus: Continuing Contradictions," Science, 158: 114, 1967. (X1)

R2. Young, Andrew T.; "Is the Four-Day 'Rotation' of Venus Illusory?" Icarus, 24:1, 1975. (X2)

R3. Hunt, Garry E.; "Super-Rotating Atmosphere of Venus," Nature, 266:15, 1977. (X1, X4)

R4. Firsoff, V. Axel; "On the Contradiction between the Radar Axial Period and the Atmospheric Circulation of Venus," British Astronomical Association, Journal, 90:464, 1980. (X3)

R5. Kerr, Richard A.; "Venus: Not Simple or Familiar, But Interesting," Science, 207:289, 1980. (X5)

R6. Fimmel, Richard O., et al; "The Atmosphere," Pioneer Venus, NASA SP-461, 1983. (X4)

AVW2 Anomalous Distribution and Scarcity of Water Vapor

Description. The great scarcity of water vapor in the Venusian atmosphere compared to the earth's. What water vapor exists seems to be concentrated at higher altitudes.

Background. Neither does Venus have surface water (AVE3), adding to the general belief that

Venus is now very deficient in water. If Venus was formed at the same time and from the same materials as the earth, there should be much more water on Venus. See AVE3.

Data Evaluation. Atmospheric water content is difficult to measure accurately. The results from the several U.S. and Russian probes are contradictory, although they all indicate very little water. Rating: 3.

Anomaly Evaluation. The current belief is that Venus was created at the same time and from the same stuff as the earth. Venus's apparent lack of water vapor flies in the face of this position. The vertical profile of water vapor is also contrary to expectations. Rating: 2.

Possible Explanations. Venus's water inventory may be normal but is locked up chemically or located below the surface. Another possibility is that the initial water complement was outgassed. Some extreme catastrophists claim that Venus is a recent addition to the inner solar system!

Similar and Related Phenomena. The lack of water on the Venusian surface (AVE3); Venus's argon anomaly (AVW4).

Examples

X1. Space probe measurements. "Unfortunately, measuring accurately the amount of water vapor in an atmosphere is very difficult. Even in connection with the Earth there persists much uncertainty about the amount of water vapor in the stratosphere. After the Venus mission of 1979 a similar state of confusion has developed concerning the amount of water vapor in the atmosphere of Venus. The neutral mass spectrometer of the large probe gave data that says there is less than 0.1% water in the atmosphere. A special optical device on the Venera probes found a small amount, too. Its measurements indicate that water decreases from 200 ppm at 50 km to 20 ppm at the surface. On the other hand, the probe gas chromatograph data show 0.52% at 42 km and 0.13% at 22 km---very much greater amounts." (p. 147) (R2) Although these results are inconsistent, the total amount of water on Venus, which has no oceans or other surface water, is much less than that present on earth. (WRC)

This anomaly is summarized in R2: "It is entirely unclear why there is so little water in the Venusian atmosphere. Has Venus formed without water? Is water hidden in the crust, or was it lost during the planet's evolution? Why is the vertical profile of water vapor concentration so extraordinary?" (p. 187) (R2)

X2. Outgassing of hydrogen suggested. "Abstract. The deuterium-hydrogen abundance ratio in the Venus atmosphere was measured while the inlets to the Pioneer Venus large probe mass spectrometer were coated with sulfuric acid from Venus' clouds. The ratio is $(1.6 \pm 0.2) \times 10^{-2}$. The hundredfold enrichment of deuterium means that at least 0.3 percent of a terrestrial ocean was outgassed on Venus, but is consistent with a much greater production." (R1) 0.3% of a terrestrial ocean is not very much water. The question of the 'missing water' is still open. (WRC)

References

R1. Donahue, T.M., et al; "Venus Was Wet: A Measurement of the Ratio of Deuterium to Hydrogen," Science, 216: 630, 1982. (X2)

R2. Fimmel, Richard O., et al; Pioneer Venus, NASA SP-461, 1983. (X1)

AVW3 Periodic Variation of CO_2 Absorption Lines

Description. The 4-day cyclic variation in the intensity of the CO_2 absorption lines in the spectrum of Venus. The fluctuations amount to about 20% and are synchronous over the entire planet.

Data Evaluation. A single series of high quality observations. Rating: 2.

Anomaly Evaluation. The cause of this variation is difficult to determine. If, as the authors of R2 suggest, it is the planet-wide expansion and contraction of the atmosphere, the anomaly is highly significant. However, if the effect is due to the 4-day masking of the CO_2 spectrum by circulating high-level clouds, the situation would be be so serious. Rating: 2.

Possible Explanations. Periodic expansion and contraction of the Venusian atmosphere; the 4-day-period passage of high-level obscuring and transparent areas in the atmosphere.

Similar and Related Phenomena. The 4-day rotation of ultraviolet features in the Venusian atmosphere.

Examples

X1. Periodic phenomenon. "Astronomers are well enough acquainted with periodic variations in the light from the stars, but a variable planet is quite a different matter. However, the planet Venus shows regular changes in the spectrum of its atmosphere, according to four scientists at Caltech's Jet Propulsion Laboratory. The strengths of carbon dioxide lines in the Venusian atmosphere swing through a four-day cycle.

Over 20 years ago Gerard Kuiper noted day-to-day fluctuations in the infrared spectrum of Venus, but no one has yet got to the bottom of the basic cause of these changes. In order to study the oscillations, A. T., L. G., and J. W. Young and J. T. Bergstrahl obtained spectra nightly during the autumn of last year. Their data on the carbon dioxide line show an unmistakable oscillation.

The observed variation is not exactly periodic, but more akin to a relaxation oscillation in which the amplitude builds up on successive cycles and then suddenly collapses. In order to produce the observed changes the cloud deck of Venus must be moving up and down as much as one kilometre, simultaneously over the entire surface of the planet. Such a large atmospheric oscillation requires a high input of mechanical energy. This condition is difficult to account for in the case of a slowly rotating planet heated uniformly by the Sun's rays. Therefore the cyclic variations point to some unexplained deep-seated property of the atmospheric dynamics." (R1) (R2)

References

R1. "Venus Breathes in Steady Fashion," New Scientist, 58:72, 1973. (X1)

R2. Young, L. G., et al; "The Planet Venus: A New Periodic Spectrum Variable," Astrophysical Journal, 181:L5, 1973. (X1)

AVW4 Noble Gas Anomalies

Description. The atmosphere of Venus contains approximately 75 times as much argon-36 and 38 as the earth's atmosphere. Argon-36 and 38 are considered primordial isotopes of argon; that is, they were present when the solar system was first formed. Neon is 45 times more abundant on Venus than earth; krypton 3 times more abundant. In contrast, radiogenic argon-40 is only one-fourth as abundant on Venus.

Background. Scientists fully expected that the inner planets would all have about the same fractions of primordial gases, since the prevailing model of solar system formation had them created at the same time from the same materials. Instead, the abundances of the primordial rare gases are much higher on Venus than on earth; and earth's abundances are much higher than those on Mars.

Data Evaluation. The data from the several U. S. and Soviet probes to Venus are emphatic that the primordial rare gases, especially the isotopes of argon and neon, are much more common on Venus. There is some argument over precise absolute abundances however. Rating: 2.

Anomaly Evaluation. Initially, planetary scientists were greatly concerned about the rare gas anomalies, but relatively modest changes in the solar system formation model may suffice to explain the new data, as described below. Of course, the data are also consistent with some more radical solar system scenarios, but these theories are generally discounted. Rating: 2.

Possible Explanations. The Venusian atmosphere may have been "salted" preferentially by primordial rare gases in the solar wind. The scarcity of argon-40 relative to earth may be due to a lower level of volcanism and erosion on Venus which keeps the argon-40 (derived from potassium-40) sealed in the crust. Another theory has the dust grains accreting to form the planets surrounded by a gas whose pressure decreases rapidly with distance from the sun; this allows the primordial gases to escape more readily from the grains farther away from the sun. Finally, Venus could in principle have formed elsewhere in the solar system and/or at a different time.

Similar and Related Phenomena. The lack of water vapor in the Venusian atmosphere (AVW2); the many isotopic anomalies found in meteorites (AYE); a few vague ancient reports that the orbit of Venus changed in historical times (AVB3); isotopic anomalies in the Martian atmosphere (AMW5).

Examples

X1. Summary of space probe results. "... a rude shock was delivered to the planetary science community when an assay of the rest of the volatiles of the Venus atmosphere was made. The case of argon illustrates the point. There are two types of argon isotopes of interest to scientists studying planetary atmospheres. Radiogenic ^{40}Ar, the most abundant kind of argon in the Earth's atmosphere is produced by radioactive decay of potassium. Its abundance tells us about the primitive concentration of potassium and outgassing conditions throughout the 4.5-billion-year history of the planet. On the other hand, ^{36}Ar and ^{38}Ar are primordial gases and they tell us about the early volatile content of planetary interiors and the early outgassing scenario. On the basis of carbon and nitrogen results, scientists expected that there would be about as much ^{36}Ar and ^{38}Ar in the atmosphere of Venus as in the atmosphere of Earth. Instead the mass spectrometers on the Pioneer and Venera landers found the concentrations of radiogenic ^{40}Ar and nonradiogenic argon to be about equal. About 30 atoms in every million atmospheric molecules (or 30 parts per million (ppm)) were ^{36}Ar. The gas chromatographs which could not distinguish among the various isotopes of argon supported the mass spectrometer results. Data from them showed the total concentration as being between 50 and 70 ppm. Since the atmosphere of Venus contains about 75 times as many molecules as that of Earth this means that it contains 75 times as much ^{36}Ar as the atmosphere of Earth. And yet the ratio of ^{38}Ar to ^{36}Ar is almost identical to the terrestrial ratio.

One discordant note has been sounded by the neutral mass spectrometer on the Bus. It could not detect argon at 130 km. By extrapolation to the lower atmosphere this result would seem to mean that there is less than 10 ppm of ^{36}Ar in the atmosphere of Venus. Even this upper limit, however, does not explude the possibility that there is 25 times as much ^{36}Ar in the atmosphere of Venus as in that of Earth.

Examination of the case of neon, another 'primordial' rare gas, confirms the argon story. The Pioneer instruments and the Venera neutral mass spectrometer place the abundance of neon between about 4 and 13 ppm---compared with 18.2 ppm for Earth. This puts the excess of neon on Venus at about 45. The ratio of ^{22}Ne to ^{20}Ne was measured as 0.07. In contrast with the argon isotopes, this ratio is lower than the value found on Earth (about 0.1), but is close to the solar ratio.

Early analysis of data from the large probe's neutral mass spectrometer did not produce any publishable values for other rare gases---krypton and xenon. Nevertheless, it was clear that the notion that Venus, Earth, and Mars were made up of materials containing the same endowment of volatiles, already shaken by the Viking results, had been completely destroyed by the data from Pioneer Venus. Why should Venus have been provided with only about twice as much carbon dioxide and nitrogen as Earth and about 50 to 100 times as much neon and non radiogenic argon?

....

Analysis of the large probe's neutral mass spectrometer data has recently produced another surprise. Although the atmosphere of Venus contains a large excess of neon and primordial argon, this is not so with two other rare gases. The absolute abundance of krypton is only about 3 times larger in the atmosphere of Venus than in that of Earth. There is much less than 30 times more xenon. In the grain accretion model there is no reason to expect the enrichment of one rare gas to be greater than another. In fact, a close look at the Mars data shows that from Mars to Earth the enrichment decreases from a factor of about 220 for neon through 165 for argon, 110 for krypton, to 30 for xenon."(R8) Some of the data reported above may also be found in R1, R2, and R4.

X2. Theoretical considerations. "One possibility suggested after the early data from the Pioneer and Venera missions were revealed was that the planets were formed from dust grains in the solar nebula which were surrounded by gas at a pressure which diminished rapidly with increasing distance from the center of the nebula. Since reactive volatiles such as carbon, nitrogen, and oxygen are chemically combined within the grains, while the rare gases are adsorbed from the surrounding gas in amounts depending on the pressure, the result would be that the grains forming the three planets would have about the same reactive volatile content but the rare gas concentration would decrease rapidly with increasing distance from the sun. This model also required that the nebula's gas temperature should be fairly constant and that for some reason, early outgassing from Mars should be less efficient by a factor of 20 than from the other two planets." (R8)

The following theory is looked upon with more favor than the preceding one. "Another way of looking at these results is to compare the ratio of 'primordial' argon to krypton on the terrestrial planets with the ratio on the Sun. In the solar atmosphere this ratio is 4000, on Venus it is 1000, on Earth 50, and on Mars 40. Thus, the ratio gets more solar-like the closer the planet is to the Sun. This fact has suggested that perhaps the material which accreted to form the planets was exposed to a strong irradiation by gas of solar composition flowing away from the sun when the Solar System was being formed. If this were true the grains and small bodies that formed the planets would have a contribution of volatiles from the Sun in addition to the contribution from the nebular gas in their neighborhood. The material forming Venus may have received a much larger share of solar gases than the other planets because in intercepting most of the solar gas it would have shielded the outer regions of the Solar System from this gas." (R8) Additional information of this theory may be found in R5, R6, and R7.

The anomalously low abundance of argon-40 on Venus may be explained in two ways: "Venus either started with considerably less potassium than Earth or is yielding up its argon from the interior more slowly than is Earth. The lack of widespread tectonism, the thicker and relatively plastic unfractured lithosphere, and the absence of surface erosion on Venus may be responsible for a slow escape of gases during the 4.5 billion years lifetime of the planet." (R8)

Proponents of Velikovsky's theories welcomed the rare gas anomalies, for Velikovsky had claimed that Venus had formed recently from Jupiter. (R3)

References

R1. Gwynne, Peter; "Venus Probes Solar System Birth," New Scientist, 80:916, 1978. (X1)

R2. Hoffman, J. H., et al; "Venus Lower Atmospheric Composition: Preliminary Results from Pioneer Venus," Science, 203:800, 1979. (X1)

R3. Greenberg, Lewis M.; "Velikovsky and Venus: A Preliminary Report on the Pioneer Probes," Kronos, 4:1, Summer 1979. (X1, X2)

R4. Kerr, Richard A.; "Venus: Not Simple or Familiar, But Interesting," Science, 207:289, 1980. (X1)

R5. McElroy, Michael B., and Prather, Michael J.; "Noble Gases in the Terrestrial Planets," Nature, 293:535, 1981. (X2)

R6. "Pioneer Data Suggests Venus' Atmosphere Received Big Input from the Sun," American Meteorological Society, Bulletin, 62:1063, 1981. (X2)

R7. Kerr, Richard A.; "Origins: A Problem with Rare Gases," Science, 215:279, 1982. (X2)

R8. Fimmel, Richard O., et al; Pioneer Venus, NASA SP-461, 1983, p. 146. (X1, X2)

AVX TRANSIT PHENOMENA

Key to Phenomena

AVX0 Introduction

Transits of Venus are much more rare than those of Mercury. The last pair took place in 1874 and 1882; the next pair occurs in 2004 and 2012! The only phenomenon of note seen during transits of Venus is the appearance of the so-called annular phase or Venus-seen-as-a-ring as the sun backlights the planet. This phenomenon provides some interesting variations during the contact phase, as detailed below.

The better-known "black drop" effect is also observed during the contact phase. It is not considered anomalous. A description may be found in AHX0.

AVX1 Ring of Light around Venus during Contact Phase

Description. The appearance of a ring of light around Venus during the precontact, contact, and post contact phases of a transit of the sun. The anomalous facets of this phenomenon, which ostensibly is due to the refraction of sunlight by Venus's atmosphere, are: (1) The rare appearance of a bright bead, similar to a Bailey's Bead, on the ring; and (2) The wide range of times during which the phenomenon first appears---from several minutes before contact to several minutes after.

Background. The basic phenomenon seems identical to Venus's "annular phase", which generally occurs near inferior conjunction; although Venus may be rather far from the sun during these apparitions. Why is the phenomenon visible (or not visible) under such a wide range of backlighting?

Data Evaluation. High quality reports from several Nineteenth Century transits. Rating: 1.

Anomaly Evaluation. Simple refraction of sunlight should account for the basic phenomenon. However, the bead effect and the variability in timing remain unexplained facets. Rating: 3.

Possible Explanations. The timing of the ring's appearance may be affected by dust and other "pollutants" in the Venusian atmosphere, much as the earth's atmosphere seems to control the darkness of lunar eclipses. Atmospheric ducting of sunlight might account for the presence of a bead on the ring.

Similar and Related Phenomena. Annular phase of Venus (AVO2); halo around Mercury during transit (AHX2); ring of light around the moon (ALO9); Bailey's Beads (ASX).

Examples

X1. December 10, 1866. Ring of light seen around Venus prior to contact. (R2) See AVO2 for details.

X2. December 12, 1874. Ring of light seen around Venus prior to contact. (R1, R3) See AVO2 for details.

X3. December 5, 1882. Observations of J. E. Keeler, Allegheny Observatory. "As soon as first contact occurred, at 20h 44m 30s, I looked for indications of the atmospheric ring around the planet, having been requested by Professor Langley to pay particular attention to the physical phenomena of the transit, but for the first three or four minutes could see nothing of it. Shortly afterwards, however, I caught a feeble glimmer of light, almost star-like in appearance, on the limb of the planet farthest from the Sun, which at 20h 49m presented the appearance of a curved streak of very faint silvery light extending for a short distance along the margin of the unimmersed portion of the planet's disc. The brightest part of this luminous arc was not directly opposite the Sun, but was situated about 20° to the west of a line joining the centres of the Sun and Venus. At the same time little horns of light, due perhaps to an optical illusion, appeared to rise from the cusps of the Sun at the margin of the planet, like the elevated rim of fluid which surrounds an immersed body through capillary action.

As the planet advanced, the arc of light gradually extended and brightened, until at 20h 54m, or nearly half way between first and second contacts, the unimmersed portion was completely surrounded by a luminous ring. The light at the place on the margin when it was first noticed, however, much exceeded in brilliance that of the adjacent portions, the brightest part extending along perhaps 30° of the planet's circumference, and on the western side the luminosity was more evident than on the eastern, when it was yet barely discernible. The juncture of the luminous arc, first observed with the western cusp of the Sun, to which it lay nearest, occurred before the eastern edge became visible. The marginal patch of light now presented the appearance of a local brightening of a continuous ring of light surrounding the planet, and according to my impression at the time, lay without its contour." (R5) Essentially the same events were observed elsewhere. (R4)

References

R1. Lyman, C. S.; "On Venus as a Luminous Ring," Philosophical Magazine,

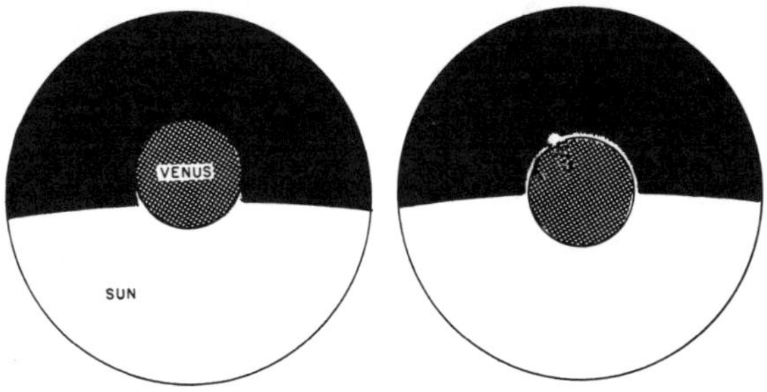

Bright ring of light develops around Venus during first contact. A brilliant "bead" was also seen. (X3)

4:49:159, 1875. (X2)
R2. Neison, E.; "The Ring of Light around Venus during the Late Transit," Astronomical Register, 13:196, 1875. (X1)
R3. Tebbutt, John; "Ring of Light around Venus," Astronomical Register, 13:222, 1875. (X2)
R4. Brashear, J. A.; "Transit of Venus," English Mechanic, 36:500, 1883. (X3)
R5. Keeler, J. E.; "The Ring of Light Surrounding Venus," Sidereal Messenger, 1:292, 1883. (X3)

AVZ MAGNETIC FIELD OF VENUS

Key to Phenomena

AVZ0 Introduction

One of the greatest surprises in scientific space exploration was the failure of spacecraft to detect an intrinsic magnetic field around Venus. This, coupled with the discovery of a magnetic field around Mercury where none was expected, has provided planetary scientists with a disturbing anomaly and a need to rethink the evolution of the inner planets.

Venus's failure to conform to expectations in the matter of magnetic field and in several other ways (viz., rare gas inventories and slow, retrograde spin) seem to make the planet "odd man out" in the otherwise cozy inner solar system. Is it possible that Venus had a different time and/or place of origin?

AVZ1 The Negligible Venusian Magnetic Field

Description. The failure of space probes to detect any appreciable intrinsic magnetic field in the vicinity of Venus.

Background. The scientific expectations prior to the direct measurements of Venus's magnetic field are detailed in X3 below. Basically, planetary scientists originally believed that Venus would have an appreciable magnetic field, while Mercury would have no intrinsic field at all. Both suppositions proved wrong.

Data Evaluation. All space probe measurements concur that the Venusian intrinsic magnetic field is negligible. Probes, however, do measure magnetic fields induced by the interaction between the solar wind and Venus's ionosphere. Rating: 1.

Anomaly Evaluation. This is another anomaly involving unfulfilled expectations. Given that Venus's sister planets, earth and Mercury, both have intrinsic fields, the implication is that Venus is substantially different internally from both. Venus with a different composition and/or evolutionary history is an important anomaly. Rating: 2.

Possible Explanations. Venus may be different compositionally from earth and Mercury. This difference may not be due simply to its different position in today's solar system and its assumed similar location when the planets accreted. Both earth and Mercury have intrinsic fields, and one would think that if Venus formed in its present position it would have an intrinsic field, too. Its composition should not be radically different from its sister planets if it was formed where it now orbits. Implication: it may have formed elsewhere in the solar system. Venus's slow spin may be a factor in suppressing dynamo action, but Mercury's spin period is also long and it does generate an intrinsic field. The core of Venus may not be hot enough for fluid convection and consequent dynamo action. The planet may be deficient in heat-generating potassium-40---a possibility underscored by Venus's low abundance of argon-40, a daughter of radioactive potassium-40. (AVW4)

Similar and Related Phenomena. The deficiency of argon-40 in the Venusian atmosphere relative to earth (AVW4); the intrinsic magnetic field of Mercury (AHZ1); the general difficulty geophysics has in accounting for the terrestrial magnetic field (EM).

Examples

X1. Early inference of a strong intrinsic magnetic field. J. Houtgast, of Utrecht, estimated in 1955 that Venus's magnetic field was roughly five times as strong as that of earth. Studying magnetic records for the earth from 1884-1953, he found that there is a marked drop in terrestrial magnetic activity as Venus passes through inferior conjunction. Venus, he surmised, provides a magnetic shield that deflects the solar wind away from the earth. (R1)

X2. Summary of U. S. and Russian space probe results. None of the several space craft sent to Venus has provided any evidence of an intrinsic magnetic field. Induced magnetic fields arising from the interaction of the solar wind with Venus's ionosphere do exist. (R2-R4) The deflection of the solar wind by virtue of this interaction with the Venusian ionosphere could, in principle, account for Houtgast's results in X1. (WRC)

X3. General observations. "One of the principal unsolved problems of geophysics is the nature of the source of the terrestrial dynamo that generates the magnetic field of the earth. Scientists hoped that a measure of a magnetic field of Venus, a planet which appears in many respects to be a twin of earth, would help clarify the effect of spin rate on the dynamo process. Venus spins on its axis much more slowly than does earth. A Venus day is 243 earth days. Dynamo theories predict that a planetary dynamo, such as the one generating the field of the earth, should depend on spin rate. If a Venus dynamo were identical to earth's, but weaker in proportion to the spin rate, the planet would have a magnetic field that would easily be detectable. But it does not. Other factors must be at work.

A planetary magnetic dynamo requires a highly electrical conducting fluid core. The absence of a conducting core may explain why earth's satellite, Moon, does not have a magnetic field, but it does not explain the absence of a field on Venus. Under the temperatures and pressures present in the core of Venus there should be a highly conducting fluid. However, the composition and electrical conductivity of the fluid may be different from that of earth. Although Venus appears to be earth's twin in size, it may not be a twin in chemical composition since it was formed at a different place in the solar nebula and presumably at a different temperature." (R4)

References

R1. "Venus' Magnetic Pull Five Times Earth's," Science News Letter, 67:281, 1955. (X1)

R2. "Why Did Mariner II Find Venus Non-Magnetic?" New Scientist, 17:10, 1963. (X2)

R3. Russell, C. T.; "Does Venus Have an Intrinsic Magnetic Field?" Nature, 275: 692, 1978. (X2)

R4. Fimmel, Richard O,: "The Intrinsic Magnetic Field," Pioneer Venus, NASA SP-461, 1983, p. 153. (X2, X3)

TIME-OF-EVENT INDEX

FIRST-AUTHOR INDEX

SOURCE INDEX

SUBJECT INDEX